AN INTRODUCTION
TO
FINITE ELEMENT ANALYSIS

AN INTRODUCTION TO FINITE ELEMENT ANALYSIS

D. H. Norrie   and   **G. de Vries**

Department of Mechanical Engineering
The University of Calgary
Calgary, Alberta, Canada

ACADEMIC PRESS   New York   San Francisco   London

A Subsidiary of Harcourt Brace Jovanovich, Publishers

ACADEMIC PRESS, INC.
111 Fifth Avenue, New York, New York 10003

United Kingdom Edition published by
ACADEMIC PRESS, INC. (LONDON) LTD.
24/28 Oval Road, London NW1

Library of Congress Cataloging in Publication Data

Norrie, D   H
      An introduction to finite element analysis.

      Includes bibliographical references.
      1.   Finite element method.   I.   d e Vries, Gerard,
joint author.   II.   Title.
TA347.F5N67   620'.001'515353   76-45996
ISBN 0–12–521660–2

PRINTED IN THE UNITED STATES OF AMERICA

To our parents

Gerardus de Vries   *Jenny de Vries*   *Gladys Norrie*

In memory of *Hector Norrie*

CONTENTS

5 Hermitian Elements, Condensation, and Coupled Boundary Conditions

6 Economization of Core Storage, Partitioning, and Tridiagonalization

7 The Variational Calculus and Its Application

8 Convergence, Completeness, and Conformity

9 Elements and Their Properties

10 Equation Solvers and Programming Techniques

11 Selected Applications of the Finite Element Method

12 Other Finite Element Methods

Appendix A Matrix Algebra

Appendix B   Matrix Calculus

Appendix C   Determinants

PREFACE

In an earlier monograph[†] on the finite element method, the authors organized the material in a progression from general analysis to particular applications. The formulation of physical problems was presented first, followed by their classification and approximate solution, using the method of trial functions. The finite element method was then shown to be a subclass of the method of trial functions and the fundamentals of the finite element method were developed. In succeeding chapters various aspects of the finite element method were considered further and the application to a wide range of physical problems was demonstrated.

The approach just outlined is suited for students in senior graduate courses as well as for researchers or practitioners who wish to broaden their present knowledge in the subject. These readers require a presentation that will "demonstrate the generality of the finite element method by providing a unified treatment of fundamentals and a broad coverage of applications."

For the student in a first course in finite elements, or for the engineer, scientist, or applied mathematician desiring an introduction to the subject, an alternative treatment is needed and the pedagogic principle "from the particular to the general" has special relevance. The present textbook is structured according to this principle and attempts to develop a basic understanding of the finite element method rather than provide a comprehensive coverage of its many applications. It is intended that the two books should be complementary, with the present treatment being more suited, as its title indicates, to introducing the finite element method to the reader.

In little over two decades, the finite element method has evolved as a technique of major importance for the solution of a wide range of scientific

[†] "The Finite Element Method—Fundamentals and Applications," D. H. Norrie and G. de Vries, Academic Press, New York, 1973.

and engineering problems. It has both stimulated new fields of mathematical research and facilitated the design of new dams, bridges, buildings, aircraft, automobiles, machine tools, and other artifacts. It has the virtues of simplicity in concept, elegance in development, and potency in application. If this book assists the reader to master the method, the authors will be well rewarded.

Since the material in this book has been drawn from many sources over a period of years, it is difficult to acknowledge all those whose contributions are in some way represented. Citations have been given whenever possible, but it would not be practical to list all persons to whom the authors are indebted. Our gratitude to all of these is therefore recorded here.

The manuscript was read at a late stage by Dr. B. M. Irons of the Department of Civil Engineering, University of Calgary, whose numerous comments and suggestions for improvement were much valued.

Particular mention must be made of the assistance received from the National Aerospace Laboratory, the Netherlands, during the period in which the second author was a Visiting Scientist. Notable among the many staff members to whom appreciation is due are T. E. Labrujere and A. A. Sanderse.

Special thanks are due to Mrs. Betty Ann Maylor for the outstanding quality of manuscript typing and for patience and equanimity through the numerous revisions.

NOTE TO THE READER

An elementary programming knowledge of FORTRAN IV is assumed in this book. The syntax used is generally that found in elementary texts and manuals on the subject. More specifically, it is that derived from the standard version known as "USA FORTRAN IV".[†] Where there is any variation from this, the syntax of the Control Data Corporation's "FORTRAN Extended Version 4" has been followed.

[†] USA Standard FORTRAN (3.9–1966), United States of America Institute of Standards, New York, 1966.

1

BASIC CONCEPTS OF THE FINITE ELEMENT METHOD

The finite element method originated in structural engineering, but within a decade the basic concepts were recognized as having wider applicability [1, 2] and were being used in several other areas. Subsequent development has been rapid, and the techniques are now well established within many scientific and engineering disciplines. Although there may be considerable diversity in the formulation, a finite element method can be distinguished by the following features:

(1) The physical region of the problem is *subdivided* into subregions or *finite elements*.

(2) One or more of the dependent variables is approximated in functional form over each element and hence over the whole domain. The parameters of this approximation subsequently become the unknowns of the problem.

(3) Substitution of the approximations into the governing equations (or their equivalent) yields a set of equations in the unknown parameters. The solution of these equations yields the parameters and hence the approximate solution to the problem.

Most commonly, the governing equations are not used directly, but an alternative formulation such as a variational principle is used instead. Sometimes a constraint is used that forces the difference between the true and approximate solutions to be small in a specified way, as in the *residual* finite element method. Since the number of unknowns in the final set of equations is often very large, it is common practice to use matrix notation, both for conciseness and to facilitate computer programming.

1.1 ASSEMBLAGES AND NETWORKS

1.1.1 Discrete and Continuous Systems

The name *finite element method*, in present usage, refers to a solution procedure for a *continuous system*, that is, a system that involves a phenomenon over a continuous region. Many problems, however, involve a *discrete system* consisting of a finite number of interconnected elements. As will be seen subsequently, there are many similarities between the analysis of a discrete system and the finite element method. From a historical point of view this is to be expected since the finite element method evolved from attempts to discretize continuum problems in structural mechanics and to apply to the resulting discrete systems those techniques that had proven so successful for structural assemblages.

Because of the similarities, the matrix formulation of discrete problems should be considered before the finite element analysis of continuous systems. In the following sections two discrete systems, *assemblages of structural elements* and *transmission networks*, will be examined.

1.1.2 Assemblages

Although structural elements are often interconnected by joints that can transmit *both* forces and moments, only a plane pin-jointed frame (see Fig. 1.1) is considered at this stage. It is assumed that the frame has been assembled without initial stress, prior to the loads being applied at the nodes as indicated. The forces $\mathbf{F}_2$ and $\mathbf{F}_3$ exerted by the pins on a typical element e_6 are shown in Fig. 1.2 in terms of their x, y components F_{x2}, F_{y2} and F_{x3}, F_{y3}, respectively. The displacements of the nodes of the element are denoted by $\bar{\boldsymbol{\delta}}_2$ and $\bar{\boldsymbol{\delta}}_3$, with components $\bar{\delta}_{x2}, \bar{\delta}_{y2}$ and $\bar{\delta}_{x3}, \bar{\delta}_{y3}$, respectively, measured from nodal positions *before* loading of the frame.

In matrix form the forces and displacements are respectively given by

$$\mathbf{F}^{e_6} = \begin{bmatrix} \mathbf{F}_2 \\ \mathbf{F}_3 \end{bmatrix}^{e_6} = \begin{bmatrix} F_{x2} \\ F_{y2} \\ F_{x3} \\ F_{y3} \end{bmatrix}^{e_6}, \tag{1.1}$$

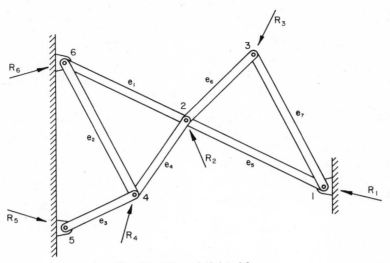

Fig. 1.1   Plane pin-jointed frame.

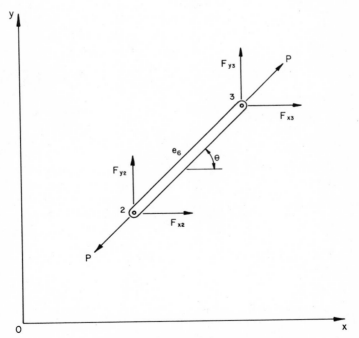

Fig. 1.2   Pin-jointed element e_6.

and

$$\delta^{e_6} = \begin{bmatrix} \bar{\delta}_2 \\ \bar{\delta}_3 \end{bmatrix}^{e_6} = \begin{bmatrix} \bar{\delta}_{x2} \\ \bar{\delta}_{y2} \\ \bar{\delta}_{x3} \\ \bar{\delta}_{y3} \end{bmatrix}^{e_6}, \tag{1.2}$$

where the superscript e_6 denotes the element under consideration. Where the context prevents confusion, this superscript will be omitted. The elongation or contraction of the bar from the unloaded length L is given by[†] $(\bar{\delta}_{x3} - \bar{\delta}_{x2}) \cos \theta + (\bar{\delta}_{y3} - \bar{\delta}_{y2}) \sin \theta$ and the strain is obtained by dividing this by L. Since the stress equals E times the strain, where E is Young's modulus, the longitudinal force P applied to the bar (Fig. 1.2) is given by

$$P = \frac{EA}{L}[(\bar{\delta}_{x3} - \bar{\delta}_{x2}) \cos \theta + (\bar{\delta}_{y3} - \bar{\delta}_{y2}) \sin \theta], \tag{1.3}$$

where A is the cross section of the bar.

The components of the longitudinal force P can be equated to the components of the pin-forces and thus Eq. (1.1) can be written as

$$\mathbf{F}^{e_6} = \begin{bmatrix} \mathbf{F}_2 \\ \mathbf{F}_3 \end{bmatrix} = \begin{bmatrix} F_{x2} \\ F_{y2} \\ F_{x3} \\ F_{y3} \end{bmatrix} = \begin{bmatrix} -P \cos \theta \\ -P \sin \theta \\ P \cos \theta \\ P \sin \theta \end{bmatrix}. \tag{1.4}$$

Substitution of P from Eq. (1.3) into the relation for F_{x2} in Eq. (1.4) yields

$$F_{x2} = \frac{EA}{L}[-(\bar{\delta}_{x3} - \bar{\delta}_{x2}) \cos^2 \theta - (\bar{\delta}_{y3} - \bar{\delta}_{y2}) \sin \theta \cos \theta], \tag{1.5a}$$

or after rearrangement

$$F_{x2} = \frac{EA}{L}[\cos^2 \theta \, \bar{\delta}_{x2} + \sin \theta \cos \theta \, \bar{\delta}_{y2} - \cos^2 \theta \, \bar{\delta}_{x3} - \sin \theta \cos \theta \, \bar{\delta}_{y3}]. \tag{1.5b}$$

In matrix form Eq. (1.5b) can be written as

$$F_{x2} = \frac{EA}{L}[\cos^2 \theta \quad \sin \theta \cos \theta \quad -\cos^2 \theta \quad -\sin \theta \cos \theta] \begin{bmatrix} \bar{\delta}_{x2} \\ \bar{\delta}_{y2} \\ \bar{\delta}_{x3} \\ \bar{\delta}_{y3} \end{bmatrix}. \tag{1.5c}$$

[†] Using the convention that the longitudinal force P is positive if tensile and negative if compressive.

The four equations of type (1.5c) for F_{x2}, F_{y2}, F_{x3}, F_{y3} can be written in matrix form as

$$
\mathbf{F}^{e_6} = \frac{EA}{L}
\begin{bmatrix}
\cos^2\theta & \sin\theta\cos\theta & -\cos^2\theta & -\sin\theta\cos\theta \\
\sin\theta\cos\theta & \sin^2\theta & -\sin\theta\cos\theta & -\sin^2\theta \\
-\cos^2\theta & -\sin\theta\cos\theta & \cos^2\theta & \sin\theta\cos\theta \\
-\sin\theta\cos\theta & -\sin^2\theta & \sin\theta\cos\theta & \sin^2\theta
\end{bmatrix}
\begin{bmatrix}
\bar{\delta}_{x2} \\
\bar{\delta}_{y2} \\
\bar{\delta}_{x3} \\
\bar{\delta}_{y3}
\end{bmatrix}.
$$

$$(1.6a)$$

With the factor EA/L taken into the square matrix, Eq. (1.6a) has the form

$$
\mathbf{F}^{e_6} =
\begin{bmatrix}
F_{x2} \\
F_{y2} \\
F_{x3} \\
F_{y3}
\end{bmatrix}
=
\left[
\begin{array}{cc|cc}
k_{x2,x2} & k_{x2,y2} & k_{x2,x3} & k_{x2,y3} \\
k_{y2,x2} & k_{y2,y2} & k_{y2,x3} & k_{y2,y3} \\
\hline
k_{x3,x2} & k_{x3,y2} & k_{x3,x3} & k_{x3,y3} \\
k_{y3,x2} & k_{y3,y2} & k_{y3,x3} & k_{y3,y3}
\end{array}
\right]
\begin{bmatrix}
\bar{\delta}_{x2} \\
\bar{\delta}_{y2} \\
\bar{\delta}_{x3} \\
\bar{\delta}_{y3}
\end{bmatrix}.
$$

$$(1.6b)$$

If the right-hand side matrices are *partitioned* by the dashed lines shown, Eq. (1.6b) can be written as

$$
\mathbf{F}^{e_6} =
\begin{bmatrix}
\mathbf{k}_{22}^{e_6} & \mathbf{k}_{23}^{e_6} \\
\mathbf{k}_{32}^{e_6} & \mathbf{k}_{33}^{e_6}
\end{bmatrix}
\begin{bmatrix}
\bar{\boldsymbol{\delta}}_2 \\
\bar{\boldsymbol{\delta}}_3
\end{bmatrix}.
$$

$$(1.6c)$$

Equation (1.6c) is the *element matrix equation* for element e_6, and its square coefficient matrix is known as the *element k matrix*. Similar equations can be obtained for the other elements. To distinguish the $\mathbf{k}_{\gamma\delta}$ submatrices of the different elements, the superscript notation has to be again introduced. Equation (1.6c) can be expanded to include a listing of all nodal displacements of the system, by an appropriate use of zeros, to give

$$
\mathbf{F}^{e_6} =
\begin{bmatrix}
0 \\
F_2^{e_6} \\
F_3^{e_6} \\
0 \\
0 \\
0
\end{bmatrix}
=
\begin{bmatrix}
0 & 0 & 0 & 0 & 0 & 0 \\
0 & \mathbf{k}_{22}^{e_6} & \mathbf{k}_{23}^{e_6} & 0 & 0 & 0 \\
0 & \mathbf{k}_{32}^{e_6} & \mathbf{k}_{33}^{e_6} & 0 & 0 & 0 \\
0 & 0 & 0 & 0 & 0 & 0 \\
0 & 0 & 0 & 0 & 0 & 0 \\
0 & 0 & 0 & 0 & 0 & 0
\end{bmatrix}
\begin{bmatrix}
\bar{\boldsymbol{\delta}}_1 \\
\bar{\boldsymbol{\delta}}_2 \\
\bar{\boldsymbol{\delta}}_3 \\
\bar{\boldsymbol{\delta}}_4 \\
\bar{\boldsymbol{\delta}}_5 \\
\bar{\boldsymbol{\delta}}_6
\end{bmatrix}.
$$

$$(1.6d)$$

Equation (1.6d) is the *expanded element matrix equation* for element e_6 and can be written in the form

$$
\mathbf{F}^{e_6} = \bar{\mathbf{k}}^{e_6}\boldsymbol{\delta},
$$

$$(1.6e)$$

where $\bar{\mathbf{k}}^{e_6}$ is the *expanded element k matrix* for element e_6, and $\boldsymbol{\delta}$ is the *system nodal vector*.

The external forces $\mathbf{R}_1, \mathbf{R}_2, \ldots, \mathbf{R}_6$ can be resolved into their x, y components $R_{x1}, R_{y1}; R_{x2}, R_{y2}; \ldots; R_{x6}, R_{y6}$ and the conditions for equilibrium at the nodes expressed in terms of these components. At node 2, for example, the condition for equilibrium in the x direction is

$$R_{x2} = F_{x2}^{e_1} + F_{x2}^{e_4} + F_{x2}^{e_5} + F_{x2}^{e_6}. \tag{1.7}$$

Although it is clear that only those elements that contain node 2 contribute to the right-hand side of Eq. (1.7), it is convenient to write the relationship more generally in either of the following ways:

$$R_{x2} = \sum_{i=1}^{7} F_{x2}^{e_i}, \qquad R_{x2} = \sum_{e=1}^{7} F_{x2}^{e}. \tag{1.8a,b}$$

A similar relationship applies to the other component R_{y2} of $\mathbf{R}_2$,

$$R_{y2} = \sum_{e=1}^{7} F_{y2}^{e}. \tag{1.8c}$$

Equations (1.8b) and (1.8c) can be summarized in matrix form as

$$\mathbf{R}_2 = \begin{bmatrix} R_{x2} \\ R_{y2} \end{bmatrix} = \sum_{e=1}^{7} \begin{bmatrix} F_{x2}^{e} \\ F_{y2}^{e} \end{bmatrix} = \sum_{e=1}^{7} \mathbf{F}_2^{e}. \tag{1.9}$$

Similar equations can be obtained for the other nodes and the resulting set of equilibrium relations written as

$$\mathbf{R} = \begin{bmatrix} \mathbf{R}_1 \\ \mathbf{R}_2 \\ \vdots \\ \mathbf{R}_6 \end{bmatrix} = \sum_{e=1}^{7} \begin{bmatrix} \mathbf{F}_1^{e} \\ \mathbf{F}_2^{e} \\ \vdots \\ \mathbf{F}_6^{e} \end{bmatrix} = \sum_{e=1}^{7} \mathbf{F}^{e}. \tag{1.10}$$

Substitution of the equations of type (1.6e) into Eq. (1.10) yields

$$\mathbf{R} = \sum_{e=1}^{7} \bar{\mathbf{k}}^{e} \boldsymbol{\delta}, \tag{1.11}$$

or

$$\mathbf{K}\boldsymbol{\delta} = \mathbf{R}, \tag{1.12}$$

where Eq. (1.12) is known as the *system matrix equation* and the *system* $\mathbf{K}$ *matrix* is given by

$$\mathbf{K} = \sum_{e=1}^{7} \bar{\mathbf{k}}^{e}. \tag{1.13}$$

The procedure used above for assembling the element matrix equations into the system matrix equation is known as *assembly by elements*.

Consideration of this assembly process (which is essentially the addition of expanded element $\mathbf{k}$ matrices) shows that the elements $\mathbf{K}_{\gamma\delta}$ of the system matrix $\mathbf{K}$ are given by

$$\mathbf{K}_{\gamma\delta} = \sum_{e=1}^{7} \mathbf{k}_{\gamma\delta}^{e}. \tag{1.14}$$

It will be noted that $\mathbf{k}_{\gamma\delta}^{e} = 0$ unless *both* γ and δ are nodes of element e. In Eq. (1.14), since γ, δ are system *node number* subscripts, $\mathbf{k}_{\gamma\delta}$ can be taken either as referring to an expanded element matrix equation [such as Eq. (1.6d)] or to an element matrix equation [such as Eq. (1.6c)].

At this point it should be recalled that the matrix elements $\mathbf{k}_{\gamma\delta}^{e}$ are submatrices [see Eqs. (1.6b) and (1.6c)]. This does not present any difficulty since summation of submatrices is carried out by summation of their corresponding elements.

Since the dimensions and properties of the bars are known, all the matrix elements $k_{\gamma\delta}^{e}$ can be computed using equations such as Eqs. (1.6), and the system matrix equation assembled via Eq. (1.14). For the frame considered in Fig. 1.1, the displacements $\bar{\delta}_{1}, \bar{\delta}_{5}, \bar{\delta}_{6}$ are known to be zero. If the applied loads $\mathbf{R}_{2}, \mathbf{R}_{3}$, and $\mathbf{R}_{4}$ are known explicitly, the set of linear algebraic equations given by Eq. (1.12) can be solved by successive substitution, matrix inversion, or iteration to give the unknown displacements $\bar{\delta}_{2}, \bar{\delta}_{3}$, and $\bar{\delta}_{4}$, and the reactions $\mathbf{R}_{1}, \mathbf{R}_{5}$, and $\mathbf{R}_{6}$.

The method of structural analysis that has been outlined above is known as the *direct stiffness method* and can be extended to allow for (a) initial strains due to assembly or temperature, (b) body forces such as those due to gravity, and (c) distributed loads applied to the bars. Three additional column vectors must then be added to Eq. (1.12), which becomes

$$\mathbf{K}\boldsymbol{\delta} + \mathbf{F}_{\varepsilon 0} + \mathbf{F}_{b} + \mathbf{F}_{d} = \mathbf{R}, \tag{1.15}$$

where $\mathbf{F}_{\varepsilon 0}, \mathbf{F}_{b}, \mathbf{F}_{d}$ are derived from the assembly/temperature stresses, the body forces, and the distributed loads, respectively.

The analysis can be extended further to three-dimensional frames and to cases where both forces and moments are transmitted through the nodes (rigid joints).

Illustrative Example 1.1 For the pin-jointed frame shown in Fig. 1.3, compute the deflections $\bar{\delta}_{x2}, \bar{\delta}_{y2}$ at node 2, assuming that the bars are each 10 in. long and have a cross section of 1 in.^{2}. Young's modulus E for the material is 30×10^{6} lb/in.^{2}.

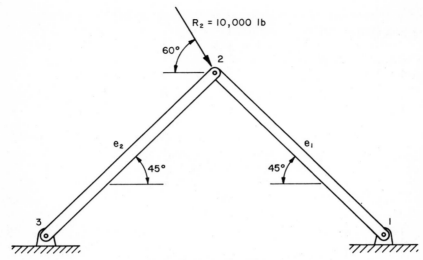

Fig. 1.3 Pin-jointed frame.

The load diagram for the problem is shown in Fig. 1.4a, and a free body diagram for a typical element e_1 in Fig. 1.4b. For element 1, from Eq. (1.3), the force applied to the bar is

$$P = \frac{EA_1}{L_1}[(\bar{\delta}_{x2} - \bar{\delta}_{x1})\cos 135 + (\bar{\delta}_{y2} - \bar{\delta}_{y1})\sin 135], \qquad (1.16)$$

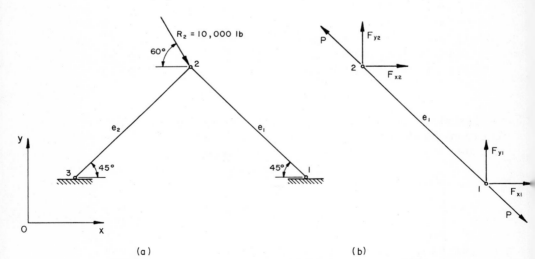

(a)   (b)

Fig. 1.4 Pin-jointed frame: (a) load diagram (b) free body diagram.

where it is assumed that a consistent set of physical units is used. Hence, using Eqs. (1.4) and (1.6a), there is obtained

$$\mathbf{F}^{e_1} = \begin{bmatrix} F_{x1} \\ F_{y1} \\ F_{x2} \\ F_{y2} \end{bmatrix} = \begin{bmatrix} -P\cos 135 \\ -P\sin 135 \\ P\cos 135 \\ P\sin 135 \end{bmatrix} = 15 \times 10^5 \begin{bmatrix} 1 & -1 & -1 & 1 \\ -1 & 1 & 1 & -1 \\ -1 & 1 & 1 & -1 \\ 1 & -1 & -1 & 1 \end{bmatrix} \begin{bmatrix} \bar{\delta}_{x1} \\ \bar{\delta}_{y1} \\ \bar{\delta}_{x2} \\ \bar{\delta}_{y2} \end{bmatrix}.$$

$$(1.17)$$

For element 2, there is similarly obtained the relations

$$P = \frac{EA_2}{L_2}[(\bar{\delta}_{x2} - \bar{\delta}_{x3})\cos 45 + (\bar{\delta}_{y2} - \bar{\delta}_{y3})\sin 45], \qquad (1.18)$$

$$\mathbf{F}^{e_2} = \begin{bmatrix} F_{x3} \\ F_{y3} \\ F_{x2} \\ F_{y2} \end{bmatrix} = \begin{bmatrix} -P\cos 45 \\ -P\sin 45 \\ P\cos 45 \\ P\sin 45 \end{bmatrix} = 15 \times 10^5 \begin{bmatrix} 1 & 1 & -1 & -1 \\ 1 & 1 & -1 & -1 \\ -1 & -1 & 1 & 1 \\ -1 & -1 & 1 & 1 \end{bmatrix} \begin{bmatrix} \bar{\delta}_{x3} \\ \bar{\delta}_{y3} \\ \bar{\delta}_{x2} \\ \bar{\delta}_{y2} \end{bmatrix}.$$

$$(1.19)$$

To illustrate the subsequent assembly process more clearly, Eq. (1.19) is rearranged so that the node numbering in its matrices follows the same ascending sequence as in Eq. (1.17), that is,

$$\mathbf{F}^{e_2} = \begin{bmatrix} F_{x2} \\ F_{y2} \\ F_{x3} \\ F_{y3} \end{bmatrix} = \begin{bmatrix} P\cos 45 \\ P\sin 45 \\ -P\cos 45 \\ -P\sin 45 \end{bmatrix} = 15 \times 10^5 \begin{bmatrix} 1 & 1 & -1 & -1 \\ 1 & 1 & -1 & -1 \\ -1 & -1 & 1 & 1 \\ -1 & -1 & 1 & 1 \end{bmatrix} \begin{bmatrix} \bar{\delta}_{x2} \\ \bar{\delta}_{y2} \\ \bar{\delta}_{x3} \\ \bar{\delta}_{y3} \end{bmatrix}.$$

$$(1.20)$$

Expanding Eqs. (1.17) and (1.20) to system size and *assembling* the resulting equations *by elements* according to Eq. (1.11), yields the system matrix equation

$$\begin{bmatrix} R_{x1} \\ R_{y1} \\ R_{x2} \\ R_{y2} \\ R_{x3} \\ R_{y3} \end{bmatrix} = 15 \times 10^5 \begin{bmatrix} 1 & -1 & -1 & 1 & 0 & 0 \\ -1 & 1 & 1 & -1 & 0 & 0 \\ -1 & 1 & 2 & 0 & -1 & -1 \\ 1 & -1 & 0 & 2 & -1 & -1 \\ 0 & 0 & -1 & -1 & 1 & 1 \\ 0 & 0 & -1 & -1 & 1 & 1 \end{bmatrix} \begin{bmatrix} \bar{\delta}_{x1} \\ \bar{\delta}_{y1} \\ \bar{\delta}_{x2} \\ \bar{\delta}_{y2} \\ \bar{\delta}_{x3} \\ \bar{\delta}_{y3} \end{bmatrix}. \qquad (1.21)$$

Since

$$
\begin{aligned}
R_{x2} &= 10{,}000 \cos 60 = 5000, \\
R_{y2} &= -10{,}000 \sin 60 = 8660, \\
\bar{\delta}_{x1} &= \bar{\delta}_{y1} = \bar{\delta}_{x3} = \bar{\delta}_{y3} = 0,
\end{aligned}
\tag{1.22}
$$

Eq. (1.21) can be written as

$$
15 \times 10^5
\begin{bmatrix}
1 & -1 & -1 & 1 & 0 & 0 \\
-1 & 1 & 1 & -1 & 0 & 0 \\
-1 & 1 & 2 & 0 & -1 & -1 \\
1 & -1 & 0 & 2 & -1 & -1 \\
0 & 0 & -1 & -1 & 1 & 1 \\
0 & 0 & -1 & -1 & 1 & 1
\end{bmatrix}
\begin{bmatrix}
0 \\
0 \\
\bar{\delta}_{x2} \\
\bar{\delta}_{y2} \\
0 \\
0
\end{bmatrix}
=
\begin{bmatrix}
R_{x1} \\
R_{y1} \\
5000 \\
-8660 \\
R_{x3} \\
R_{y3}
\end{bmatrix}.
\tag{1.23}
$$

Partitioning the matrices in Eq. (1.23) as indicated by the dashed lines then allows $\bar{\delta}_{x2}$ and $\bar{\delta}_{y2}$ to be solved from

$$
15 \times 10^5
\begin{bmatrix}
2 & 0 \\
0 & 2
\end{bmatrix}
\begin{bmatrix}
\bar{\delta}_{x2} \\
\bar{\delta}_{y2}
\end{bmatrix}
=
\begin{bmatrix}
5000 \\
-8660
\end{bmatrix},
\tag{1.24}
$$

as

$$
\bar{\delta}_{x2} = \frac{5000}{30 \times 10^5} = 1.667 \times 10^{-3} \text{ in.},
$$

$$
\bar{\delta}_{y2} = -\frac{8660}{30 \times 10^5} = -2.889 \times 10^{-3} \text{ in.}
\tag{1.25}
$$

Substituting Eqs. (1.25) into Eq. (1.23) finally results in

$$
\begin{bmatrix}
R_{x1} \\
R_{y1} \\
R_{x3} \\
R_{y3}
\end{bmatrix}
=
\begin{bmatrix}
-1 & 1 \\
1 & -1 \\
-1 & -1 \\
-1 & -1
\end{bmatrix}
\begin{bmatrix}
2500 \\
-4330
\end{bmatrix},
\tag{1.26}
$$

which can be solved to give

$$
\begin{aligned}
R_{x1} &= -6830 \text{ lb}, & R_{y1} &= 6830 \text{ lb}, \\
R_{x3} &= 1830 \text{ lb}, & R_{y3} &= 1830 \text{ lb}.
\end{aligned}
\tag{1.27}
$$

These results may be checked by noting that

$$
\sum_{i=1}^{3} R_{xi} = -6830 + 5000 + 1830 = 0,
$$

$$
\sum_{i=1}^{3} R_{yi} = 6830 - 8660 + 1830 = 0,
\tag{1.28}
$$

thus verifying that the conditions of overall equilibrium for the frame are satisfied.

1.1.3 Networks

The matrix equations for a network of interconnected elements are similar in form to those derived in the previous section for structural assemblages. To illustrate, consider the hydraulic pipe network shown in Fig. 1.5. For slow flows (that is, laminar conditions), the flow Q through a pipe is proportional to the pressure difference between the ends of the pipe. Thus for element e_6 (Fig. 1.6) the flows *into* this pipe section at nodes 2 and 3, respectively, are

$$Q_2^{e_6} = c^{e_6}(\bar{p}_2 - \bar{p}_3), \qquad Q_3^{e_6} = -c^{e_6}(\bar{p}_2 - \bar{p}_3), \qquad (1.29)$$

where $\bar{p}_2$ and $\bar{p}_3$ are the pressures at nodes 2 and 3, Q_2 and Q_3 are the flows at the same nodes, and c is a constant dependent on the fluid properties, the pipe diameter, and its length.

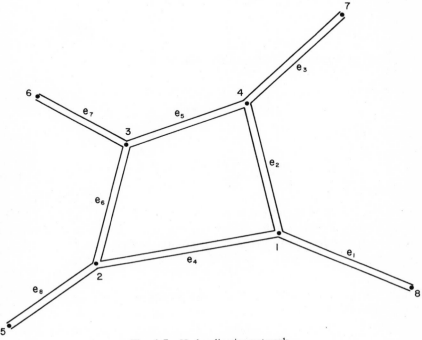

Fig. 1.5 Hydraulic pipe network.

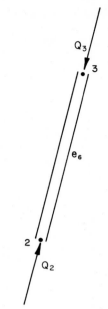

Fig. 1.6 Pipe element e_6.

In matrix form Eqs. (1.29) can be written as

$$\mathbf{Q}^{e_6} = \begin{bmatrix} Q_2 \\ Q_3 \end{bmatrix}^{e_6} = \begin{bmatrix} c^{e_6} & -c^{e_6} \\ -c^{e_6} & c^{e_6} \end{bmatrix} \begin{bmatrix} \bar{p}_2 \\ \bar{p}_3 \end{bmatrix}, \tag{1.30a}$$

or alternatively as

$$\mathbf{Q}^{e_6} = \begin{bmatrix} k_{22}^{e_6} & k_{23}^{e_6} \\ k_{32}^{e_6} & k_{33}^{e_6} \end{bmatrix} \begin{bmatrix} \bar{p}_2 \\ \bar{p}_3 \end{bmatrix}. \tag{1.30b}$$

Equation (1.30b) is the *element matrix equation* for element e_6, [cf. Eq. (1.6c)]. It can also be written as the *expanded element matrix equation* identical in form to Eq. (1.6d) or in the form corresponding to Eq. (1.6e)

$$\mathbf{Q}^{e_6} = \bar{\mathbf{k}}^{e_6}\mathbf{p}, \tag{1.31}$$

where $\mathbf{p}$ is the system nodal vector of the pressures $\bar{p}_1, \bar{p}_2, \ldots, \bar{p}_8$.

Suppose fluid is supplied *to* the network at the nodes $1, 2, \ldots, 8$, with the respective injected flows being $R_1, R_2, \ldots, R_8$. The continuity condition at node 2, for example, becomes

$$R_2 = \sum_{i=1}^{8} Q_2^{e_i} = 0 + 0 + 0 + Q_2^{e_4} + 0 + Q_2^{e_6} + 0 + Q_2^{e_8}. \quad (1.32)$$

For each node, the set of equations of type (1.32) can be expressed [cf. Eq. (1.10)] as

$$\mathbf{R} = \begin{bmatrix} R_1 \\ R_2 \\ \vdots \\ R_8 \end{bmatrix} = \sum_{e=1}^{8} \begin{bmatrix} Q_1^{e} \\ Q_2^{e} \\ \vdots \\ Q_8^{e} \end{bmatrix} = \sum_{e=1}^{8} \mathbf{Q}^e. \quad (1.33)$$

Substitution of the equations of type (1.31) into Eq. (1.33) leads, as in the previous section, to the *system matrix equation*

$$\mathbf{Kp} = \mathbf{R}. \quad (1.34)$$

If $\mathbf{p}$ is substituted for $\boldsymbol{\delta}$, Eqs. (1.10)–(1.14) are also applicable to the present case.

For known values of the injected flows, the nodal pressures can be obtained by solution of Eq. (1.34). The flows through each pipe can then be computed using equations such as Eqs. (1.30).

Exercise 1.1   The pipe network shown in Fig. 1.5 represents a portion of the water distribution system for a village. Because of damage to the supply dam, pressure in the system is so low that the local engineer assumes laminar flow will exist throughout. On this basis, calculate the outflows at the pipe nodes 6, 7, and 8 for the following data:

element e_8 (the supply main) $d = 2$ in., $l = 4000$ ft

elements e_1, e_3, e_5: $d = 1$ in., $l = 3000$ ft

elements e_2, e_4, e_6, e_7: $d = 1$ in., $l = 1500$ ft

pressure at node 5   30 lb/in.^{2}

pressure at nodes 6, 7, 8:   atmospheric (14.7 lb/in.^{2})

The pressure drop Δp for laminar flow in a pipe of diameter d and length l is given by the Hagen–Poiseuille equation $\Delta p = 32\mu l v/d^2$ where μ is the dynamic viscosity and v is the average velocity, with consistent units being used. Assume that the water in the system is at a constant temperature of 60°F and the dynamic viscosity is 2.35×10^{-5} lb sec/ft^2. All pipes are horizontal and junction losses are to be neglected.

Experiment has shown that laminar flow can exist in a pipe only up to a Reynolds number $\mathrm{Re} = \rho v d/\mu$ of approximately 2000. Above this value, the relationship between pressure and flow becomes nonlinear. Taking the density ρ of water as 1.94 slugs/ft³, check whether the engineer's assumption of laminar flow in all pipe sections was correct. (*Answer:* Outflows at nodes 6, 7, 8 are .03059, .01276, .01786 ft³/sec; Reynolds numbers range from 3198 in e_5 to 41,903 in e_6 and the laminar flow assumption is incorrect.)

Exercise 1.2   Given the dc circuit with constant voltage source V shown in Fig. 1.7, develop element equations (as in Section 1.2.3) and assemble these to form the system matrix equation. Hence find the nodal voltages V_1 and V_2, and thence the currents I_1, I_2, and I_3. (*Answer:* $V_1 = 29.268$ V, $V_2 = 4.878$ V, $I_1 = 3.537$ A, $I_2 = 0.610$ A, $I_3 = 2.927$ A).

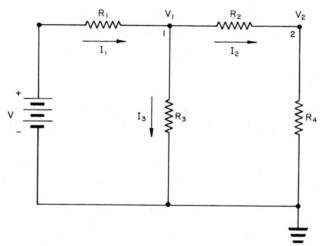

Fig. 1.7   DC circuit with constant voltage source V: $R_1 = 20\,\Omega$, $R_2 = 40\,\Omega$, $R_3 = 10\,\Omega$, $R_4 = 8\,\Omega$, $V = 100$ V.

Exercise 1.3   A simple pin-jointed frame is shown in Fig. 1.8, where

element e_1: $A = 2$ in.², $L = 6$ ft

element e_2: $A = 1$ in.², $L = $ (to be determined)

element e_3: $A = 1$ in.², $L = 8$ ft

element e_4: $A = 2$ in.², $L = 6$ ft.

E is the same for all elements and has the value 30×10^6 lb/in.².

$R_{x2} = 200$ lb, $R_{y2} = 0$ lb

$R_{x3} = 0$ lb, $R_{y3} = 400$ lb

Determine the unknown deflections and reactions.

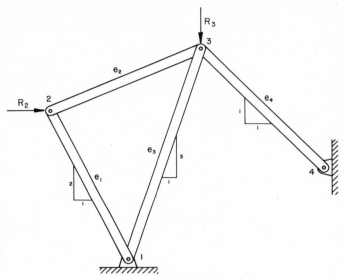

Fig. 1.8 Simple pin-jointed frame.

Exercise 1.4 Given the unbalanced bridge network with a source of known current I shown in Fig. 1.9, find the current I_4, through the conductance G_4 using node 3 as the reference point for voltage.

This problem should first be solved by assembling element equations as in Section 1.2.3. As a check, it can then be reworked by writing down directly the component (nodal) equations of the system matrix equation, using Kirchhoff's law. Choosing a node as a reference point is equivalent to taking the datum voltage as zero at that point. (*Answer:* $I_4 = 0.792$ A).

Exercise 1.5 For the structure shown in Fig. 1.10, determine the unknown reactions and deflections given the following data:

element e_1: $A = 1$ in.^{2}, $L = 7$ ft.
element e_2: $A = 2$ in.^{2}, $L = 8$ ft.
element e_3: $A = 1$ in.^{2}, $L = 4$ ft.
element e_4: $A = 2$ in.^{2}, $L = $ (to be determined)
element e_5: $A = 1$ in.^{2}, $L = $ (to be determined)
element e_6: $A = 2$ in.^{2}, $L = 5$ ft,
element e_7: $A = 3$ in.^{2}, $L = 8$ ft,

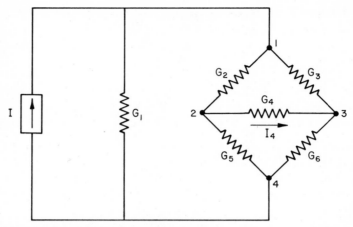

Fig. 1.9   Unbalanced bridge network with constant current source I: $G_1 = 3$ mhos, $G_2 = 2$ mhos, $G_3 = 1$ mho, $G_4 = 1$ mho, $G_5 = 1$ mho, $G_6 = 1$ mho, $I = 80$ A.

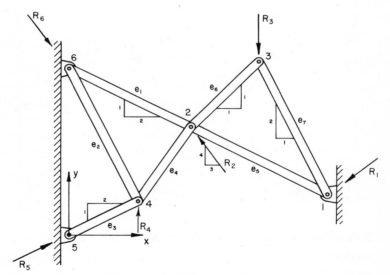

Fig. 1.10   Plane pin-jointed structure.

E is the same for all elements and has the value 30×10^6 lb/in.^{2}.

$R_{x2} = -300$ lb, $R_{y2} = 400$ lb

$R_{x3} = 0$ lb, $R_{y3} = -400$ lb

$R_{x4} = 0$ lb, $R_{y4} = 200$ lb

(*Answer:* $R_{x1} = -132.63$ lb, $R_{y1} = -130.05$ lb, $R_{x5} = 368.75$ lb, $R_{y5} = 184.38$ lb, $R_{x6} = 63.88$ lb, $R_{y6} = -254.33$ lb, $\bar{\delta}_{x2} = -0.0023398$ in., $\bar{\delta}_{y2} = -0.0001321$ in., $\bar{\delta}_{x4} = -0.0002844$ in., $\bar{\delta}_{y4} = -0.0009063$ in.)

Exercise 1.6   Given the network with a source of known current I shown in Fig. 1.11, find the currents I_1, I_2, and I_4 through the conductances G_1, G_2, and G_4, using point 7 as reference point. This problem should be solved using both of the approaches indicated in Exercise 1.4. (*Answer:* $I_1 = 100$ A, $I_2 = 65$ A, $I_4 = 35$ A.)

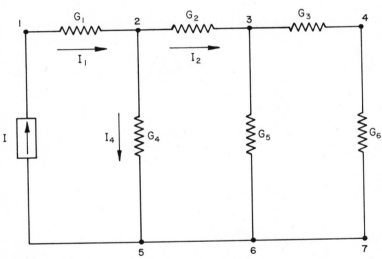

Fig. 1.11   Network with constant current source I: $G_1 = 8$ mhos, $G_2 = 2$ mhos, $G_3 = 4$ mhos, $G_4 = 2$ mhos, $G_5 = 1$ mho, $G_6 = 2$ mhos, $I = 100$ A.

1.2   EVOLUTION OF THE FINITE ELEMENT METHOD

In Section 1.1 the direct stiffness method of structural analysis was used to formulate the system matrix equation for a simple pin-jointed assembly. The method can be extended to structures where the elements are other than simple bars, in which case the force–deflection relationships in Eqs. (1.5) will be more complex but still of similar form to Eqs. (1.6). The extension to the three-dimensional case is quite straightforward, but there is a corresponding increase in the size of the matrices. Nonetheless, for even the most complicated case, the general form of the system matrix equation is that of Eq. (1.15). By combining the $\mathbf{F}$ and $\mathbf{R}$ vectors, the final system matrix equations can be reduced to the standard form of Eq. (1.12).

The direct stiffness method, however, was not the first matrix method to be devised for structural analysis. The inverse formulation involving *flexibilities*, with forces as primary unknowns instead of displacements, was already highly developed by the early 1950s when stiffness approaches first began to appear. A significant contribution was made in 1954/5 by Argyris [3] who showed that the system matrix equation for either the flexibility or stiffness method could be obtained by minimizing the potential energy of the system. Subsequently, the emphasis swung to energy (variational) formulations, using stiffness matrices, with a natural alliance being made with the then rapidly evolving computer, allowing the solution of many large-scale practical problems with an accuracy not previously possible.

The story of the finite element method in engineering begins back in the early 1950s. Attempts were then being made to apply the matrix methods that were successful with discrete structures to continuous structures by subdividing the structure into a finite number of hypothetical elements. In 1956 the group formed by Turner at the Boeing Aircraft Company described [4] a procedure of this type embodying several of the key features of the finite element method.[†] Rapid development of this approach to cover a wide range of problems in structural engineering and solid mechanics then followed. The extension of the finite element method to nonstructural problems began in the early 1960s, using a variational approach. More recently, in addition to the variational finite element method, which might be called the "classical" formulation, other finite element methods have come into use. The best known of these are the *Galerkin method* which is, itself, a particular case of the *weighted-residual method*, the *least-squares method*, and a procedure known variously as the *direct method*, the *global balance method*, or *Oden's method*.

Some of the areas in which the finite element method has been applied are: aircraft, automotive, and ship structures; steel and reinforced-concrete bridges; building structures; earthquake response of reservoir–dam systems; rock mechanics; plasticity design; fracture mechanics; dynamics of sub-

---

[†] Separately, in applied mathematics, the variational finite element method (though not known by that name) was already under development. In 1943 Courant [5] described a solution procedure based on the (variational) principle of minimum potential energy and using linear approximations over triangular elements. Some of the basic concepts in the finite element method were subsequently used by Pólya, Prager, Synge, and others, but it was not until the 1960s that the mathematical and engineering streams of development began to intermingle.

merged structures; fibrous composites; viscous, subsonic, and supersonic flows; oscillating aerofoils; sonar transducers; acoustic fields; electromagnetic fields; magnet design; plasma flows; nuclear reactor fluxes; glacier ice flow; tectonic plate movements; seepage and ground water flows; oil and gas reservoir engineering; biomechanics; sap movement in trees; pollutant dispersion in tidal estuaries; surface waves; spontaneous ignition; statistics.

The rapid growth of the finite element method is shown by the increase in citations listed in finite element bibliographies over the past decade:

Compiler(s)	Year of publication	Number of citations	Reference
Singhal	1969	775	6
Akin, Fenton, Stoddart	1972	1096	7
Norrie, de Vries	1974	2800	8
Whiteman	1975	2166	9
Norrie, de Vries	1975	3800	10
Norrie, de Vries	1976	7115	11

1.3   A ONE-DIMENSIONAL EXAMPLE OF THE VARIATIONAL FINITE ELEMENT METHOD

Only the variational, residual, and direct finite element methods will be considered in this textbook, although other formulations do exist [12, 13]. To begin, a simple one-dimensional example using the variational approach is presented.

Consider a unit mass moving under the influence of a force that varies proportionally with the distance y traveled. If the proportionality constant is unity, the motion is governed by the equation

$$\frac{d^2 y}{dt^2} - y = 0. \tag{1.35}$$

It is required to find the distance traveled at various times between $t = 0$ and $t = 2$ given that

$$y|_{t=0} = y(0) = 1, \qquad y|_{t=2} = y(2) = e^2 = 7.389. \tag{1.36}$$

1.3.1   Basic Formulation

In the variational finite element method, the governing equation is not used directly, but an equivalent *variational formulation* is used instead. For the present problem, it can be shown using the variational calculus that the solution y to Eq. (1.35) is identical to the function that causes the *functional*

$$\chi = \frac{1}{2} \int_0^2 \left[\hat{y}^2 + \left(\frac{d\hat{y}}{dt} \right)^2 \right] dt, \tag{1.37}$$

to assume its minimum value, provided that the *trial functions* $\hat{y}$ used in Eq. (1.37) are continuous, have piecewise continuous first derivatives, and satisfy the principal boundary conditions [Eqs. (1.36)].

The region of the problem, $0 \le t \le 2$, is subdivided into *n finite elements* $e_1, e_2, \ldots, e_n$ and y is approximated by a trial function $\hat{y}^{e_i}$ within each element (Fig. 1.12). The functional χ [Eq. (1.37)] can be written as the following sum of element integrals

$$\chi = \frac{1}{2} \int_{t_1}^{t_2} \left[(\hat{y})^2 + \left(\frac{d\hat{y}}{dt}\right)^2 \right]^{e_1} dt + \frac{1}{2} \int_{t_2}^{t_3} \left[(\hat{y})^2 + \left(\frac{d\hat{y}}{dt}\right)^2 \right]^{e_2} dt + \cdots$$

$$+ \frac{1}{2} \int_{t_i}^{t_{i+1}} \left[(\hat{y})^2 + \left(\frac{d\hat{y}}{dt}\right)^2 \right]^{e_i} dt + \cdots + \frac{1}{2} \int_{t_{n-1}}^{t_n} \left[(\hat{y})^2 + \left(\frac{d\hat{y}}{dt}\right)^2 \right]^{e_{n-1}} dt$$

$$+ \frac{1}{2} \int_{t_n}^{t_{n+1}} \left[(\hat{y})^2 + \left(\frac{d\hat{y}}{dt}\right)^2 \right]^{e_n} dt, \tag{1.38}$$

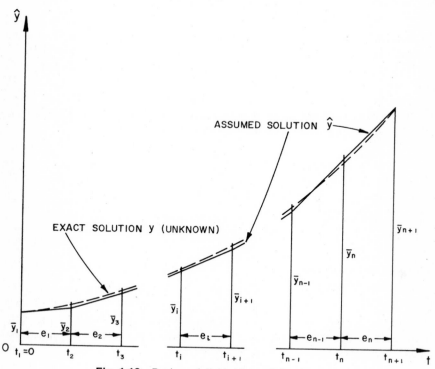

Fig. 1.12   Region subdivided into *n* finite elements.

or simply as

$$\chi = \sum_{i=1}^{n} \frac{1}{2} \int_{t_i}^{t_{i+1}} \left[(\hat{y})^2 + \left(\frac{d\hat{y}}{dt}\right)^2 \right]^{e_i} dt. \tag{1.39}$$

Defining an *element contribution* χ^{e_i} as

$$\chi^{e_i} = \frac{1}{2} \int_{t_i}^{t_{i+1}} \left[(\hat{y})^2 + \left(\frac{d\hat{y}}{dt}\right)^2 \right]^{e_i} dt \tag{1.40}$$

allows Eq. (1.39) to be written as

$$\chi = \sum_{i=1}^{n} \chi^{e_i}. \tag{1.41}$$

Choosing a linear trial function $\hat{y}^{e_i}$ gives

$$\hat{y}^{e_i} = \alpha_1^{e_i} + \alpha_2^{e_i} t, \qquad t_i \le t \le t_{i+1}, \tag{1.42}$$

where $\alpha_1^{e_i}$, $\alpha_2^{e_i}$ are constants that can be determined from the nodal conditions

$$\hat{y}^{e_i}(t_i) = \bar{y}_i, \qquad \hat{y}^{e_i}(t_{i+1}) = \bar{y}_{i+1}, \tag{1.43}$$

as

$$\alpha_1^{e_i} = \frac{t_{i+1}\bar{y}_i - t_i\bar{y}_{i+1}}{t_{i+1} - t_i}, \qquad \alpha_2^{e_i} = \frac{\bar{y}_{i+1} - \bar{y}_i}{t_{i+1} - t_i}. \tag{1.44}$$

Substitution of Eqs. (1.44) into Eq. (1.42) now yields the element approximation for y in the form

$$\hat{y}^{e_i} = \left(\frac{t_{i+1} - t}{t_{i+1} - t_i}\right)\bar{y}_i + \left(\frac{t - t_i}{t_{i+1} - t_i}\right)\bar{y}_{i+1}, \qquad t_i \le t \le t_{i+1}. \tag{1.45}$$

Defining the coefficients of $\bar{y}_i$ and $\bar{y}_{i+1}$ in Eq. (1.45) as *shape functions* $N_i^{e_i}$ and $N_{i+1}^{e_i}$ allows Eq. (1.45) to be written as the *shape function equation*

$$\hat{y}^{e_i} = N_i^{e_i}\bar{y}_i + N_{i+1}^{e_i}\bar{y}_{i+1}, \qquad t_i \le t \le t_{i+1}. \tag{1.46a}$$

It should be noted that the shape functions $N_i^{e_i}$ and $N_{i+1}^{e_i}$ are functions of *only* the *independent variable t* and the *nodal coordinates*.

Outside the element e_i, $\hat{y}^{e_i}$ is given by

$$\hat{y}^{e_i} = 0, \qquad t_{i+1} < t < t_i. \tag{1.46b}$$

In matrix form Eq. (1.46a) can be written as

$$\hat{y}^{e_i} = \mathbf{N}^{e_i}\mathbf{y}^{e_i}, \qquad t_i \le t \le t_{i+1}, \tag{1.47}$$

where

$$\mathbf{N}^{e_i} = \begin{bmatrix} N_i^{e_i} & N_{i+1}^{e_i} \end{bmatrix} = \begin{bmatrix} N_i & N_{i+1} \end{bmatrix}^{e_i}, \qquad (1.48)$$

$$\mathbf{y}^{e_i} = \begin{bmatrix} \bar{y}_i \\ \bar{y}_{i+1} \end{bmatrix}^{e_i}. \qquad (1.49)$$

$\mathbf{N}^{e_i}$ and $\mathbf{y}^{e_i}$ are respectively known as the *shape function matrix* and the *nodal vector* for element e_i.

If there were more than two nodes per element, with only the nodal values of the function specified at each node, the shape function equation (1.47) would retain the same form. The size of the matrices $\mathbf{N}_i^e$ and $\mathbf{y}_i^e$ however would increase to equal the number of nodal values used for the element.

If the elements $e_1, e_2, \ldots, e_n$ are chosen to be of the same length,[†] then

$$t_{i+1} - t_i = \frac{t_{n+1} - t_1}{n} \qquad \text{for} \quad i = 1, 2, \ldots, n. \qquad (1.50)$$

Since $t_1 = 0$ and $t_{n+1} = 2$, Eq. (1.50) reduces to

$$t_{i+1} - t_i = 2/n. \qquad (1.51)$$

Insertion of Eq. (1.51) into Eq. (1.45) allows the element trial function $\hat{y}^{e_i}$ to be written as

$$y^{e_i} = \frac{n}{2} [(t_{i+1} - t)\bar{y}_i + (t - t_i)\bar{y}_{i+1}], \qquad t_i \le t \le t_{i+1}. \qquad (1.52)$$

Substituting Eq. (1.52) into Eq. (1.40) yields the element contribution

$$\chi^{e_i} = \frac{n^2}{8} \int_{t_i}^{t_{i+1}} \{[(t_{i+1} - t)\bar{y}_i + (t - t_i)\bar{y}_{i+1}]^2 + (-\bar{y}_i + \bar{y}_{i+1})^2\} \, dt. \qquad (1.53)$$

Integrating Eq. (1.53) results in

$$\chi^{e_i} = \frac{n^2}{8} \left\{ \frac{[(t_{i+1} - t)\bar{y}_i + (t - t_i)\bar{y}_{i+1}]^3}{3(\bar{y}_{i+1} - \bar{y}_i)} + (-\bar{y}_i + \bar{y}_{i+1})^2 t \right\} \Bigg|_{t_i}^{t_{i+1}}, \qquad (1.54)$$

which, by virtue of Eq. (1.51), reduces to

$$\chi^{e_i} = \frac{n}{4} \left[\left(\frac{4}{3n^2} + 1 \right) \bar{y}_i^2 + \left(\frac{4}{3n^2} - 2 \right) \bar{y}_i \bar{y}_{i+1} + \left(\frac{4}{3n^2} + 1 \right) \bar{y}_{i+1}^2 \right]. \qquad (1.55)$$

[†] This reduces the computation, but is not a necessary requirement (see, for example, Exercise 1.8).

If the appropriate node numbers for each element are used successively in Eq. (1.55) and the resulting equations substituted into Eq. (1.41), χ will be seen to be a *function* of the nodal values $\bar{y}_1, \bar{y}_2, \ldots, \bar{y}_{n+1}$, that is,

$$\chi = \chi(\bar{y}_1, \bar{y}_2, \ldots, \bar{y}_i, \ldots, \bar{y}_n, \bar{y}_{n+1}). \tag{1.56}$$

From the calculus of variations, conditions for χ to be a minimum are

$$\frac{\partial \chi}{\partial \bar{y}_p} = 0, \qquad p = 2, 3, \ldots, n. \tag{1.57}$$

The *system node numbers* $p = 1$ and $n + 1$ do not appear in Eq. (1.57) since the *nodal values* $\bar{y}_1$ and $\bar{y}_{n+1}$ can be substituted as constants from the boundary conditions [Eqs. (1.36)]. There are $n - 1$ equations of the type (1.57), which, together with the boundary conditions

$$\bar{y}_1 = 1, \tag{1.58a}$$

$$\bar{y}_{n+1} = 7.389, \tag{1.58b}$$

give a set of $n + 1$ equations that can be solved for the $n + 1$ nodal values $\bar{y}_1, \bar{y}_2, \ldots, \bar{y}_{n+1}$. A common practice, however, is to obtain the minimization conditions for all nodes, that is,

$$\frac{\partial \chi}{\partial \bar{y}_p} = 0, \qquad p = 1, 2, \ldots, n + 1, \tag{1.59}$$

and to replace the appropriate equations by the prescribed boundary conditions subsequently. This approach will be followed here. In terms of element contributions Eqs. (1.59) become from Eq. (1.41)

$$\frac{\partial \chi}{\partial \bar{y}_p} = \sum_{i=1}^{n} \frac{\partial \chi^{e_i}}{\partial \bar{y}_p} = 0, \qquad p = 1, 2, \ldots, n + 1. \tag{1.60}$$

The summation in Eq. (1.60) is taken over all elements, although the summation only needs to be taken over the elements *surrounding*, that is, adjacent to p, since the contributions from all other elements are zero. For example, for the nodal value $\bar{y}_i$, the only contributing elements in Eq. (1.60) are e_{i-1} and e_i, since these are the only elements that contain the nodal variable $\bar{y}_i$; see Eq. (1.55) and Fig. 1.12.

For a typical element e_i, there is obtained from Eq. (1.55)

$$\frac{\partial \chi^{e_i}}{\partial \bar{y}_i} = \frac{n}{4}\left[2\left(\frac{4}{3n^2} + 1\right)\bar{y}_i + \left(\frac{4}{3n^2} - 2\right)\bar{y}_{i+1}\right]. \tag{1.61}$$

$$\frac{n}{4}\begin{bmatrix}
2\left(\frac{4}{3n^2}+1\right) & \left(\frac{4}{3n^2}-2\right) & & & & & \\
\left(\frac{4}{3n^2}-2\right) & 4\left(\frac{4}{3n^2}+1\right) & \left(\frac{4}{3n^2}-2\right) & & & & \\
 & \left(\frac{4}{3n^2}-2\right) & 4\left(\frac{4}{3n^2}+1\right) & \left(\frac{4}{3n^2}-2\right) & & & \\
 & & \left(\frac{4}{3n^2}-2\right) & 4\left(\frac{4}{3n^2}+1\right) & \left(\frac{4}{3n^2}-2\right) & & \\
 & & & & \ddots & & \\
 & & & & \left(\frac{4}{3n^2}-2\right) & 4\left(\frac{4}{3n^2}+1\right) & \left(\frac{4}{3n^2}-2\right) \\
 & & & & & \left(\frac{4}{3n^2}-2\right) & 4\left(\frac{4}{3n^2}+1\right) & \left(\frac{4}{3n^2}-2\right) \\
 & & & & & & \left(\frac{4}{3n^2}-2\right) & 2\left(\frac{4}{3n^2}+1\right)
\end{bmatrix}
\begin{bmatrix}
\bar{y}_1 \\ \bar{y}_2 \\ \bar{y}_3 \\ \bar{y}_4 \\ \cdots \\ \bar{y}_{n-1} \\ \bar{y}_n \\ \bar{y}_{n+1}
\end{bmatrix}
=
\begin{bmatrix}
0 \\ 0 \\ 0 \\ 0 \\ \cdots \\ 0 \\ 0 \\ 0
\end{bmatrix}
\qquad (1.66)$$

Similarly it can be shown that for element e_{i-1}

$$\frac{\partial \chi^{e_{i-1}}}{\partial \bar{y}_i} = \frac{n}{4}\left[\left(\frac{4}{3n^2} - 2\right)\bar{y}_{i-1} + 2\left(\frac{4}{3n^2} + 1\right)\bar{y}_i\right]. \qquad (1.62)$$

Substituting Eqs. (1.61) and (1.62) into Eq. (1.60) and noting that e_i and e_{i-1} are the only elements contributing to the summation of Eq. (1.60), allows the minimization condition Eq. (1.59) for $p = i$ to be obtained as

$$\frac{\partial \chi}{\partial \bar{y}_i} = \frac{n}{4}\left[\left(\frac{4}{3n^2} - 2\right)\bar{y}_{i-1} + 4\left(\frac{4}{3n^2} + 1\right)\bar{y}_i + \left(\frac{4}{3n^2} - 2\right)\bar{y}_{i+1}\right] = 0. \qquad (1.63)$$

It can simply be shown that Eq. (1.63) holds for $i = 2, 3, \ldots, n$; whereas for $i = 1$ and $i = n + 1$,

$$\frac{\partial \chi}{\partial \bar{y}_1} = \frac{n}{4}\left[2\left(\frac{4}{3n^2} + 1\right)\bar{y}_1 + \left(\frac{4}{3n^2} - 2\right)\bar{y}_2\right] = 0, \qquad (1.64)$$

$$\frac{\partial \chi}{\partial \bar{y}_{n+1}} = \frac{n}{4}\left[\left(\frac{4}{3n^2} - 2\right)\bar{y}_n + 2\left(\frac{4}{3n^2} + 1\right)\bar{y}_{n+1}\right] = 0, \qquad (1.65)$$

since the only element contributing to $\partial \chi / \partial \bar{y}_1$ is e_1 and to $\partial \chi / \partial \bar{y}_{n+1}$ is e_n. Substituting Eqs. (1.63)–(1.65) into Eq. (1.60) yields a set of $n + 1$ simultaneous linear algebraic equations, which may be written in matrix form as shown in Eq. (1.66), p. 24. To insert the prescribed boundary conditions, the first and $(n + 1)$th equation need to be replaced by Eqs. (1.58a) and (1.58b), respectively. This can be accomplished[†] by first overwriting a one in the diagonal position in the first and $(n + 1)$th rows of the first left-hand side matrix of Eq. (1.66), secondly overwriting zeros in the remaining positions of these two rows, and thirdly replacing the first and $(n + 1)$th elements of the right-hand side matrix by the prescribed values of y at nodal points 1 and $n + 1$. The resulting *system matrix equation* is

$$\begin{bmatrix} 1 & & & & & & & & \\ a & b & a & & & & & & \\ & a & b & a & & & & & \\ & & a & b & a & & & & \\ & & & & \ddots & & & & \\ & & & & & a & b & a & \\ & & & & & & a & b & a \\ & & & & & & & a & b & a \\ & & & & & & & & & 1 \end{bmatrix} \begin{bmatrix} \bar{y}_1 \\ \bar{y}_2 \\ \bar{y}_3 \\ \bar{y}_4 \\ \vdots \\ \bar{y}_{n-2} \\ \bar{y}_{n-1} \\ \bar{y}_n \\ \bar{y}_{n+1} \end{bmatrix} = \begin{bmatrix} 1 \\ 0 \\ 0 \\ 0 \\ \vdots \\ 0 \\ 0 \\ 0 \\ 7.389 \end{bmatrix} \qquad (1.67)$$

[†] See also Note 1 following this section.

where

$$a = \frac{1}{3n} - \frac{n}{2}, \qquad b = \frac{4}{3n} + n. \qquad (1.68)$$

Equation (1.67) can also be written as

$$\mathbf{Ky} = \mathbf{R}, \qquad (1.69)$$

and can be seen to have the same general form as Eqs. (1.12) and (1.34). Equation (1.69) can be solved by successive substitution, matrix inversion, iteration, or more conveniently, by one of the standard library matrix solution procedures available on most computer installations.

Insertion of the Dirichlet conditions as outlined above (see also Note 1) causes the *symmetric* system $\mathbf{K}$ matrix in Eq. (1.66) to *lose* its symmetry, as can be seen from Eq. (1.67). This symmetry can be retained if the Payne–Irons method [12] for inserting Dirichlet conditions is adopted. Both approaches are summarized below.

Note 1
Rule for Dirichlet Nodes—I If p is a node for which the nodal value is specified explicitly as g_p, that is, $\bar{y}_p = g_p$, then the procedure for inserting this Dirichlet condition into the system matrix equation can be summarized as follows:

Enter zeros in the pth row of the system $\mathbf{K}$ matrix except for a 1 in the diagonal position, and enter g_p in the pth row of the matrix $\mathbf{R}$.

Rule for Dirichlet Nodes—II The Payne–Irons procedure for insertion of a Dirichlet condition into the system matrix equation is the following:

If $\bar{y}_p$ is specified by the Dirichlet condition $\bar{y}_p = g_p$, the diagonal element in the pth row of the $\mathbf{K}$ matrix is multiplied by a very large number, and at the same time the pth element in the $\mathbf{R}$ matrix is replaced by the same large number multiplied by g_p and by the diagonal element.

It should be obvious that the accuracy of the previous formulation and indeed of any finite element solution can be improved by either increasing the number of elements or by using a trial function that approximates the exact solution more closely. It is therefore not surprising that the two most commonly used approaches when a high degree of accuracy is required are (1) the use of many elements of a simple type and (2) the adoption of a more complex[†] element based on a higher order polynomial.

[†] Such an element may have only nodal values as nodal parameters, or have both nodal values and nodal derivatives as nodal parameters.

Note 2

For any element e, the derivatives of the element contribution χ^e with respect to the nodal values can be written as the *element matrix equation*. Thus, in the present example, the relation for $\partial\chi^{e_i}/\partial\bar{y}_i$ from Eq. (1.61) and a similar equation for $\partial\chi^{e_i}/\partial\bar{y}_{i+1}$ can be put in the form

$$\frac{\partial\chi^{e_i}}{\partial\mathbf{y}^{e_i}} = \begin{bmatrix} \partial\chi^{e_i}/\partial\bar{y}_i \\ \partial\chi^{e_i}/\partial\bar{y}_{i+1} \end{bmatrix} = \begin{bmatrix} k^{e_i}_{i,i} & k^{e_i}_{i,i+1} \\ k^{e_i}_{i+1,i} & k^{e_i}_{i+1,i+1} \end{bmatrix} \begin{bmatrix} \bar{y}_i \\ \bar{y}_{i+1} \end{bmatrix} \tag{1.70a}$$

or

$$\frac{\partial\chi^{e_i}}{\partial\mathbf{y}^{e_i}} = \mathbf{k}^{e_i}\mathbf{y}^{e_i} \tag{1.70b}$$

where $\mathbf{y}^{e_i}$ is the *element nodal vector* for the element e_i.

The formulation outlined in this section is often known as the *Ritz finite element method*. When this approach is used, the system $\mathbf{K}$ matrix is symmetric and the system matrix equation is linear provided that the functional for the problem is of quadratic or quadratic-linear[†] form [13].

Exercise 1.7   Show that for the particular problem considered in Section 1.3.1, the matrix elements[‡] $k^e_{\alpha\beta}$ can be expressed in the form

$$k^e_{\alpha\beta} = \int_{D_e}\left(N^e_\alpha N^e_\beta + \frac{dN^e_\alpha}{dt}\frac{dN^e_\beta}{dt}\right)dD_e, \tag{1.71}$$

where $\alpha = 1, 2, \beta = 1, 2$ are the *node identifiers* of element e.

Exercise 1.8   For the problem of Section 1.3.1, show that if the nodes of an element are *identified* as 1 and 2, the coefficients in the element matrix equation (Eq. 1.70) are

$$k_{11} = k_{22} = \frac{h}{3} + \frac{1}{h}, \tag{1.72a}$$

$$k_{12} = k_{21} = \frac{h}{6} - \frac{1}{h} \tag{1.72b}$$

for every element. The length h need not be the same for every element.

Exercise 1.9   Using four elements of equal size in the problem considered in Section 1.3.1, evaluate the element matrix equations with the aid of Eq. (1.70). Assemble these matrix equations *by elements* via Eq. (1.60), and show that

---

† See Appendix A.
‡ For simplicity and where no confusion is possible, e will be used in place of e_i to denote the element number.

the system $\mathbf{K}$ matrix, before insertion of the Dirichlet conditions, is

$$\mathbf{K} = \begin{bmatrix} 52 & -46 & 0 & 0 & 0 \\ -46 & 104 & -46 & 0 & 0 \\ 0 & -46 & 104 & -46 & 0 \\ 0 & 0 & -46 & 104 & -46 \\ 0 & 0 & 0 & -46 & 52 \end{bmatrix}. \qquad (1.73)$$

Alternatively using the Rule for Dirichlet Nodes–I, show that the *final* system matrix $\mathbf{K}$ is given by

$$\mathbf{K} = \begin{bmatrix} 1 & 0 & 0 & 0 & 0 \\ -46 & 104 & -46 & 0 & 0 \\ 0 & -46 & 104 & -46 & 0 \\ 0 & 0 & -46 & 104 & -46 \\ 0 & 0 & 0 & 0 & 1 \end{bmatrix}, \qquad (1.74a)$$

and the corresponding matrix $\mathbf{R}$ by

$$\mathbf{R} = \begin{bmatrix} 1 \\ 0 \\ 0 \\ 0 \\ e^2 \end{bmatrix}. \qquad (1.74b)$$

Using the Rule for Dirichlet Nodes–II, insert the boundary conditions of this problem and demonstrate that the system $\mathbf{K}$ matrix remains symmetric.

1.3.2   Computer Solution

If only a few elements are considered, the finite element solution of the one-dimensional problem of the previous section can be worked through on a hand calculator. If many elements are used, particularly if they are of unequal length so that the element $\mathbf{k}$ matrix is not common to all elements, a computer solution becomes necessary.

The following program for the problem of Section 1.3.1 uses *node identifiers* 1 and 2 in the element matrix equation [see Eq. (1.70)], and the computation of the element k_{mn} is based on Eqs. (1.72). To obtain the system $\mathbf{K}$ matrix the procedure outlined earlier, *assembly by elements*, is used, that is, each element is taken in turn and its element $\mathbf{k}$ matrix added into the system $\mathbf{K}$ matrix immediately after computation. This assembly procedure is equivalent to adding the expanded element matrix equations [see, for example, Eq. (1.6e)] via an assembly equation of the form of Eq. (1.11) or (1.60). Since for this illustrative example the exact solution, $y = e^t$, is known, the percentage errors for each nodal value are printed out as part of the output.

The program was run in time-sharing mode and no attempt was made to optimize the procedure.

1.3.2.1 Computer Program

```
C      .....FINITE ELEMENT METHOD,PROGRAM 1.....
C      SOLUTION OF AN ORDINARY DIFFERENTIAL EQUATION.
C      THE RESULTING SYSTEM MATRIX EQUATION IS SOLVED USING
C      THE STANDARD LIBRARY SUBROUTINE LEQT1F.
C
C      .....THE FOLLOWING IS A LIST OF SYMBOLS USED.....
C
C      NPOIN      -      TOTAL NUMBER OF NODES
C      NELEM      -      TOTAL NUMBER OF ELEMENTS
C      X(I)       -      X COORDINATE OF NODE I
C      STE(M,N)   -      THE M-N TH ELEMENT OF THE ELEMENT K
C                        MATRIX,M AND N BEING NODE IDENTIFIERS
C      ST(I,J)    -      THE I-J TH ELEMENT OF THE ASSEMBLED
C                        SYSTEM K MATRIX
C      RHS(I,1)   -      RIGHT-HAND SIDE MATRIX IN THE
C                        SYSTEM MATRIX EQUATION
C      LEQT1F     -      STANDARD LIBRARY SUBROUTINE USED
C                        TO SOLVE THE SYSTEM MATRIX EQUATION
C      PER(I)     -      PERCENTAGE ERROR IN THE SOLUTION
C                        AT NODE I
C
C      .....THE SYSTEM MATRIX EQUATION TO BE SOLVED IS.....
C
C                   ST * SOLUTION = RHS
C
       PROGRAM PRGM1(INPUT,OUTPUT,TAPE5,TAPE6)
       DIMENSION X(20),ST(20,20),RHS(20,1),STE(2,2),WKAREA(20),PER(20)
C
C      .....THE TOTAL NUMBER OF ELEMENTS IS READ IN.....
       READ(5,10) NELEM
10     FORMAT(I3)
C
C      .....THE TOTAL NUMBER OF NODES IS DETERMINED.....
       NPOIN=NELEM+1
C
C      .....THE X COORDINATES ARE READ IN FOR ALL NODES.....
       READ(5,20) (I,X(I),I=1,NPOIN)
20     FORMAT(9(I3,F4.1))
C
C      .....THE TOTAL NUMBER OF NODES IS PRINTED OUT.....
       WRITE(6,30) NPOIN
30     FORMAT(//////,1X,22H TOTAL NUMBER OF NODES,I3)
C
C      .....THE TOTAL NUMBER OF ELEMENTS IS PRINTED OUT.....
       WRITE(6,40) NELEM
40     FORMAT(1X,25H TOTAL NUMBER OF ELEMENTS,I3)
C
C      .....THE X COORDINATES ARE PRINTED OUT FOR ALL NODES.....
       WRITE(6,50)
50     FORMAT(//,1X,34H THE NODES AND THEIR X COORDINATES,/)
       WRITE(6,60)
60     FORMAT(1X,10H NODE    X,4(14H      NODE    X))
       WRITE(6,70) (I,X(I),I=1,NPOIN)
70     FORMAT((1X,I4,F7.2,4(I7,F7.2)))
```

```
C
C       .....THE SYSTEM K MATRIX AND THE RIGHT-HAND SIDE
C            MATRIX ARE INITIALIZED TO ZERO.....
        DO 90 I=1,NPOIN
        DO 80 J=1,NPOIN
80      ST(I,J)=0.0
90      RHS(I,1)=0.0
C
C       .....THE ELEMENT K MATRICES ARE OBTAINED AND ASSEMBLED
C            FOR ALL ELEMENTS AND THE SYSTEM K MATRIX IS
C            OBTAINED.....
        DO 100 I=1,NELEM
        COEF=X(I+1)-X(I)
        STE(1,1)=COEF/3.0+1.0/COEF
        STE(1,2)=COEF/6.0-1.0/COEF
        STE(2,1)=STE(1,2)
        STE(2,2)=STE(1,1)
        ST(I,I)=ST(I,I)+STE(1,1)
        ST(I,I+1)=ST(I,I+1)+STE(1,2)
        ST(I+1,I)=ST(I+1,I)+STE(2,1)
100     ST(I+1,I+1)=ST(I+1,I+1)+STE(2,2)
C
C       .....THE DIRICHLET BOUNDARY CONDITIONS ARE INSERTED
C            IN THE RIGHT-HAND SIDE MATRIX AND THE SYSTEM
C            K MATRIX IS CORRECTED.....
        INCR=NPOIN-1
        DO 120 I=1,NPOIN,INCR
        DO 110 J=1,NPOIN
110     ST(I,J)=0.0
        ST(I,I)=1.0
120     RHS(I,1)=EXP(X(I))
C
C       .....THE SYSTEM MATRIX EQUATION IS SOLVED USING
C            THE STANDARD LIBRARY SUBROUTINE LEQT1F.....
        MM=1
        IDGT=0
        NN=20
        CALL LEQT1F(ST,MM,NPOIN,NN,RHS,IDGT,WKAREA,IER)
C
C       .....THE SOLUTION HAS BEEN OBTAINED AND IS PRINTED OUT.....
        WRITE(6,130)
130     FORMAT(//,1X,31H THE SOLUTION HAS BEEN OBTAINED)
        WRITE(6,140)
140     FORMAT(//,1X,36H THE NODES AND THEIR FUNCTION VALUES,/)
        WRITE(6,150)
150     FORMAT(1X,14H NODE    VALUE,3(18H     NODE    VALUE))
        WRITE(6,160) (I,RHS(I,1),I=1,NPOIN)
160     FORMAT((1X,I4,F10.3,3(I8,F10.3)))
C
C       .....THE PERCENTAGE ERRORS ARE CALCULATED AND PRINTED OUT.....
        DO 170 I=1,NPOIN
170     PER(I)=ABS((RHS(I,1)-EXP(X(I)))/EXP(X(I)))*100.0
        WRITE(6,180)
180     FORMAT(//,1X,26H THE PERCENTAGE ERRORS ARE,/)
        WRITE(6,190)
190     FORMAT(1X,13H NODE    PER,3(16H     NODE    PER))
        WRITE(6,200) (I,PER(I),I=1,NPOIN)
200     FORMAT((1X,I4,F9.2,3(I7,F9.2)))
        STOP
        END
```

To illustrate the use of this program, the result for six unequally spaced elements is given below.

1.3.2.2   Input Data

```
6
1  0.0    2  0.1    3  0.3    4  0.6    5  1.0    6  1.5    7  2.0
```

1.3.2.3   Results

```
THE TOTAL NUMBER OF NODES   7
THE TOTAL NUMBER OF ELEMENTS   6
```

THE NODES AND THEIR X COORDINATES

NODE	X	NODE	X	NODE	X	NODE	X	NODE	X
1	0.00	2	.10	3	.30	4	.60	5	1.00
6	1.50	7	2.00						

THE SOLUTION HAS BEEN OBTAINED

THE NODES AND THEIR FUNCTION VALUES

NODE	VALUE	NODE	VALUE	NODE	VALUE	NODE	VALUE
1	1.000	2	1.103	3	1.344	4	1.810
5	2.699	6	4.462	7	7.389		

THE PERCENTAGE ERRORS ARE

NODE	PER	NODE	PER	NODE	PER	NODE	PER
1	0.00	2	.19	3	.46	4	.67
5	.70	6	.44	7	0.00		

1.3.3   Local Coordinate Formulation

In the previous section many of the basic concepts of the variational finite element method were presented. The matrix formulation was based upon the *fixed Ot* coordinate system shown in Fig. 1.12 and the shape function representation for each element (and subsequent equations) used the same reference frame. Such a *common* system of reference is known as a *global coordinate system.*

An alternative approach allowing more concise formulation uses a *local coordinate system* specific to each element when developing the element matrix equations. The previous problem will be reworked using local coordinates so that the mechanism and the advantages of this approach may become clear.

Figure 1.13 shows both the local coordinate system $\bar{O}\xi$ and the global system Ot for element e_i. The relationship between the two coordinate systems for element e_i is

$$\xi = t - t_i, \qquad t_i \le t \le t_{i+1}. \tag{1.75}$$

where $i = 1, 2, \ldots, n$. Such a relationship between two coordinate descriptions is often referred to as a *transformation.*

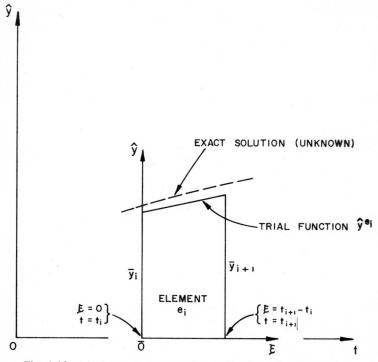

Fig. 1.13   Local coordinate system for the one-dimensional element e_i.

It will be noted that in this case the shift of the origin from O to $\bar{O}$ does not affect the nodal parameters $\bar{y}_i$. In general, the *nodal values* of the function remain unchanged during both a translation and a rotation of coordinate axes. Although *nodal derivatives* remain unchanged for a translation, they *do* change for a rotation.

The length of the element e_i will be denoted by h^{e_i}, where

$$h^{e_i} = t_{i+1} - t_i. \tag{1.76}$$

Choosing a linear approximation for $\hat{y}$ within element e_i allows $\hat{y}^{e_i}$ to be written as

$$\hat{y}^{e_i} = \alpha_1^{e_i} + \alpha_2^{e_i}\xi, \qquad 0 \le \xi \le h^{e_i}, \tag{1.77}$$

where $\alpha_1^{e_i}$, $\alpha_2^{e_i}$ are constants, different for each element, which can be determined in terms of the nodal values $\bar{y}_i$ and $\bar{y}_{i+1}$ at $\xi = 0$ and $\xi = h^{e_i}$, respectively. Applying Eq. (1.77) at $\xi = 0$ and $\xi = h^{e_i}$, gives

$$\bar{y}_i = \alpha_1^{e_i}, \qquad \bar{y}_{i+1} = \alpha_1^{e_i} + \alpha_2^{e_i}h^{e_i}. \tag{1.78}$$

In matrix form Eqs. (1.78) can be written as

$$\begin{bmatrix} 1 & 0 \\ 1 & h \end{bmatrix} \begin{bmatrix} \alpha_1 \\ \alpha_2 \end{bmatrix} = \begin{bmatrix} \bar{y}_i \\ \bar{y}_{i+1} \end{bmatrix}, \tag{1.79}$$

where the superscripts e_i have been dropped for convenience. Equation (1.79) can be written more simply as

$$\mathbf{A}\alpha = \mathbf{y}^e, \tag{1.80}$$

where

$$\mathbf{A} = [a_{mn}] = \begin{bmatrix} 1 & 0 \\ 1 & h \end{bmatrix}, \qquad m = 1, 2, \quad n = 1, 2, \tag{1.81a}$$

$$\alpha = [\alpha_m] = \begin{bmatrix} \alpha_1 \\ \alpha_2 \end{bmatrix}, \qquad m = 1, 2, \tag{1.81b}$$

$$\mathbf{y}^e = [y_m] = \begin{bmatrix} \bar{y}_i \\ \bar{y}_{i+1} \end{bmatrix}, \qquad m = 1, 2, \tag{1.81c}$$

The subscripts m, n correspond to the node identifiers of the element. The determinant of the coefficient matrix $\mathbf{A}$ can be seen to equal h, the length of the element. Since this never equals zero, the matrix is nonsingular and hence can be inverted to give

$$\mathbf{B} = \mathbf{A}^{-1} = [b_{mn}] = \frac{1}{h} \begin{bmatrix} h & 0 \\ -1 & 1 \end{bmatrix}, \qquad m = 1, 2, \quad n = 1, 2. \tag{1.82}$$

Premultiplying Eq. (1.80) by $\mathbf{B}$ results in the *unique* solution

$$\alpha = \mathbf{A}^{-1}\mathbf{y}^e = \mathbf{B}\mathbf{y}^e. \tag{1.83}$$

Using the standard *summation convention* that repeated subscripts denote a sum,[†] a typical element of α can be written as

$$\alpha_m = b_{mn}y_n, \qquad m = 1, 2, \quad n = 1, 2. \tag{1.84}$$

† For example, if $q = 1, 2, \ldots, 4$,

$$C_{pq}d_q = \sum_{q=1}^{4} C_{pq}d_q = C_{p1}d_1 + C_{p2}d_2 + C_{p3}d_3 + C_{p4}d_4$$

and if $k = 1, 2, 3$,

$$a_{kk} = \sum_{k=1}^{3} a_{kk} = a_{11} + a_{22} + a_{33}.$$

Substituting the appropriate matrix elements from Eqs. (1.81) into Eq. (1.84) allows α_1 and α_2 to be determined as

$$\alpha_1 = \bar{y}_i, \qquad \alpha_2 = -\frac{1}{h}\bar{y}_i + \frac{1}{h}\bar{y}_{i+1}. \tag{1.85}$$

The trial function $\hat{y}$ within element e_i in the local system becomes, from Eqs. (1.77) and (1.85),

$$\hat{y}^{e_i} = \bar{y}_i + \left(-\frac{1}{h}\bar{y}_i + \frac{1}{h}\bar{y}_{i+1}\right)\xi, \qquad 0 \le \xi \le h, \tag{1.86a}$$

or

$$\hat{y}^{e_i} = \left(1 - \frac{\xi}{h}\right)\bar{y}_i + \frac{\xi}{h}\bar{y}_{i+1}, \qquad 0 \le \xi \le h. \tag{1.86b}$$

It is, of course, possible to obtain Eq. (1.86b) by simple algebra, but the matrix procedure introduced above is more convenient when there are more than two nodal parameters for each element. Equation (1.86b) can also be written in the shape function form of Eq. (1.46a) but with the shape functions in terms of the local coordinates.

Having obtained the trial function over an element in terms of its nodal parameters [Eqs. (1.86)], this can be substituted into the element form of the functional to obtain the element contribution χ^{e_i}. When the trial function is a polynomial, however, the element contribution χ^{e_i} and the element matrix equation $\partial\chi^{e_i}/\partial\bar{y}$ can be obtained more directly by retaining a series representation for $\hat{y}^{e_i}$ in the analysis. This approach bypasses the explicit determination of the shape functions in Eq. (1.86b) and provides a simple integration for the element contributions χ^{e_i}. Because of these advantages,[†] this approach is adopted in the following; but it should be noted that equivalent element matrix equations are obtained from alternative procedures.

The series representation for $\hat{y}^{e_i}$ [Eq. (1.77)] can be written in the form

$$\hat{y}^{e_i} = \sum_{i=1}^{2} \alpha_i \xi^{m_i}, \tag{1.87}$$

where

$$m_1 = 0, \qquad m_2 = 1, \tag{1.88}$$

and where the superscripts for α_1 and α_2 have been dropped for convenience.

[†] These will become evident if Exercise 1.12 is worked through.

Equation (1.87) can be differentiated with respect to ξ to give

$$\frac{d\hat{y}^{ei}}{d\xi} = \sum_{i=1}^{2} \alpha_i m_i \xi^{m_i - 1}. \tag{1.89}$$

Using the transformation in Eq. (1.75), it is seen that the element contribution χ^{ei} [Eq. (1.40)] becomes in the local system

$$\chi^{ei} = \frac{1}{2} \int_0^h \left[(\hat{y})^2 + \left(\frac{d\hat{y}}{d\xi} \right)^2 \right]^{ei} d\xi. \tag{1.90}$$

Substituting Eqs. (1.87) and (1.89) into Eq. (1.90) yields the typical element contribution

$$\chi^{ei} = \frac{1}{2} \int_0^h \sum_{i=1}^{2} \sum_{j=1}^{2} \left[\alpha_i \alpha_j \xi^{m_i + m_j} + \alpha_i \alpha_j m_i m_j \xi^{m_i + m_j - 2} \right] d\xi, \tag{1.91}$$

which can be written as a summation of integrals by interchanging the order of integration and summation

$$\chi^{ei} = \frac{1}{2} \sum_{i=1}^{2} \sum_{j=1}^{2} \alpha_i \alpha_j \left[\int_0^h \xi^{m_i + m_j} d\xi + \int_0^h m_i m_j \xi^{m_i + m_j - 2} d\xi \right]. \tag{1.92}$$

Introducing the notation

$$g_{ij} = \int_0^h \xi^{m_i + m_j} d\xi + m_i m_j \int_0^h \xi^{m_i + m_j - 2} d\xi, \tag{1.93}$$

allows Eq. (1.92) to be written as

$$\chi^{ei} = \frac{1}{2} \sum_{i=1}^{2} \sum_{j=1}^{2} \alpha_i g_{ij} \alpha_j, \tag{1.94}$$

which is seen to be a *quadratic form* (Appendix A) which can also be put in the form

$$\chi^{ei} = \tfrac{1}{2} \boldsymbol{\alpha}^{\mathrm{T}} \mathbf{G} \boldsymbol{\alpha}, \tag{1.95}$$

where $\boldsymbol{\alpha}$ has been defined in Eq. (1.81b) and $\mathbf{G} = [g_{ij}]$, defined through Eq. (1.93), is known as the *element integration matrix*.

Carrying out the integration in Eq. (1.93) gives a typical element of the matrix $\mathbf{G}$ as

$$g_{ij} = \frac{h^{m_i + m_j + 1}}{m_i + m_j + 1} + m_i m_j \frac{h^{m_i + m_j - 1}}{m_i + m_j - 1}, \tag{1.96}$$

which is seen to be symmetric, and where, of course, the second right-hand side term vanishes whenever m_i or m_j equals zero. Since the values for h,

m_i, and m_j are known, the integration matrix $\mathbf{G}$ can be computed directly from Eq. (1.96).

Substituting Eq. (1.83) into Eq. (1.95) yields the element contribution in terms of its element nodal vector $\mathbf{y}^e$ as

$$\chi^{e_i} = \tfrac{1}{2}(\mathbf{y}^e)^{\mathrm{T}}\mathbf{B}^{\mathrm{T}}\mathbf{GB}(\mathbf{y}^e). \tag{1.97}$$

Differentiating this equation with respect to the nodal vector $\mathbf{y}^e$ (see Appendix B) gives

$$\frac{\partial \chi^{e_i}}{\partial \mathbf{y}^e} = (\mathbf{B}^{\mathrm{T}}\mathbf{GB})\mathbf{y}^e, \tag{1.98}$$

where

$$\frac{\partial \chi^{e_i}}{\partial \mathbf{y}^e} = \begin{bmatrix} \partial \chi^{e_i}/\partial \bar{y}_i \\ \partial \chi^{e_i}/\partial \bar{y}_{i+1} \end{bmatrix}. \tag{1.99}$$

More conveniently, Eq. (1.98) can be written in the form

$$\frac{\partial \chi^{e_i}}{\partial \mathbf{y}^e} = \mathbf{k}\mathbf{y}^e, \tag{1.100}$$

where the element $\mathbf{k}$ matrix is defined as

$$\mathbf{k} = \mathbf{B}^{\mathrm{T}}\mathbf{GB}, \tag{1.101}$$

and where Eq. (1.100) is recognized as the *element matrix equation*.

To illustrate the above procedure, let the region $0 \le t \le 2$ be divided into n equal intervals, so that the length of each element is

$$h = h^{e_i} = 2/n, \qquad i = 1, 2, \ldots, n. \tag{1.102}$$

Substituting Eqs. (1.88) and (1.102) into Eq. (1.96) gives the integration matrix $\mathbf{G}$ for an element as

$$\mathbf{G} = [g_{ij}] = \begin{bmatrix} h & \dfrac{h^2}{2} \\ \dfrac{h^2}{2} & \dfrac{h^3}{3} + h \end{bmatrix} = \frac{h}{6}\begin{bmatrix} 6 & 3h \\ 3h & 2h^2 + 6 \end{bmatrix}. \tag{1.103}$$

Since the length of each element is the same, the evaluated integration matrix $\mathbf{G}$ is common to all elements. Substituting Eqs. (1.82) and (1.103) into Eq. (1.101) gives the element $\mathbf{k}$ matrix as

$$\begin{aligned}
\mathbf{k} &= \frac{1}{6h}\begin{bmatrix} h & -1 \\ 0 & 1 \end{bmatrix}\begin{bmatrix} 6 & 3h \\ 3h & 6 + 2h^2 \end{bmatrix}\begin{bmatrix} h & 0 \\ -1 & 1 \end{bmatrix} \\
&= \frac{1}{6h}\begin{bmatrix} 2h^2 + 6 & h^2 - 6 \\ h^2 - 6 & 2h^2 + 6 \end{bmatrix}. \tag{1.104}
\end{aligned}$$

With the aid of Eq. (1.102), Eq. (1.104) can also be written in terms of n as

$$\mathbf{k} = \frac{1}{6n}\begin{bmatrix} 4 + 3n^2 & 2 - 3n^2 \\ 2 - 3n^2 & 4 + 3n^2 \end{bmatrix}. \tag{1.105}$$

The element matrix equation thus becomes, from Eq. (1.100),

$$\frac{\partial \chi^{e_i}}{\partial y^e} = \frac{1}{6n}\begin{bmatrix} 4 + 3n^2 & 2 - 3n^2 \\ 2 - 3n^2 & 4 + 3n^2 \end{bmatrix}\begin{bmatrix} \bar{y}_i \\ \bar{y}_{i+1} \end{bmatrix}, \tag{1.06}$$

or, if expanded to system size,

$$
\begin{array}{c}
\\
\\
\\
\\
\text{row } i \rightarrow \\
\\
\text{row } i+1 \rightarrow \\
\\
\\
\\
\end{array}
\begin{bmatrix}
0 \\
0 \\
\vdots \\
0 \\
\dfrac{\partial \chi^{e_i}}{\partial \bar{y}_i} \\
\dfrac{\partial \chi^{e_i}}{\partial \bar{y}_{i+1}} \\
0 \\
\vdots \\
0 \\
0
\end{bmatrix}
= \frac{1}{6n}
\begin{bmatrix}
0 & 0 & \cdots & 0 & 0 & 0 & 0 & \cdots & 0 & 0 \\
0 & 0 & \cdots & 0 & 0 & 0 & 0 & \cdots & 0 & 0 \\
\vdots & & & \vdots & \vdots & \vdots & & & \vdots & \\
0 & 0 & \cdots & 0 & 0 & 0 & 0 & \cdots & 0 & 0 \\
0 & 0 & \cdots & 0 & 4 + 3n^2 & 2 - 3n^2 & 0 & \cdots & 0 & 0 \\
0 & 0 & \cdots & 0 & 2 - 3n^2 & 4 + 3n^2 & 0 & \cdots & 0 & 0 \\
0 & 0 & \cdots & 0 & 0 & 0 & 0 & \cdots & 0 & 0 \\
\vdots & \vdots & & \vdots & \vdots & \vdots & & & \vdots & \\
0 & 0 & \cdots & 0 & 0 & 0 & 0 & \cdots & 0 & 0 \\
0 & 0 & \cdots & 0 & 0 & 0 & 0 & \cdots & 0 & 0
\end{bmatrix}
\begin{bmatrix}
\bar{y}_1 \\
\bar{y}_2 \\
\vdots \\
\bar{y}_{i-1} \\
\bar{y}_i \\
\bar{y}_{i+1} \\
\bar{y}_{i+2} \\
\vdots \\
\bar{y}_n \\
\bar{y}_{n+1}
\end{bmatrix}.
\tag{1.107}
$$

(Column i and Column $i+1$ indicated above the matrix.)

Similar results can be obtained for the other elements. In fact, since the $\mathbf{k}$ matrix is identical for every element, the four entries in each expanded element $\mathbf{k}$ matrix will be respectively the same, although appearing in different rows and columns.

Assembling by elements according to Eq. (1.60), which is the same as adding the expanded element matrix equations for all elements, yields the system matrix equation

$$
\frac{1}{6n}
\begin{bmatrix}
4 + 3n^2 & 2 - 3n^2 & & & & & & \\
2 - 3n^2 & 8 + 6n^2 & 2 - 3n^2 & & & & & \\
& 2 - 3n^2 & 8 + 6n^2 & 2 - 3n^2 & & & & \\
& & 2 - 3n^2 & 8 + 6n^2 & 2 - 3n^2 & & & \\
& & & & \ddots & & & \\
& & & & 8 + 6n^2 & 2 - 3n^2 & \\
& & & & 2 - 3n^2 & 8 + 6n^2 & 2 - 3n^2 \\
& & & & & 2 - 3n^2 & 4 + 3n^2
\end{bmatrix}
\begin{bmatrix}
\bar{y}_1 \\
\bar{y}_2 \\
\bar{y}_3 \\
\bar{y}_4 \\
\vdots \\
\bar{y}_{n-1} \\
\bar{y}_n \\
\bar{y}_{n+1}
\end{bmatrix}
=
\begin{bmatrix}
0 \\
0 \\
0 \\
0 \\
\vdots \\
0 \\
0 \\
0
\end{bmatrix}.
\tag{1.108}
$$

Inserting the Dirichlet boundary conditions as before finally gives the *corrected* system matrix equation (1.67).

To further illustrate, let the region be divided into two equal elements, that is, $n = 2$ and $h = 1$. The corrected system matrix equation can be evaluated as

$$\begin{bmatrix} 1 & 0 & 0 \\ -5 & 16 & -5 \\ 0 & 0 & 1 \end{bmatrix} \begin{bmatrix} \bar{y}_1 \\ \bar{y}_2 \\ \bar{y}_3 \end{bmatrix} = \begin{bmatrix} 1 \\ 0 \\ 7.389 \end{bmatrix}, \qquad (1.109)$$

and solved by inversion to give

$$\begin{bmatrix} \bar{y}_1 \\ \bar{y}_2 \\ \bar{y}_3 \end{bmatrix} = \frac{1}{16} \begin{bmatrix} 16 & 0 & 0 \\ 5 & 1 & 5 \\ 0 & 0 & 16 \end{bmatrix} \begin{bmatrix} 1 \\ 0 \\ 7.389 \end{bmatrix} = \begin{bmatrix} 1 \\ 2.622 \\ 7.389 \end{bmatrix}. \qquad (1.110)$$

Increasing the number of elements increases the accuracy of the solution, as is shown in Table 1.1.

TABLE 1.1

Accuracy as a Function of Number of Elements

Number of elements of equal length	Percentage errors in nodal values (internal nodes only)			
2			3.53		
3		1.64		1.11	
4	0.85		0.81		0.47

Exercise 1.10   The shape functions for the trial function $\hat{y}^e$ are seen, from Eq. (1.86b), to be

$$N_1{}^e = 1 - \frac{\xi}{h}, \qquad N_2{}^e = \frac{\xi}{h}. \qquad (1.111)$$

Using 1 and 2 as node identifiers, show that the entries of the element **k** matrix in the local system are

$$k_{ij}^e = \int_0^h \left(N_i^e N_j^e + \frac{dN_i^e}{d\xi} \frac{dN_j^e}{d\xi} \right) d\xi, \qquad (1.112)$$

for the problem considered in Section 1.3.3 and verify that for n equal elements of length h, the entries are the same as those given in Eq. (1.72).

Exercise 1.11   Show that with four elements of equal size, the corrected system matrix equation for the problem considered in Section 1.3.3 becomes

$$\frac{1}{24}\begin{bmatrix} 1 & 0 & 0 & 0 & 0 \\ -23 & 52 & -23 & 0 & 0 \\ 0 & -23 & 52 & -23 & 0 \\ 0 & 0 & -23 & 52 & -23 \\ 0 & 0 & 0 & 0 & 1 \end{bmatrix}\begin{bmatrix} \bar{y}_1 \\ \bar{y}_2 \\ \bar{y}_3 \\ \bar{y}_4 \\ \bar{y}_5 \end{bmatrix} = \begin{bmatrix} 1 \\ 0 \\ 0 \\ 0 \\ 7.389 \end{bmatrix}. \tag{1.113}$$

Solve this equation by successive substitution or other standard technique and verify that the solution is given by

$$\begin{bmatrix} \bar{y}_1 \\ \bar{y}_2 \\ \bar{y}_3 \\ \bar{y}_4 \\ \bar{y}_5 \end{bmatrix} = \begin{bmatrix} 1 \\ 1.649 \\ 2.718 \\ 4.482 \\ 7.389 \end{bmatrix}. \tag{1.114}$$

Exercise 1.12   Using the *quadratic* trial function

$$\hat{y}^{ei} = \alpha_1^{ei} + \alpha_2^{ei}\xi + \alpha_3^{ei}\xi^2, \tag{1.115}$$

and the local coordinate system $\bar{O}\xi$ shown in Fig. 1.14, show that $\mathbf{A}$ and $\mathbf{B}$ corresponding to Eqs. (1.81a) and (1.82), respectively, become

$$\mathbf{A} = \begin{bmatrix} 1 & 0 & 0 \\ 1 & \dfrac{h}{2} & \dfrac{h^2}{4} \\ 1 & h & h^2 \end{bmatrix}, \qquad \mathbf{B} = \frac{1}{h^2}\begin{bmatrix} h^2 & 0 & 0 \\ -3h & 4h & -h \\ 2 & -4 & 2 \end{bmatrix}. \tag{1.116}$$

Also, by deriving the analogous equation to Eq. (1.93), show that the integration matrix $\mathbf{G}$ becomes

$$\mathbf{G} = \frac{h}{60}\begin{bmatrix} 60 & 30h & 20h^2 \\ 30h & 20h^2 + 60 & 15h^3 + 60h \\ 20h^2 & 15h^3 + 60h & 12h^4 + 80h^2 \end{bmatrix}. \tag{1.117}$$

Hence, verify that the element $\mathbf{k}$ matrix is given by

$$\mathbf{k} = \mathbf{B}^T\mathbf{G}\mathbf{B} = \frac{1}{30h}\begin{bmatrix} 4h^2 + 70 & 2h^2 - 80 & -h^2 + 10 \\ 2h^2 - 80 & 16h^2 + 160 & 2h^2 - 80 \\ -h^2 + 10 & 2h^2 - 80 & 4h^2 + 70 \end{bmatrix}. \tag{1.118}$$

Exercise 1.13   Modify the computer program of Section 1.3.2 to use the quadratic element of Exercise 1.12 in place of the linear element. Solve the

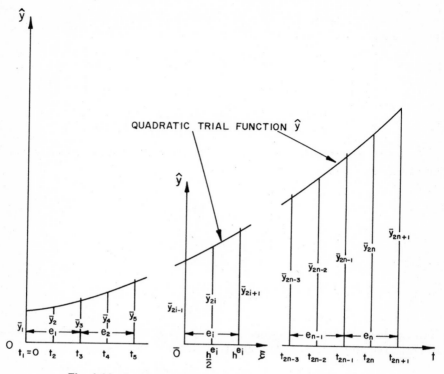

Fig. 1.14 Local coordinate system for the three-node element e_i.

previous problem using two quadratic elements of equal length and verify that the nodal values are given by

$$\begin{bmatrix} \bar{y}_1 \\ \bar{y}_2 \\ \bar{y}_3 \\ \bar{y}_4 \\ \bar{y}_5 \end{bmatrix} = \begin{bmatrix} 1 \\ 1.649 \\ 2.720 \\ 4.480 \\ 7.389 \end{bmatrix}. \tag{1.119}$$

REFERENCES

1. O. C. Zienkiewicz and Y. K. Cheung, Finite elements in the solution of field problems, *The Engineer* pp, 507–510 (September 1965).
2. H. C. Martin and G. F. Carey, A brief history of Finite Element Theory, *in* "Introduction to Finite Element Analysis." McGraw-Hill, New York, 1973.
3. J. H. Argyris, Energy theorems and structural analysis, *Aircraft Engrg.* **26** (October 1954), 347–356, (November 1954), 383–394; **27** (February 1955), 42–58, (March 1955), 80–94,

(April 1955), 125–134, (May 1955), 145–158; these papers formed the basis of the later publication by J. H. Argyris and S. Kelsey, "Energy Theorems and Structural Analysis." Butterworth, London, 1960.

4. M. J. Turner, R. W. Clough, H. C. Martin, and L. J. Topp, Stiffness and deflection analysis of complex structures, *J. Aeronaut. Sci.* **23**, No. 9, 805–824 (September 1956).

5. R. Courant, Variational methods for the solution of problems of equilibrium and vibrations, *Bull. Amer. Math. Soc.* **49**, 1–23 (1943).

6. A. C. Singhal, 775 Selected References on the Finite Element Method and Matrix Methods of Structural Analysis. Rep. S-12, Civil Engrg. Dept., Laval Univ., Quebec (January 1969).

7. J. E. Akin, D. L. Fenton, and W. C. T. Stoddart, The Finite Element Method—A Bibliography of Its Theory and Application. Rep. EM 72-1, Dept. of Engrg. Science and Mech., Univ. of Tennessee, Knoxsville, Tennessee (February 1972).

8. D. H. Norrie and G. de Vries, A Finite Element Bibliography. Part I—Rep. 57, Part II— Rep. 58, Part III—Rep. 59, Dept. of Mech. Engrg., Univ. of Calgary, Alberta (June 1974).

9. J. R. Whiteman, "A Bibliography for Finite Elements." Academic Press, New York, 1975.

10. D. H. Norrie and G. de Vries, A Finite Element Bibliography. Part I—Rep. 57, Part II— Rep. 58, Part III—Rep. 59, Dept. of Mech. Engrg., Univ. of Calgary, Alberta, 2nd ed. (1975).

11. D. H. Norrie and G. de Vries, "A Finite Element Bibliography." Plenum Press, New York, 1976.

12. O. C. Zienkiewicz, "The Finite Element Method in Engineering Science," 2nd ed. McGraw-Hill, New York, 1971.

13. D. H. Norrie and G. de Vries, "The Finite Element Method—Fundamentals and Applications." Academic Press, New York 1973.

2

THE VARIATIONAL
FINITE ELEMENT METHOD

In the previous chapter a one-dimensional example illustrated several basic concepts of the variational finite element method. The extension to two-dimensional problems is now described, taking as a particular case two-dimensional heat flow through a square block. The problem is first formulated for a global frame of reference and subsequently reworked using a local coordinate system.

2.1 A TWO-DIMENSIONAL HEAT FLOW EXAMPLE

2.1.1 Formulation in Global Coordinates

Consider the problem of two-dimensional heat flow through the square block shown in Fig. 2.1. The upper surface of the block is maintained at 100°C, the lower surface at 50°C, and the side walls are perfectly insulated. It is required to determine the temperature distribution in the block and, in particular, the temperature at point A shown in Fig. 2.1. The variational finite element method is used, with the formulation being based on the global coordinates Oxy.

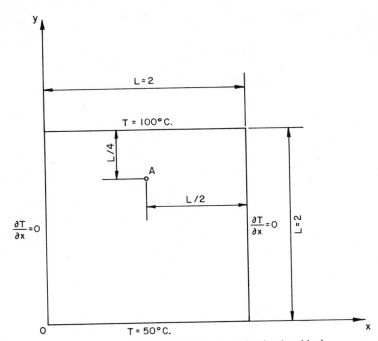

Fig. 2.1   Two-dimensional thermal conduction in a block.

The exact solution, for comparison, can be obtained by inspection since the problem can be regarded as one dimensional. The following formulation is, however, two dimensional and the problem will be solved on a two-dimensional basis. This formulation will also be used in the next chapter to solve a *truly* two-dimensional heat-flow problem.

The governing equation in the present case is Laplace's equation [1, 2]

$$\nabla^2 T = \frac{\partial^2 T}{\partial x^2} + \frac{\partial^2 T}{\partial y^2} = 0 \quad \text{in } D, \tag{2.1}$$

with Dirichlet boundary conditions being prescribed on part of the boundary:

$$T = 50, \qquad y = 0, \tag{2.2a}$$
$$T = 100, \qquad y = L, \tag{2.2b}$$

and Neuman conditions on the remaining part of the boundary:

$$\partial T/\partial x = 0, \qquad x = 0, \tag{2.3a}$$
$$\partial T/\partial x = 0, \qquad x = L, \tag{2.3b}$$

The region R of the problem consists of the domain D and the boundary S, that is, $R = D + S$.

Equations (2.1)–(2.3) will not be employed directly, but instead the *equivalent* variational formulation will be used. It can be shown from the calculus of variations (see Chapter 7) that the solution $T(x, y)$ satisfying Eqs. (2.1)–(2.3) is identical to that function which minimizes the *functional*

$$\chi = \frac{1}{2} \int\!\!\int_D \left[\left(\frac{\partial \hat{T}}{\partial x}\right)^2 + \left(\frac{\partial \hat{T}}{\partial y}\right)^2 \right] dx\,dy, \tag{2.4}$$

where $\hat{T}(x, y)$ are *admissible trial functions* over the domain D. For this problem, the trial functions $\hat{T}(x, y)$ are admissible if they are continuous and have piecewise continuous first derivatives in the domain D. In addition, the trial functions are required to satisfy the *principal boundary conditions* [Eqs. (2.2)]. The Neuman boundary conditions [Eqs. (2.3)] will be satisfied *automatically* by that function minimizing the functional in Eq. (2.4) as a natural consequence of the variational procedure and are therefore referred to as *natural boundary conditions*.

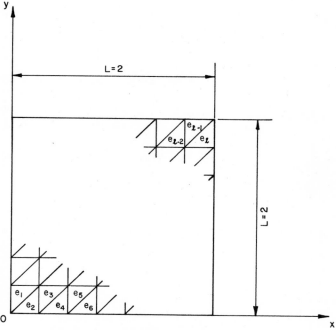

Fig. 2.2 Region subdivided into l finite elements.

The region is subdivided into l finite elements, which in this example are chosen to be triangular; see Fig. 2.2. The total number of nodes is denoted by n. In two- and three-dimensional problems, there is no obvious relationship between the total number of elements and the total number of nodes as was the case in the previous chapter. The subdivision of the region and the continuity conditions imposed on the trial functions allow Eq. (2.4) to be written as

$$\chi = \sum_{i=1}^{l} \chi^{e_i}, \tag{2.5}$$

where the *element contribution* χ^{e_i} is defined as

$$\chi^{e_i} = \frac{1}{2} \int\int_{e_i} \left[\left(\frac{\partial \hat{T}^{e_i}}{\partial x} \right)^2 + \left(\frac{\partial \hat{T}^{e_i}}{\partial y} \right)^2 \right] dx\, dy. \tag{2.6}$$

Consider a typical element e_i shown in Fig. 2.3, where the *node identifiers* i, j, and m have been assigned to the nodes in a counterclockwise sense. For the arbitrary element e_i, the trial function $\hat{T}^{e_i}(x, y)$ in this example is chosen

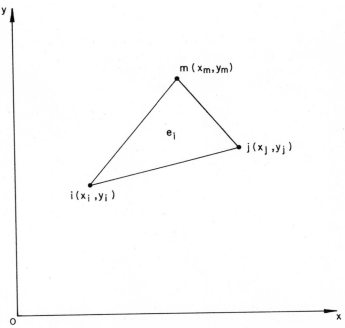

Fig. 2.3 A typical triangular element e_i.

to be linear, that is,

$$\hat{T}^{e_i}(x, y) = \alpha_1^{e_i} + \alpha_2^{e_i}x + \alpha_3^{e_i}y, \qquad x, y \text{ in } e_i, \tag{2.7}$$

where $\alpha_1^{e_i}$, $\alpha_2^{e_i}$, and $\alpha_3^{e_i}$ are constants, in general, different for each element.

To determine these constants, Eq. (2.7) is applied successively to each of the nodes i, j, and m to give

$$\bar{T}_i = \alpha_1 + \alpha_2 x_i + \alpha_3 y_i, \tag{2.8a}$$

$$\bar{T}_j = \alpha_1 + \alpha_2 x_j + \alpha_3 y_j, \tag{2.8b}$$

$$\bar{T}_m = \alpha_1 + \alpha_2 x_m + \alpha_3 y_m, \tag{2.8c}$$

$\bar{T}_i$, $\bar{T}_j$, and $\bar{T}_m$ are the values of T at the nodes i, j, and m, respectively, and the superscripts e_i have been dropped for convenience. The system of equations (2.8) will yield a unique solution for the constants α_1, α_2, and α_3, provided that the determinant of the coefficient matrix does not vanish, that is,

$$2\Delta = \begin{vmatrix} 1 & x_i & y_i \\ 1 & x_j & y_j \\ 1 & x_m & y_m \end{vmatrix} \neq 0. \tag{2.9}$$

It can be simply shown by trigonometry that this determinant equals twice the area Δ of the triangle, as indicated in Eq. (2.9). Since the area of the triangle never equals zero, that is, $\Delta \neq 0$, the solution for α_1, α_2, and α_3 *exists* and is *unique*. Solving Eqs. (2.8) yields α_1, α_2, α_3 as

$$\alpha_1 = \frac{1}{2\Delta}(a_i\bar{T}_i + a_j\bar{T}_j + a_m\bar{T}_m), \tag{2.10a}$$

$$\alpha_2 = \frac{1}{2\Delta}(b_i\bar{T}_i + b_j\bar{T}_j + b_m\bar{T}_m), \tag{2.10b}$$

$$\alpha_3 = \frac{1}{2\Delta}(c_i\bar{T}_i + c_j\bar{T}_j + c_m\bar{T}_m), \tag{2.10c}$$

where

$$a_i = x_j y_m - x_m y_j, \qquad b_i = y_j - y_m, \qquad c_i = x_m - x_j, \tag{2.11}$$

and where $a_j, a_m, b_j, b_m, c_j, c_m$, can be obtained by cyclic permutation of the indices. In the above equations the superscript e_i on the a's, b's, c's, and α's has again been dropped for convenience. Substitution of Eqs. (2.10) into

(2.7) now yields the shape function representation

$$\hat{T}^{ei}(x, y) = \frac{1}{2\Delta}[(a_i + b_i x + c_i y)\bar{T}_i + (a_j + b_j x + c_j y)\bar{T}_j$$

$$+ (a_m + b_m x + c_m y)\bar{T}_m], \tag{2.12a}$$

or

$$\hat{T}^{ei} = N_i\bar{T}_i + N_j\bar{T}_j + N_m\bar{T}_m = \mathbf{N}^e\mathbf{T}^e, \tag{2.12b}$$

where the shape function matrix is defined as

$$\mathbf{N}^e = [N_i \quad N_j \quad N_m], \tag{2.13a}$$

and the element nodal vector as

$$\mathbf{T}^e = \begin{bmatrix} \bar{T}_i \\ \bar{T}_j \\ \bar{T}_m \end{bmatrix}. \tag{2.13b}$$

The required derivatives can be obtained from Eq. (2.12a) as

$$\frac{\partial T^{ei}}{\partial x} = \frac{1}{2\Delta}[b_i\bar{T}_i + b_j\bar{T}_j + b_m\bar{T}_m], \tag{2.14a}$$

$$\frac{\partial T^{ei}}{\partial y} = \frac{1}{2\Delta}[c_i\bar{T}_i + c_j\bar{T}_j + c_m\bar{T}_m]. \tag{2.14b}$$

Substitution of Eqs. (2.14) into the element contribution [Eq. (2.6)] yields

$$\chi^{ei} = \frac{1}{8\Delta^2} \int\int_{ei} [(b_i\bar{T}_i + b_j\bar{T}_j + b_m\bar{T}_m)^2$$

$$+ (c_i\bar{T}_i + c_j\bar{T}_j + c_m\bar{T}_m)^2]\,dx\,dy. \tag{2.15}$$

Since the integrand in Eq. (2.15) is independent of the variables x and y, and since

$$\int\int_{ei} dx\,dy = \Delta, \tag{2.16}$$

Eq. (2.15) may be written as

$$\chi^{ei} = \frac{1}{8\Delta}[(b_i\bar{T}_i + b_j\bar{T}_j + b_m\bar{T}_m)^2 + (c_i\bar{T}_i + c_j\bar{T}_j + c_m\bar{T}_m)^2]. \tag{2.17}$$

An expression of the form of Eq. (2.17) can be obtained for each element. Substituting all these element contributions into Eq. (2.5) transforms the functional in Eq. (2.4) into a *function* of all the nodal values $\bar{T}_1, \bar{T}_2, \ldots, \bar{T}_n$, that is,

$$\chi = \chi(\bar{T}_1, \bar{T}_2, \ldots, \bar{T}_n). \qquad (2.18)$$

For the present, as in the previous chapter, all of the nodal values $\bar{T}_1$, $\bar{T}_2, \ldots, \bar{T}_n$ are treated as variables, including those that are prescribed. The conditions for χ to be a minimum can therefore be written as

$$\frac{\partial \chi}{\partial \bar{T}_p} = 0, \qquad p = 1, 2, \ldots, n. \qquad (2.19)$$

Substitution of Eq. (2.5) into Eq. (2.19) allows this equation to be written as

$$\frac{\partial \chi}{\partial \bar{T}_p} = \sum_{i=1}^{l} \frac{\partial \chi^{e_i}}{\partial \bar{T}_p} = 0, \qquad p = 1, 2, \ldots, n. \qquad (2.20)$$

It is evident that although the summation in Eq. (2.20) is taken over all the elements, only those elements that have node p *in common* have a *nonzero* contribution. Differentiating Eq. (2.17) with respect to the proper $\bar{T}_p$ allows the contribution $\partial \chi^{e_i}/\partial \bar{T}_p$ of element e_i in Eq. (2.20) to be determined. Thus, if the node identifiers i, j, and m of element e_i refer to the system node numbers p, q, and r, respectively, differentiation of Eq. (2.17) with respect to $\bar{T}_p$ yields

$$\frac{\partial \chi^{e_i}}{\partial \bar{T}_p} = \frac{1}{4\Delta} [b_p (b_p \bar{T}_p + b_q \bar{T}_q + b_r \bar{T}_r) + c_p (c_p \bar{T}_p + c_q \bar{T}_q + c_r \bar{T}_r)]. \qquad (2.21)$$

The assembly of the component element equations, prescribed by Eq. (2.20), is an *assembly by nodes* since the assembly process must be carried out separately for each of the nodes of the system. This contrasts with *assembly by elements* used in Section 1.3.1.

The generation of the contributions $\partial \chi^{e_i}/\partial \bar{T}_p$ and their assembly will be illustrated further by subdividing the region of the present problem into 16 elements having a total of 15 nodes, as shown in Fig. 2.4.

Table 2.1 relates the node identifiers i, j, m of each element to the system node numbers. Reference to Fig. 2.4 shows that the node identifiers are allocated counterclockwise, according to a systematic pattern. All elements are of the same size and the assignment of identifiers is such that there are only two types of elements. These are *not* requirements, but simplify the computation, as will become apparent. The x and y coordinates of the nodes are listed in Table 2.2. The parameters b_i, b_j, b_m, c_i, c_j, and c_m, calculated according to Eqs. (2.11), are given in Table 2.3. Consider now, for illustration, element 5. Table 2.1 shows that the node identifiers i, j, and m for this element correspond, respectively, to the system node numbers 7, 4, and 8. Substituting this information and the appropriate parameters for

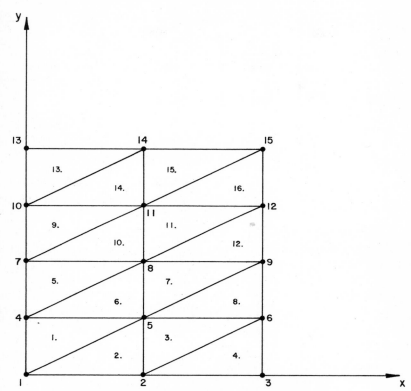

Fig. 2.4   Region subdivided into 16 finite elements.

TABLE 2.1

Relationship between System Node Numbers and Element Node Identifiers

Element	Node Number			Element	Node Number		
	i	j	m		i	j	m
1	4	1	5	9	10	7	11
2	2	5	1	10	8	11	7
3	5	2	6	11	11	8	12
4	3	6	2	12	9	12	8
5	7	4	8	13	13	10	14
6	5	8	4	14	11	14	10
7	8	5	9	15	14	11	15
8	6	9	5	16	12	15	11

TABLE 2.2

Nodal Coordinates

Node	Coordinate x	Coordinate y	Node	Coordinate x	Coordinate y	Node	Coordinate x	Coordinate y
1	0	0	6	2	0.5	11	1	1.5
2	1	0	7	0	1	12	2	1.5
3	2	0	8	1	1	13	0	2
4	0	0.5	9	2	1	14	1	2
5	1	0.5	10	0	1.5	15	2	2

TABLE 2.3

Element Characteristic Dimensions

Element	b_i	b_j	b_m	c_i	c_j	c_m
1	-0.5	0	0.5	1	-1	0
2	0.5	0	-0.5	-1	1	0
3	-0.5	0	0.5	1	-1	0
4	0.5	0	-0.5	-1	1	0
5	-0.5	0	0.5	1	-1	0
6	0.5	0	-0.5	-1	1	0
7	-0.5	0	0.5	1	-1	0
8	0.5	0	-0.5	-1	1	0
9	-0.5	0	0.5	1	-1	0
10	0.5	0	-0.5	-1	1	0
11	-0.5	0	0.5	1	-1	0
12	0.5	0	-0.5	-1	1	0
13	-0.5	0	0.5	1	-1	0
14	0.5	0	-0.5	-1	1	0
15	-0.5	0	0.5	1	-1	0
16	0.5	0	-0.5	-1	1	0

element 5 from Table 2.3 into Eq. (2.17) yields the element contribution as

$$\chi^5 = \frac{1}{8\Delta^5}[(-\tfrac{1}{2}\bar{T}_7 + 0\bar{T}_4 + \tfrac{1}{2}\bar{T}_8)^2 + (1\bar{T}_7 - 1\bar{T}_4 + 0\bar{T}_8)^2]. \quad (2.22)$$

The superscript on the area Δ denotes which element is involved and can be deleted since for this problem the areas are the same for all elements, namely,

$$\Delta^{e_i} = 0.25, \qquad i = 1, 2, \ldots, 16. \quad (2.23)$$

From Eq. (2.22) it is seen that $\partial \chi^{e_i}/\partial \bar{T}_p$ is nonzero only when $p = 7, 4,$ or 8. In other words element 5 is one of the *contributing elements* for the system equations involving $\bar{T}_7$, $\bar{T}_4$, and $\bar{T}_8$. From Eq. (2.22) the only nonzero derivatives for element 5 are

$$\frac{\partial \chi^5}{\partial \bar{T}_7} = \frac{1}{4\Delta}\left[-\tfrac{1}{2}(-\tfrac{1}{2}\bar{T}_7 + 0\bar{T}_4 + \tfrac{1}{2}\bar{T}_8) + 1(1\bar{T}_7 - 1\bar{T}_4 + O\bar{T}_8)\right], \qquad (2.24a)$$

$$\frac{\partial \chi^5}{\partial \bar{T}_4} = \frac{1}{4\Delta}\left[0(-\tfrac{1}{2}\bar{T}_7 + 0\bar{T}_4 + \tfrac{1}{2}\bar{T}_8) - 1(1\bar{T}_7 - 1\bar{T}_4 + 0\bar{T}_8)\right], \qquad (2.24b)$$

$$\frac{\partial \chi^5}{\partial \bar{T}_8} = \frac{1}{4\Delta}\left[\tfrac{1}{2}(-\tfrac{1}{2}\bar{T}_7 + 0\bar{T}_4 + \tfrac{1}{2}\bar{T}_8) + 0(1\bar{T}_7 - 1\bar{T}_4 + 0\bar{T}_8)\right]. \qquad (2.24c)$$

In matrix form Eqs. (2.24) become the *element matrix equation*

$$\begin{bmatrix} \partial \chi^5/\partial \bar{T}_7 \\ \partial \chi^5/\partial \bar{T}_4 \\ \partial \chi^5/\partial \bar{T}_8 \end{bmatrix} = \frac{1}{16\Delta}\begin{bmatrix} 5 & -4 & -1 \\ -4 & 4 & 0 \\ -1 & 0 & 1 \end{bmatrix}\begin{bmatrix} \bar{T}_7 \\ \bar{T}_4 \\ \bar{T}_8 \end{bmatrix}. \qquad (2.25)$$

The three derivatives, with respect to $\bar{T}_7$, $\bar{T}_4$, and $\bar{T}_8$, in Eqs. (2.24) were deliberately written in the order corresponding to the node identifiers i, j, and m for this element (see Table 2.1). Equation (2.25) is thus the particular case, for element 5, of the general form of the element matrix equation

$$\begin{bmatrix} \partial \chi^e/\partial \bar{T}_i \\ \partial \chi^e/\partial \bar{T}_j \\ \partial \chi^e/\partial \bar{T}_m \end{bmatrix} = \begin{bmatrix} k_{ii}^e & k_{ij}^e & k_{im}^e \\ k_{ji}^e & k_{jj}^e & k_{jm}^e \\ k_{mi}^e & k_{mj}^e & k_{mm}^e \end{bmatrix}\begin{bmatrix} \bar{T}_i \\ \bar{T}_j \\ \bar{T}_m \end{bmatrix}, \qquad (2.26a)$$

or

$$\partial \chi^e/\partial \mathbf{T}^e = \mathbf{k}^e\mathbf{T}^e, \qquad (2.26b)$$

where

$$\mathbf{k}^e = \begin{bmatrix} k_{ii}^e & k_{ij}^e & k_{im}^e \\ k_{ji}^e & k_{jj}^e & k_{jm}^e \\ k_{mi}^e & k_{mj}^e & k_{mm}^e \end{bmatrix}, \qquad (2.27)$$

is the *element* **k** *matrix* for element e, and $\mathbf{T}^e$ is the *element nodal vector* defined in Eq. (2.13b). For later convenience, the superscript e_i has been replaced in Eqs. (2.26) and (2.27). This convention will be used from this point onward.

Since the elements $1, 3, 5, 7, 9, 11, 13, 15$ are of identical size and orientation with respect to the frame of reference Oxy, it can be shown that the *element* **k** *matrix* is the *same* for these elements *provided* the node identifiers of Eq. (226a) are replaced by their proper corresponding node numbers. Thus, for element

9, the element matrix equation is

$$
\begin{bmatrix} \partial \chi^9 / \partial \bar{T}_{10} \\ \partial \chi^9 / \partial \bar{T}_7 \\ \partial \chi^9 / \partial \bar{T}_{11} \end{bmatrix} = \frac{1}{16\Delta} \begin{bmatrix} 5 & -4 & -1 \\ -4 & 4 & 0 \\ -1 & 0 & 1 \end{bmatrix} \begin{bmatrix} \bar{T}_{10} \\ \bar{T}_7 \\ \bar{T}_{11} \end{bmatrix}. \tag{2.28}
$$

The remaining elements 2, 4, 6, 8, 10, 12, 14, 16 also have a common element **k** matrix, which for this problem, as may be verified, is identical to that obtained above for the odd-numbered elements. Thus, for element 10, the element matrix equation is

$$
\begin{bmatrix} \partial \chi^{10} / \partial \bar{T}_8 \\ \partial \chi^{10} / \partial \bar{T}_{11} \\ \partial \chi^{10} / \partial \bar{T}_7 \end{bmatrix} = \frac{1}{16\Delta} \begin{bmatrix} 5 & -4 & -1 \\ -4 & 4 & 0 \\ -1 & 0 & 1 \end{bmatrix} \begin{bmatrix} \bar{T}_8 \\ \bar{T}_{11} \\ \bar{T}_7 \end{bmatrix}. \tag{2.29}
$$

Having obtained the contributions $\partial \chi^e / \partial \bar{T}_p$ for all elements, the next step is to assemble these into the system matrix equation. In the previous chapter, assembly by elements was demonstrated and it was seen that this procedure involved calculating the *complete* element **k** matrix for each element in turn. The entries of each element **k** matrix were added (assembled) into the system **K** matrix *before* proceeding to the next element. The alternative procedure, assembly by nodes, is used in the following and will be seen to be distinctly different. Assembly by elements corresponds to assembly of the system matrix by submatrices, whereas assembly by nodes corresponds to assembly by rows.

For assembly by nodes, the basic assembly equation is Eq. (2.20). For node 7, for example, the relevant assembly relation is

$$
\frac{\partial \chi}{\partial \bar{T}_7} = \sum_{e=1}^{l} \frac{\partial \chi^e}{\partial \bar{T}_7}, \qquad l = 16. \tag{2.30a}
$$

Figure 2.4 shows that for node 7 the only contributing elements are 5, 9, and 10 and, consequently, Eq. (2.30a) reduces to

$$
\frac{\partial \chi}{\partial \bar{T}_7} = \frac{\partial \chi^5}{\partial \bar{T}_7} + \frac{\partial \chi^9}{\partial \bar{T}_7} + \frac{\partial \chi^{10}}{\partial \bar{T}_7}. \tag{2.30b}
$$

The contributions of elements 5, 9, and 10, to the right-hand side of Eq. (2.30b) have already been evaluated in Eqs. (2.25), (2.28), and (2.29), respectively. From these equations, substituting the relevant expressions for $\partial \chi^5 / \partial \bar{T}_7$, $\partial \chi^9 / \partial \bar{T}_7$, and $\partial \chi^{10} / \partial \bar{T}_7$ into Eq. (2.30b) gives

$$
\frac{\partial \chi}{\partial \bar{T}_7} = \frac{1}{16\Delta} \left\{ \begin{bmatrix} 5 & -4 & -1 \end{bmatrix} \begin{bmatrix} \bar{T}_7 \\ \bar{T}_4 \\ \bar{T}_8 \end{bmatrix} + \begin{bmatrix} -4 & 4 & 0 \end{bmatrix} \begin{bmatrix} \bar{T}_{10} \\ \bar{T}_7 \\ \bar{T}_{11} \end{bmatrix} \right.
$$

$$
\left. + \begin{bmatrix} -1 & 0 & 1 \end{bmatrix} \begin{bmatrix} \bar{T}_8 \\ \bar{T}_{11} \\ \bar{T}_7 \end{bmatrix} \right\} = 0, \tag{2.31}
$$

which after some manipulation reduces to

$$\frac{\partial \chi}{\partial \bar{T}_7} = \frac{1}{16\Delta}[-4\bar{T}_4 + 10\bar{T}_7 - 2\bar{T}_8 - 4\bar{T}_{10}] = 0. \tag{2.32}$$

Equation (2.32) may also be written in its expanded form as

$$\frac{\partial \chi}{\partial \bar{T}_7} = \frac{1}{16\Delta}[0 \quad 0 \quad 0 \quad -4 \quad 0 \quad 0 \quad 10 \quad -2 \quad 0 \quad -4 \quad 0 \quad 0 \quad 0 \quad 0 \quad 0]\mathbf{T} = 0, \tag{2.33}$$

where the system nodal vector T is defined by

$$\mathbf{T} = \begin{bmatrix} \bar{T}_1 \\ \bar{T}_2 \\ \vdots \\ \bar{T}_n \end{bmatrix}. \tag{2.34}$$

For each nodal parameter in the system, an equation similar to Eq. (2.33) can be obtained. Assembling these equations into a single matrix equation gives

$$\frac{\partial \chi}{\partial \mathbf{T}} = \begin{bmatrix} \partial \chi / \partial \bar{T}_1 \\ \partial \chi / \partial \bar{T}_2 \\ \vdots \\ \partial \chi / \partial \bar{T}_n \end{bmatrix} = \mathbf{KT} = \mathbf{0}, \tag{2.35}$$

which will be recognized as the system matrix equation. As in Chapter 1, this must now be corrected for the Dirichlet boundary conditions. It is left as an exercise to show that, if the Dirichlet boundary conditions are inserted according to Rule I, the final system matrix equation becomes

$$\begin{bmatrix} 1 & 0 & 0 & 0 & 0 & 0 & 0 & 0 & 0 & 0 & 0 & 0 & 0 & 0 & 0 \\ 0 & 1 & 0 & 0 & 0 & 0 & 0 & 0 & 0 & 0 & 0 & 0 & 0 & 0 & 0 \\ 0 & 0 & 1 & 0 & 0 & 0 & 0 & 0 & 0 & 0 & 0 & 0 & 0 & 0 & 0 \\ -4 & 0 & 0 & 10 & -2 & 0 & -4 & 0 & 0 & 0 & 0 & 0 & 0 & 0 & 0 \\ 0 & -8 & 0 & -2 & 20 & -2 & 0 & -8 & 0 & 0 & 0 & 0 & 0 & 0 & 0 \\ 0 & 0 & -4 & 0 & -2 & 10 & 0 & 0 & -4 & 0 & 0 & 0 & 0 & 0 & 0 \\ 0 & 0 & 0 & -4 & 0 & 0 & 10 & -2 & 0 & -4 & 0 & 0 & 0 & 0 & 0 \\ 0 & 0 & 0 & 0 & -8 & 0 & -2 & 20 & -2 & 0 & -8 & 0 & 0 & 0 & 0 \\ 0 & 0 & 0 & 0 & 0 & -4 & 0 & -2 & 10 & 0 & 0 & -4 & 0 & 0 & 0 \\ 0 & 0 & 0 & 0 & 0 & 0 & -4 & 0 & 0 & 10 & -2 & 0 & -4 & 0 & 0 \\ 0 & 0 & 0 & 0 & 0 & 0 & 0 & -8 & 0 & -2 & 20 & -2 & 0 & -8 & 0 \\ 0 & 0 & 0 & 0 & 0 & 0 & 0 & 0 & -4 & 0 & -2 & 10 & 0 & 0 & -4 \\ 0 & 0 & 0 & 0 & 0 & 0 & 0 & 0 & 0 & 0 & 0 & 0 & 1 & 0 & 0 \\ 0 & 0 & 0 & 0 & 0 & 0 & 0 & 0 & 0 & 0 & 0 & 0 & 0 & 1 & 0 \\ 0 & 0 & 0 & 0 & 0 & 0 & 0 & 0 & 0 & 0 & 0 & 0 & 0 & 0 & 1 \end{bmatrix} \begin{bmatrix} \bar{T}_1 \\ \bar{T}_2 \\ \bar{T}_3 \\ \bar{T}_4 \\ \bar{T}_5 \\ \bar{T}_6 \\ \bar{T}_7 \\ \bar{T}_8 \\ \bar{T}_9 \\ \bar{T}_{10} \\ \bar{T}_{11} \\ \bar{T}_{12} \\ \bar{T}_{13} \\ \bar{T}_{14} \\ \bar{T}_{15} \end{bmatrix} = \begin{bmatrix} 50 \\ 50 \\ 50 \\ 0 \\ 0 \\ 0 \\ 0 \\ 0 \\ 0 \\ 0 \\ 0 \\ 0 \\ 100 \\ 100 \\ 100 \end{bmatrix}. \tag{2.36}$$

Using any standard solution procedure, the solution to Eq. (2.36) will be found to be

$$
\begin{aligned}
&\bar{T}_1 = 50, &&\bar{T}_2 = 50, &&\bar{T}_3 = 50, \\
&\bar{T}_4 = 62.5, &&\bar{T}_5 = 62.5, &&\bar{T}_6 = 62.5, \\
&\bar{T}_7 = 75, &&\bar{T}_8 = 75, &&\bar{T}_9 = 75, \\
&\bar{T}_{10} = 87.5, &&\bar{T}_{11} = 87.5, &&\bar{T}_{12} = 87.5, \\
&\bar{T}_{13} = 100, &&\bar{T}_{14} = 100, &&\bar{T}_{15} = 100.
\end{aligned} \tag{2.37}
$$

The required solution at point A is therefore given by

$$
T|_{\text{at } A} \simeq \bar{T}_{11} = 87.5^\circ\text{C}. \tag{2.38}
$$

It will be noted that no special account was taken of the Neumann conditions [Eqs. (2.3a) and (2.3b)] applying at nodes 4, 7, 10 and 6, 9, 12, respectively. As previously stated, minimization at these boundary points is sufficient to impose these conditions as a *natural* consequence.

The question as to how to program assembly by nodes may well be asked at this point. To answer this, first consider Eq. (2.36) and note that Eq. (2.33), corresponding to the seventh node, forms the seventh individual equation in Eq. (2.36). The elements of the row matrix in Eq. (2.33) are thus $K_{7\delta}$, $\delta = 1, 2, \ldots, n$. More generally, it is seen that the γth component equation in the system matrix equation is the nodal equation

$$
\frac{\partial \chi}{\partial \bar{T}_\gamma} = [K_{\gamma 1} \quad K_{\gamma 2} \quad \cdots \quad K_{\gamma \delta} \quad \cdots \quad K_{\gamma n}]\mathbf{T} = 0. \tag{2.39}
$$

Hence, the problem reduces to that of computing the row matrices $\partial \chi / \partial \bar{T}_\gamma$, $\gamma = 1, 2, \ldots, n$.

A study of the development between Eqs. (2.24) and Eq. (2.33) shows that for the row matrix deriving from the nodal value $\bar{T}_\gamma$ [see Eq. (2.39)]

$$
K_{\gamma \delta} = \sum_{e=1}^{l} k_{\gamma \delta}^e, \tag{2.40}
$$

where the summation needs only include those elements surrounding node γ since all other elements yield zero contributions. With an explicit formula[†] by which to calculate $k_{\gamma \delta}^e$, the programming strategy will be to compute a row (in the system matrix equation) corresponding to each node taken in turn, by means of Eq. (2.40).

Alternatively, if assembly by elements were used, the assembly procedure would (conceptually) be the expansion of the element $\mathbf{k}$ matrices and their successive addition into the system $\mathbf{K}$ matrix. Thus, when the element "DO LOOP" reaches element 9, the element $\mathbf{k}$ matrix of Eq. (2.28) would be

[†] Obtainable by substituting Eq. (2.21) into (2.26b). Also see Exercise 2.1.

computed and assembled directly into the system $\mathbf{K}$ matrix using the expanded element $\mathbf{k}$ matrix for that element given by

$$
\begin{bmatrix} \partial\chi^9/\partial\bar{T}_1 \\ \partial\chi^9/\partial\bar{T}_2 \\ \cdot \\ \cdot \\ \cdot \\ \cdot \\ \partial\chi^9/\partial\bar{T}_7 \\ \cdot \\ \cdot \\ \partial\chi^9/\partial\bar{T}_{10} \\ \partial\chi^9/\partial\bar{T}_{11} \\ \cdot \\ \cdot \\ \cdot \\ \partial\chi^9/\partial\bar{T}_{15} \end{bmatrix}
\begin{bmatrix}
0 & 0 & 0 & 0 & 0 & 0 & 0 & 0 & 0 & 0 & 0 & 0 & 0 & 0 & 0 \\
0 & 0 & 0 & 0 & 0 & 0 & 0 & 0 & 0 & 0 & 0 & 0 & 0 & 0 & 0 \\
0 & 0 & 0 & 0 & 0 & 0 & 0 & 0 & 0 & 0 & 0 & 0 & 0 & 0 & 0 \\
0 & 0 & 0 & 0 & 0 & 0 & 0 & 0 & 0 & 0 & 0 & 0 & 0 & 0 & 0 \\
0 & 0 & 0 & 0 & 0 & 0 & 0 & 0 & 0 & 0 & 0 & 0 & 0 & 0 & 0 \\
0 & 0 & 0 & 0 & 0 & 0 & 0 & 0 & 0 & 0 & 0 & 0 & 0 & 0 & 0 \\
0 & 0 & 0 & 0 & 0 & 0 & 4 & 0 & 0 & -4 & 0 & 0 & 0 & 0 & 0 \\
0 & 0 & 0 & 0 & 0 & 0 & 0 & 0 & 0 & 0 & 0 & 0 & 0 & 0 & 0 \\
0 & 0 & 0 & 0 & 0 & 0 & 0 & 0 & 0 & 0 & 0 & 0 & 0 & 0 & 0 \\
0 & 0 & 0 & 0 & 0 & 0 & -4 & 0 & 0 & 5 & -1 & 0 & 0 & 0 & 0 \\
0 & 0 & 0 & 0 & 0 & 0 & 0 & 0 & 0 & -1 & 1 & 0 & 0 & 0 & 0 \\
0 & 0 & 0 & 0 & 0 & 0 & 0 & 0 & 0 & 0 & 0 & 0 & 0 & 0 & 0 \\
0 & 0 & 0 & 0 & 0 & 0 & 0 & 0 & 0 & 0 & 0 & 0 & 0 & 0 & 0 \\
0 & 0 & 0 & 0 & 0 & 0 & 0 & 0 & 0 & 0 & 0 & 0 & 0 & 0 & 0 \\
0 & 0 & 0 & 0 & 0 & 0 & 0 & 0 & 0 & 0 & 0 & 0 & 0 & 0 & 0
\end{bmatrix}
\begin{bmatrix} \bar{T}_1 \\ \bar{T}_2 \\ \cdot \\ \cdot \\ \cdot \\ \cdot \\ \bar{T}_7 \\ \cdot \\ \cdot \\ \bar{T}_{10} \\ \bar{T}_{11} \\ \cdot \\ \cdot \\ \cdot \\ \bar{T}_{15} \end{bmatrix}
$$

(the arrows above columns point at positions 7, 10, 11)

$$(2.41)$$

Exercise 2.1   If 1, 2, and 3 are the node identifiers of the three-node linear triangular element e, show that for the Laplace problem considered in Section 2.1.1 the entries of the element $\mathbf{k}$ matrix are given by

$$k_{\alpha\beta} = \frac{1}{4\Delta}(b_\alpha b_\beta + c_\alpha c_\beta), \qquad \alpha = 1, 2, 3, \quad \beta = 1, 2, 3, \qquad (2.42)$$

where the b's and c's are defined by Eqs. (2.11).

Exercise 2.2   The functional associated with the Poisson equation

$$\frac{\partial^2\phi}{\partial x^2} + \frac{\partial^2\phi}{\partial y^2} + \frac{\partial^2\phi}{\partial z^2} = f(x, y, z) \qquad \text{in } D, \qquad (2.43)$$

subject to the Dirichlet boundary condition

$$\phi = g(x, y, z) \qquad \text{on } S_1 \qquad (2.44)$$

and the Neumann condition[†]

$$\frac{\partial\phi}{\partial n} = n_x \frac{\partial\phi}{\partial x} + n_y \frac{\partial\phi}{\partial y} + n_z \frac{\partial\phi}{\partial z} = 0 \qquad \text{on } S_2, \qquad (2.45)$$

[†] The complete boundary S comprises $S_1 + S_2$. The unit outward normal to the boundary is n and n_x, n_y, n_z are the direction cosines of the unit outward normal.

is given by

$$\chi = \frac{1}{2} \int_D \left[\left(\frac{\partial \phi}{\partial x} \right)^2 + \left(\frac{\partial \phi}{\partial y} \right)^2 + \left(\frac{\partial \phi}{\partial z} \right)^2 + 2f\phi \right] dx\, dy\, dz. \qquad (2.46)$$

For the linear triangular element, show that the element matrix equation is given by

$$\frac{\partial \chi^e}{\partial \mathbf{T}^e} = \mathbf{k}^e \mathbf{T}^e + \mathbf{F}^e, \qquad (2.47)$$

where the elements of the $\mathbf{k}^e$ and $\mathbf{F}^e$ matrices are, respectively, given by

$$k_{\alpha\beta}^e = \int_{D_e} \left(\frac{\partial N_\alpha}{\partial x} \frac{\partial N_\beta}{\partial x} + \frac{\partial N_\alpha}{\partial y} \frac{\partial N_\beta}{\partial y} + \frac{\partial N_\alpha}{\partial z} \frac{\partial N_\beta}{\partial z} \right) dx\, dy\, dz,$$

$$\alpha = 1, 2, 3, \quad \beta = 1, 2, 3, \qquad (2.48)$$

$$F_\alpha^e = \int_{D_e} f N_\alpha\, dx\, dy\, dz, \qquad \alpha = 1, 2, 3, \qquad (2.49)$$

and where 1, 2, and 3 are the node identifiers of element e. (Note: The Dirichlet boundary condition [Eq. (2.44)] is the *principal* boundary condition for this problem and must be used to correct the system matrix equation in the usual way. The Neumann boundary condition [Eq. (2.45)] is a natural boundary condition and is satisfied automatically in the final solution, provided that the minimization conditions are applied at the nodes where this condition is prescribed.)

2.1.2   Formulation in Local Coordinates

The two-dimensional problem of Fig. 2.1 is now reformulated using a local coordinate system in place of the global system adopted previously. Instead of the linear trial function, the quadratic approximation function is chosen. Six-node triangular elements with three vertex and three midside nodes are used (Fig. 2.5) with the nodal value of the function being the only variable at each node. The numbers 1, 2, ..., 6 are used as node identifiers for these elements, assigned in a counterclockwise sense.

In terms of the global coordinates x and y, the second-order polynomial trial function for the six-node triangular element is

$$\hat{T}^e = \alpha_1 + \alpha_2 x + \alpha_3 y + \alpha_4 x^2 + \alpha_5 xy + \alpha_6 y^2, \qquad (2.50)$$

where $\alpha_1, \alpha_2, \ldots, \alpha_6$ are constants, in general different for each element. For convenience, the superscripts e have been dropped on the right-hand side of Eq. (2.50).

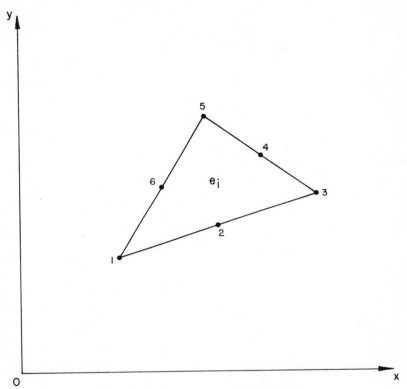

Fig. 2.5 The six-node triangular element.

If the local coordinate system $\bar{O}\xi\eta$ shown in Fig. 2.6 is used, the quadratic trial function is written as

$$\hat{T}^e = \alpha_1 + \alpha_2\xi + \alpha_3\eta + \alpha_4\xi^2 + \alpha_5\xi\eta + \alpha_6\eta^2, \qquad (2.51)$$

where again the superscripts on the α_i have been dropped.

It will be seen subsequently that the characteristic dimensions a, b, c of triangle e (see Fig. 2.6) play a useful role in the local coordinate formulation. These dimensions can be calculated from the known global coordinates of the nodes 1, 3, 5 in the following way. From trigonometry, there can be written

$$r = [(x_3 - x_1)^2 + (y_3 - y_1)^2]^{1/2}, \qquad (2.52)$$

$$\cos\theta = (x_3 - x_1)/r, \qquad (2.53a)$$

$$\sin\theta = (y_3 - y_1)/r, \qquad (2.53b)$$

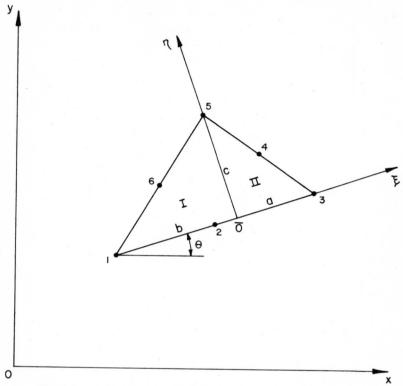

Fig. 2.6   Local coordinate system for the six-node triangular element.

where r is the distance from point 1 to point 3. The characteristic dimensions a, b, and c are given by

$$a = (x_3 - x_5)\cos\theta - (y_5 - y_3)\sin\theta, \qquad (2.54a)$$

$$b = (x_5 - x_1)\cos\theta + (y_5 - y_1)\sin\theta, \qquad (2.54b)$$

$$c = (y_5 - y_3)\cos\theta + (x_3 - x_5)\sin\theta. \qquad (2.54c)$$

Substitution of Eqs. (2.53) into Eqs. (2.54) then yields the characteristic dimensions in the form

$$a = [(x_3 - x_5)(x_3 - x_1) - (y_5 - y_3)(y_3 - y_1)]/r, \qquad (2.55a)$$

$$b = [(x_5 - x_1)(x_3 - x_1) + (y_5 - y_1)(y_3 - y_1)]/r, \qquad (2.55b)$$

$$c = [(y_5 - y_3)(x_3 - x_1) + (x_3 - x_5)(y_3 - y_1)]/r, \qquad (2.55c)$$

which are the required relationships.

To determine the constants α_i, Eq. (2.51) is applied to each of the nodes 1, 2, ..., 6 in turn. After substituting the ξ, η nodal coordinates in terms of the dimensions a, b, c, the resulting equations can be combined to form the matrix equation

$$
\begin{bmatrix}
1 & -b & 0 & b^2 & 0 & 0 \\
1 & (a-b)/2 & 0 & [(a-b)/2]^2 & 0 & 0 \\
1 & a & 0 & a^2 & 0 & 0 \\
1 & a/2 & c/2 & (a/2)^2 & ac/4 & (c/2)^2 \\
1 & 0 & c & 0 & 0 & c^2 \\
1 & -b/2 & c/2 & (-b/2)^2 & -bc/4 & (c/2)^2
\end{bmatrix}
\begin{bmatrix}
\alpha_1 \\ \alpha_2 \\ \alpha_3 \\ \alpha_4 \\ \alpha_5 \\ \alpha_6
\end{bmatrix}
=
\begin{bmatrix}
\bar{T}_1 \\ \bar{T}_2 \\ \bar{T}_3 \\ \bar{T}_4 \\ \bar{T}_5 \\ \bar{T}_6
\end{bmatrix},
\tag{2.56}
$$

Equation (2.56) can be solved for the α_i provided that the determinant of the coefficient matrix does not vanish. The determinant can be shown to be $-c^4(a+b)^4/64$, which never equals zero since the area of the triangle equals $\frac{1}{2}c(a+b)$ and never vanishes. Hence, the coefficient matrix of Eq. (2.56) is nonsingular and can be inverted.

Denoting the coefficient matrix in Eq. (2.56) by $\mathbf{A}$ allows this equation to be written as

$$
\mathbf{A}\boldsymbol{\alpha} = \mathbf{T}^e,
\tag{2.57}
$$

where

$$
\boldsymbol{\alpha} = [\alpha_i] =
\begin{bmatrix}
\alpha_1 \\ \alpha_2 \\ \vdots \\ \alpha_6
\end{bmatrix},
\qquad
\mathbf{T}^e =
\begin{bmatrix}
\bar{T}_1 \\ \bar{T}_2 \\ \vdots \\ \bar{T}_6
\end{bmatrix},
\tag{2.58}
$$

and where $\mathbf{T}^e$ is the *nodal vector* for element e. Premultiplying Eq. (2.57) by the inverse of $\mathbf{A}$ gives

$$
\boldsymbol{\alpha} = \mathbf{A}^{-1}\mathbf{T}^e = \mathbf{B}\mathbf{T}^e,
\tag{2.59}
$$

where

$$
\mathbf{A}^{-1} = \mathbf{B} = [b_{ij}].
\tag{2.60}
$$

The next step in the formulation is to express the element contribution given in Eq. (2.6) in terms of the local coordinates ξ and η. Reference to Eq. (2.6) shows that equivalent local coordinate expressions are required for the derivatives $\partial \hat{T}/\partial x$ and $\partial \hat{T}/\partial y$.

In the previous chapter it was noted that a simple translation of coordinate axes does not change a nodal value nor a nodal derivative; that is, these have the same numeric value with respect to *either* global or local coordinates. If there is a rotation between the two sets of axes, the nodal values still remain unchanged but derivatives (such as $\partial \hat{T}/\partial x$, $\partial \hat{T}/\partial \xi$) are no longer respectively the same in the two coordinate systems. Since the trial function in Eq. (2.51) is an *interpolation* between the nodal values $\bar{T}_1$, $\bar{T}_2$, ..., $\bar{T}_6$, [see Eq. (2.56)] and does not involve any of the derivatives of T, the element nodal vector T^e is the same in both the global and local systems. In this example, it is not necessary, therefore, to use sub- or superscripts to indicate the frame of reference being used.

Consider now the relationships between the x, y and ξ, η coordinates of a point. Simple trigonometry shows that these relationships, due to the rotation θ, here summarized in matrix form, are

$$\begin{bmatrix} x \\ y \end{bmatrix} = \begin{bmatrix} \cos \theta & -\sin \theta \\ \sin \theta & \cos \theta \end{bmatrix} \begin{bmatrix} \xi \\ \eta \end{bmatrix}, \tag{2.61a}$$

with the inverse relationships being

$$\begin{bmatrix} \xi \\ \eta \end{bmatrix} = \begin{bmatrix} \cos \theta & \sin \theta \\ -\sin \theta & \cos \theta \end{bmatrix} \begin{bmatrix} x \\ y \end{bmatrix}. \tag{2.61b}$$

The square matrix on the right-hand side of Eq. (2.61b) will be denoted by $\mathbf{R}$ and is known as the *rotation matrix*. The inverse of the rotation matrix $\mathbf{R}^{-1}$ appears in Eq. (2.61a). It is easily seen that the *inverse of* $\mathbf{R}$ *is equal to its transpose*, a property that is true for all *orthogonal transformations* [3].

If T is regarded as $T = T(\xi, \eta)$, there can be written from the calculus

$$\frac{\partial T}{\partial x} = \frac{\partial T}{\partial \xi} \frac{\partial \xi}{\partial x} + \frac{\partial T}{\partial \eta} \frac{\partial \eta}{\partial x}, \tag{2.62a}$$

$$\frac{\partial T}{\partial y} = \frac{\partial T}{\partial \xi} \frac{\partial \xi}{\partial y} + \frac{\partial T}{\partial \eta} \frac{\partial \eta}{\partial y}. \tag{2.62b}$$

Differentiation of Eq. (2.61b) yields the relations

$$\frac{\partial \xi}{\partial x} = \cos \theta, \qquad \frac{\partial \xi}{\partial y} = \sin \theta,$$

$$\frac{\partial \eta}{\partial x} = -\sin \theta, \qquad \frac{\partial \eta}{\partial y} = \cos \theta, \tag{2.63}$$

and substitution of Eqs. (2.63) into Eqs. (2.62) gives

$$\frac{\partial T}{\partial x} = \cos \theta \frac{\partial T}{\partial \xi} - \sin \theta \frac{\partial T}{\partial \eta}, \tag{2.64a}$$

$$\frac{\partial T}{\partial y} = \sin \theta \frac{\partial T}{\partial \xi} + \cos \theta \frac{\partial T}{\partial \eta}, \tag{2.64b}$$

or in matrix notation

$$\begin{bmatrix} \partial T/\partial x \\ \partial T/\partial y \end{bmatrix} = \begin{bmatrix} \cos\theta & -\sin\theta \\ \sin\theta & \cos\theta \end{bmatrix} \begin{bmatrix} \partial T/\partial\xi \\ \partial T/\partial\eta \end{bmatrix}. \tag{2.65a}$$

The inverse of Eq. (2.65a) is

$$\begin{bmatrix} \partial T/\partial\xi \\ \partial T/\partial\eta \end{bmatrix} = \begin{bmatrix} \cos\theta & \sin\theta \\ -\sin\theta & \cos\theta \end{bmatrix} \begin{bmatrix} \partial T/\partial x \\ \partial T/\partial y \end{bmatrix}. \tag{2.65b}$$

Equations (2.65) allow conversion of global first derivatives to local first derivatives and vice versa. The appearance of the rotation matrix in Eq. (2.65b) and its inverse in Eq. (2.65a) should be noted.

The integrand of the functional in Eq. (2.6) can now be converted to local coordinates by substitution from Eqs. (2.64) or (2.65a), when it will be found that

$$\left(\frac{\partial \hat{T}}{\partial x}\right)^2 + \left(\frac{\partial \hat{T}}{\partial y}\right)^2 = \left(\frac{\partial \hat{T}}{\partial\xi}\right)^2 + \left(\frac{\partial \hat{T}}{\partial\eta}\right)^2. \tag{2.66}$$

From the calculus [4, 5],

$$\int_D f(x, y)\,dx\,dy = \int_D g(\xi, \eta)|\mathbf{J}|\,d\xi\,d\eta, \tag{2.67}$$

where $f(x, y)$ and $g(\xi, \eta)$ are equivalent expressions for a function in two alternative coordinate systems Oxy and $\bar{O}\xi\eta$, and $\mathbf{J}$ is the *Jacobian of transformation* $\partial(x, y)/\partial(\xi, \eta)$. From Eq. (2.67), the replacement for $dx\,dy$ in Eq. (2.6) is given by

$$dx\,dy = |\mathbf{J}|\,d\xi\,d\eta = \left|\frac{\partial(x, y)}{\partial(\xi, \eta)}\right|\,d\xi\,d\eta = \begin{vmatrix} \partial x/\partial\xi & \partial x/\partial\eta \\ \partial y/\partial\xi & \partial y/\partial\eta \end{vmatrix}\,d\xi\,d\eta$$

$$= \begin{vmatrix} \cos\theta & -\sin\theta \\ \sin\theta & \cos\theta \end{vmatrix}\,d\xi\,d\eta = d\xi\,d\eta, \tag{2.68}$$

making use of Eq. (2.61a) for the derivatives $\partial x/\partial\xi$, $\partial x/\partial\eta$, $\partial y/\partial\xi$, $\partial y/\partial\eta$. Substitution of Eqs. (2.66) and (2.68) into Eq. (2.6) yields the element contribution as

$$\chi^e = \frac{1}{2}\int_e \left[\left(\frac{\partial \hat{T}^e}{\partial\xi}\right)^2 + \left(\frac{\partial \hat{T}^e}{\partial\eta}\right)^2\right]\,d\xi\,d\eta. \tag{2.69}$$

Writing the trial function [Eq. (2.51)] in the form

$$\hat{T}^e = \sum_{i=1}^{6} \alpha_i \xi^{m_i}\eta^{n_i}, \tag{2.70}$$

where $m_1, m_2, \ldots, m_6$ and $n_1, n_2, \ldots, n_6$ are defined in Table 2.4, allows the derivatives $\partial \hat{T}^e/\partial \xi$, $\partial \hat{T}^e/\partial \eta$ to be written as

$$\frac{\partial \hat{T}^e}{\partial \xi} = \sum_{i=1}^{6} \alpha_i m_i \xi^{m_i - 1} \eta^{n_i}, \tag{2.71a}$$

$$\frac{\partial \hat{T}^e}{\partial \eta} = \sum_{i=1}^{6} \alpha_i n_i \xi^{m_i} \eta^{n_i - 1}. \tag{2.71b}$$

TABLE 2.4

Series Indices For Six-Node
Quadratic Element

i	m_i	n_i	i	m_i	n_i
1	0	0	4	2	0
2	1	0	5	1	1
3	0	1	6	0	2

Substitution of Eqs. (2.71) into Eq. (2.69) yields the element contribution in the form

$$\chi^e = \frac{1}{2} \int_e \left\{ \sum_{i=1}^{6} \sum_{j=1}^{6} \left[\alpha_i \alpha_j m_i m_j \xi^{m_i + m_j - 2} \eta^{n_i + n_j} + \alpha_i \alpha_j n_i n_j \xi^{m_i + m_j} \eta^{n_i + n_j - 2} \right] \right\} d\xi \, d\eta. \tag{2.72}$$

Introducing the notation

$$g_{ij} = \int_e \left[m_i m_j \xi^{m_i + m_j - 2} \eta^{n_i + n_j} + n_i n_j \xi^{m_i + m_j} \eta^{n_i + n_j - 2} \right] d\xi \, d\eta. \tag{2.73}$$

allows Eq. (2.72) to be written as

$$\chi^e = \frac{1}{2} \sum_{i=1}^{6} \sum_{j=1}^{6} \alpha_i g_{ij} \alpha_j. \tag{2.74}$$

In matrix form (see Appendix A.3), Eq. (2.74) is

$$\chi^e = \tfrac{1}{2} \boldsymbol{\alpha}^T \mathbf{G} \boldsymbol{\alpha}, \tag{2.75}$$

where $\boldsymbol{\alpha}$ is defined in Eqs. (2.58), and

$$\mathbf{G} = [g_{ij}], \tag{2.76}$$

is the *integration matrix*. Substitution of Eq. (2.59) into (2.75) then yields the element contribution as

$$\chi^e = \tfrac{1}{2} (\mathbf{T}^e)^T \mathbf{B}^T \mathbf{G} \mathbf{B} (\mathbf{T}^e). \tag{2.77}$$

As was mentioned previously, the nodal values $\bar{T}_1, \bar{T}_2, \ldots, \bar{T}_6$ in the local system are the same as the nodal values $\bar{T}_1, \bar{T}_2, \ldots, \bar{T}_6$ in the global system. Equation (2.77) is therefore applicable in either system. Equation (2.77) can also be written as

$$\chi^e = \tfrac{1}{2}(\mathbf{T}^e)^{\mathrm{T}}\mathbf{k}(\mathbf{T}^e), \tag{2.78}$$

where

$$\mathbf{k} = \mathbf{B}^{\mathrm{T}}\mathbf{GB}, \tag{2.79}$$

is recognized as the *element* **k** *matrix* for element *e*.

Differentiating Eq. (2.78) (see Appendix B.2) gives the element matrix equation

$$\frac{\partial \chi^e}{\partial \mathbf{T}^e} = \mathbf{kT}^e. \tag{2.80}$$

The integration matrix **G** has yet to be determined explicitly. To evaluate the entries g_{ij}, it is necessary [see Eq. (2.73)] to evaluate integrals of the form

$$h(m, n) = \int_e \xi^m \eta^n \, d\xi \, d\eta, \tag{2.81}$$

where m and n are integers. Reference to Fig. 2.6 shows that this integral can be written as

$$h(m, n) = \int_{\xi=-b}^{\xi=0} \int_{\eta=0}^{\eta=(c/b)\xi+c} \xi^m \eta^n \, d\eta \, d\xi + \int_{\xi=0}^{\xi=a} \int_{\eta=0}^{\eta=-(c/a)\xi+c} \xi^m \eta^n \, d\eta \, d\xi, \tag{2.82}$$

where the first and second right-hand side integrals are the contributions over subtriangles I and II, respectively. Introducing the change of variable

$$u = \xi/b, \tag{2.83}$$

allows the first right-hand side integral to be written as

$$I_1 = \int_{u=-1}^{u=0} \int_{\eta=0}^{\eta=c(u+1)} b(ub)^m \eta^n \, du \, d\eta, \tag{2.84a}$$

or, after integrating with respect to η, as

$$I_1 = \frac{c^{n+1}b^{m+1}}{n+1} \int_{u=-1}^{u=0} u^m (u + 1)^{n+1} \, du. \tag{2.84b}$$

Equations (2.84) may also be written as

$$I_1 = -\frac{c^{n+1}(-b)^{m+1}}{n+1} \int_{\bar{u}=0}^{\bar{u}=1} \bar{u}^{n+1}(1 - \bar{u})^m \, d\bar{u}, \tag{2.85}$$

using the change of variable

$$\bar{u} = u + 1. \tag{2.86}$$

The integral in expression (2.85) is recognized as the *beta function* $\beta(n + 2, m + 1)$ expressible [5], successively, in terms of *gamma functions* and *factorials* as

$$\int_{\bar{u}=0}^{\bar{u}=1} \bar{u}^{n+1}(1 - \bar{u})^m \, d\bar{u} = \frac{\Gamma(n + 2)\Gamma(m + 1)}{\Gamma(m + n + 3)} = \frac{(n + 1)!m!}{(m + n + 2)!}. \quad (2.87)$$

Substitution of Eq. (2.87) into Eq. (2.85) gives

$$I_1 = \int_{\xi=-b}^{\xi=0} \int_{\eta=0}^{\eta=(c/b)\xi+c} \xi^m \eta^n \, d\eta \, d\xi = -c^{n+1}(-b)^{m+1}\left[\frac{m!n!}{(m + n + 2)!}\right]. \quad (2.88a)$$

Similarly, it can be shown that

$$I_2 = \int_{\xi=0}^{\xi=a} \int_{\eta=0}^{\eta=-(c/2)\xi+c} \xi^m \eta^n \, d\eta \, d\xi = c^{n+1}a^{m+1}\left[\frac{m!n!}{(m + n + 2)!}\right]. \quad (2.88b)$$

Substituting Eqs. (2.88) into Eq. (2.81) gives

$$h(m, n) = \int_e \xi^m \eta^n \, d\xi \, d\eta = \frac{c^{n+1}[a^{m+1} - (-b)^{m+1}]m!n!}{(m + n + 2)!}, \quad (2.89)$$

which can be used to evaluate Eq. (2.73) as

$$g_{ij} = m_i m_j h(m_i + m_j - 2, n_i + n_j) + n_i n_j h(m_i + m_j, n_i + n_j - 2). \quad (2.90)$$

To further illustrate the local coordinate formulation, consider the problem of two-dimensional heat flow through the square block shown in Fig. 2.1. Let the region be divided into eight six-node triangular elements as shown in Fig. 2.7. As can be seen from this figure, the total number of nodes n equals 25.

In Table 2.5, the x and y coordinates for the nodes are listed. Table 2.6 summarizes the relationship between node identifiers, system node numbers, and the characteristic dimensions a, b, c, for all elements.

Substitution of the dimensions a, b, and c from Table 2.6 into Eq. (2.56) yields the coefficient matrix $\mathbf{A}$ as

$$\mathbf{A} = \begin{bmatrix} 1 & 0 & 0 & 0 & 0 & 0 \\ 1 & \frac{1}{2} & 0 & \frac{1}{4} & 0 & 0 \\ 1 & 1 & 0 & 1 & 0 & 0 \\ 1 & \frac{1}{2} & \frac{1}{2} & \frac{1}{2} & \frac{1}{2} & \frac{1}{2} \\ 1 & 0 & 1 & 0 & 0 & 1 \\ 1 & 0 & \frac{1}{2} & 0 & 0 & \frac{1}{4} \end{bmatrix}. \quad (2.91)$$

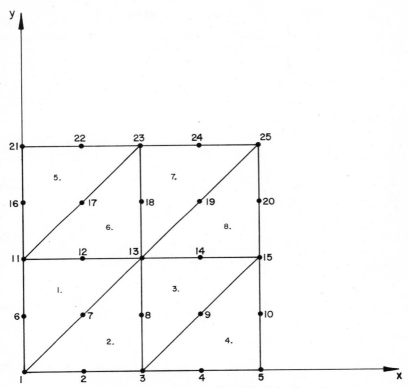

Fig. 2.7 Region subdivided into eight finite elements.

TABLE 2.5

Nodal Coordinates

Node	Coordinate x	Coordinate y	Node	Coordinate x	Coordinate y	Node	Coordinate x	Coordinate y
1	0	0	10	2	0.5	19	1.5	1.5
2	0.5	0	11	0	1	20	2	1.5
3	1	0	12	0.5	1	21	0	2
4	1.5	0	13	1	1	22	0.5	2
5	2	0	14	1.5	1	23	1	2
6	0	0.5	15	2	1	24	1.5	2
7	0.5	0.5	16	0	1.5	25	2	2
8	1	0.5	17	0.5	1.5			
9	1.5	0.5	18	1	1.5			

TABLE 2.6

Relationship between System Node Numbers
and Element Node Identifiers;
Element Characteristic Dimensions

Element	Node Number						Dimension		
	1	2	3	4	5	6	a	b	c
1	11	6	1	7	13	12	1	0	1
2	3	8	13	7	1	2	1	0	1
3	13	8	3	9	15	14	1	0	1
4	5	10	15	9	3	4	1	0	1
5	21	16	11	17	23	22	1	0	1
6	13	18	23	17	11	12	1	0	1
7	23	18	13	19	25	24	1	0	1
8	15	20	25	19	13	14	1	0	1

Inversion of Eq. (2.91) gives the matrix **B** as

$$\mathbf{B} = \mathbf{A}^{-1} = \begin{bmatrix} 1 & 0 & 0 & 0 & 0 & 0 \\ -3 & 4 & -1 & 0 & 0 & 0 \\ -3 & 0 & 0 & 0 & -1 & 4 \\ 2 & -4 & 2 & 0 & 0 & 0 \\ 4 & -4 & 0 & 4 & 0 & -4 \\ 2 & 0 & 0 & 0 & 2 & -4 \end{bmatrix}. \quad (2.92)$$

The integration matrix **G** is next required, the elements of which are given by Eq. (2.90). Since Table 2.6 shows that a, b, c are respectively the same for all elements, insertion of these into Eq. (2.89) results in

$$h(m, n) = m!n!/(m + n + 2)!, \qquad e = 1, 2, \ldots, 8. \quad (2.93)$$

Substitution of Eq. (2.93) into Eq. (2.90) finally yields the required elements g_{ij}.

In a computer program, $h(m, n)$ would be evaluated directly using Eq. (2.89) and the known values for m, n, a, b, and c. Zero values for m, n, a, b, c can lead to error unless care is taken in the programming. Consider, for example, the case when $i = 1$ and $j = 1$. From Table 2.4, $m_i = 0$, $m_j = 0$, $n_i = 0$, and $n_j = 0$. Substituting these values, together with $a = 1$, $b = 0$, and $c = 1$, into Eq. (2.89), for $m = m_i + m_j - 2$ and $n = n_i + n_j$, results in

$$h(m_i + m_j - 2, n_i + n_j)$$

$$= \frac{1^{n_i+n_j+1}[1^{m_i+m_j-1} - (-0)^{m_i+m_j-1}](m_i + m_j - 2)!(n_i + n_j)!}{(m_i + m_j + n_i + n_j)!}. \quad (2.94)$$

Any attempt to compute Eq. (2.94) directly would result in an error message since the term $(-0)^{m_i + m_j - 1}$ becomes, in the present case, $(-0)^{-1}$, which is *indefinite*. The problem can only arise when $m_i + m_j - 1$ has its minimum value of -1, that is, when *both* m_i and m_j are zero. As Eq. (2.90) shows, the expression in Eq. (2.94) need not be evaluated when the product $m_i m_j = 0$, that is, when *either* m_i or m_j equals zero. The difficulty of indefinite arguments in the first term of Eq. (2.90) can thus be neatly avoided by testing the coefficient $m_i m_j$ against zero and *only* evaluating the term $h(m_i + m_j - 2, n_i + n_j)$ if this coefficient does not vanish. Similarly, $h(m_i + m_j, n_i + n_j - 2)$ is evaluated only if $n_i n_j$ is nonzero.

Applying the above procedure to Eq. (2.90) results in

$$
\mathbf{G} = \begin{bmatrix}
0 & 0 & 0 & 0 & 0 & 0 \\
0 & h(0,0) & 0 & 2h(1,0) & h(0,1) & 0 \\
0 & 0 & h(0,0) & 0 & h(1,0) & 2h(0,1) \\
0 & 2h(1,0) & 0 & 4h(2,0) & 2h(1,1) & 0 \\
0 & h(0,1) & h(1,0) & 2h(1,1) & h(0,2) + h(2,0) & 2h(1,1) \\
0 & 0 & 2h(0,1) & 0 & 2h(1,1) & 4h(0,2)
\end{bmatrix}, \quad (2.95)
$$

which on account of Eq. (2.93) reduces to

$$
\mathbf{G} = \frac{1}{12} \begin{bmatrix}
0 & 0 & 0 & 0 & 0 & 0 \\
0 & 6 & 0 & 4 & 2 & 0 \\
0 & 0 & 6 & 0 & 2 & 4 \\
0 & 4 & 0 & 4 & 1 & 0 \\
0 & 2 & 2 & 1 & 2 & 1 \\
0 & 0 & 4 & 0 & 1 & 4
\end{bmatrix}. \quad (2.96)
$$

Forming the product $\mathbf{B}^T\mathbf{G}\mathbf{B}$ from Eqs. (2.92) and (2.96) gives, from Eq. (2.79), the element $\mathbf{k}$ matrix

$$
\mathbf{k} = \mathbf{B}^T\mathbf{G}\mathbf{B} = \frac{1}{6} \begin{bmatrix}
6 & -4 & 1 & 0 & 1 & -4 \\
-4 & 16 & -4 & -8 & 0 & 0 \\
1 & -4 & 3 & 0 & 0 & 0 \\
0 & -8 & 0 & 16 & 0 & -8 \\
1 & 0 & 0 & 0 & 3 & -4 \\
-4 & 0 & 0 & -8 & -4 & 16
\end{bmatrix}, \quad (2.97)
$$

which is seen to be symmetric. It is left as an exercise to show that $\mathbf{k}$, given in *node identifier* form in Eq. (2.97), is the *same* for all elements. Converting from node identifiers to node numbers allows the element matrix equations to be assembled, either by nodes or elements, into the system matrix equation.

$$
\mathbf{K}\,\overline{T} =
\begin{bmatrix}
\overline{T}_1 \\
\overline{T}_2 \\
\overline{T}_3 \\
\overline{T}_4 \\
\overline{T}_5 \\
\overline{T}_6 \\
\vdots \\
\vdots \\
\overline{T}_{20} \\
\overline{T}_{21} \\
\overline{T}_{22} \\
\overline{T}_{23} \\
\overline{T}_{24} \\
\overline{T}_{25}
\end{bmatrix}
=
\begin{bmatrix}
50 \\
50 \\
50 \\
50 \\
50 \\
0 \\
\vdots \\
\vdots \\
0 \\
100 \\
100 \\
100 \\
100 \\
100
\end{bmatrix}
\tag{2.98}
$$

It may be verified that the system matrix equation, after it is corrected for the Dirichlet boundary conditions according to Rule 1 of Chapter 1, is given by Eq. (2.98). The elements not shown in the coefficient matrix of this equation are all zeros.

Solution of Eq. (2.98) by any standard technique then gives the nodal values as

$$
\begin{aligned}
&\bar{T}_1 = 50, \quad &&\bar{T}_2 = 50, \quad &&\bar{T}_3 = 50, \quad &&\bar{T}_4 = 50, \quad &&\bar{T}_5 = 50, \\
&\bar{T}_6 = 62.5, \quad &&\bar{T}_7 = 62.5, \quad &&\bar{T}_8 = 62.5, \quad &&\bar{T}_9 = 62.5, \quad &&\bar{T}_{10} = 62.5, \\
&\bar{T}_{11} = 75, \quad &&\bar{T}_{12} = 75, \quad &&\bar{T}_{13} = 75, \quad &&\bar{T}_{14} = 75, \quad &&\bar{T}_{15} = 75, \quad (2.99) \\
&\bar{T}_{16} = 87.5, \quad &&\bar{T}_{17} = 87.5, \quad &&\bar{T}_{18} = 87.5, \quad &&\bar{T}_{19} = 87.5, \quad &&\bar{T}_{20} = 87.5, \\
&\bar{T}_{21} = 100, \quad &&\bar{T}_{22} = 100, \quad &&\bar{T}_{23} = 100, \quad &&\bar{T}_{24} = 100, \quad &&\bar{T}_{25} = 100.
\end{aligned}
$$

The required solution at point A (see Fig. 2.7) is finally obtained as

$$
T|_{\text{at } A} \simeq \bar{T}_{18} = 87.5°\text{C}. \tag{2.100}
$$

Note 1   In the preceding formulation, the shape function interpolation for the element can be used in place of the polynomial series [Eq. (2.70)]. For the shape function approach, Eq. (2.51) is written as

$$
\hat{T}^e = [1 \quad \xi \quad \eta \quad \xi^2 \quad \xi\eta \quad \eta^2]\boldsymbol{\alpha}. \tag{2.101}
$$

Substitution for $\boldsymbol{\alpha}$ from Eq. (2.59) then yields

$$
\hat{T}^e = [1 \quad \xi \quad \eta \quad \xi^2 \quad \xi\eta \quad \eta^2]\mathbf{B}\mathbf{T}^e, \tag{2.102}
$$

which may be written in the shape function form

$$
\hat{T}^e = \mathbf{N}^e\mathbf{T}^e, \tag{2.103}
$$

where

$$
\mathbf{N}^e = [1 \quad \xi \quad \eta \quad \xi^2 \quad \xi\eta \quad \eta^2]\mathbf{B}. \tag{2.104}
$$

Equation (2.104) gives the shape function matrix in local coordinate form. A similar derivation can be used to obtain the shape function matrix in global coordinate form.

Equation (2.104) may be substituted into Eq. (2.106) following to yield the element $\mathbf{k}$ matrix, or alternatively Eq. (2.103) may be substituted into Eq. (2.69) to give χ^e as a function of the nodal values of the element. For *assembly by nodes*, the derivatives $\partial\chi/\partial\bar{T}_p$ can then be obtained for use in Eq. (2.20), whereas for *assembly by elements* the element matrix equation $\partial\chi/\partial\mathbf{T}^e$ can be obtained in similar form to Eq. (2.80).

The shape function approach is often used in practice, but it will be found that the required integrations cannot be evaluated as simply as in the polynomial procedure, at least when local coordinates are adopted.

There are some types of elements (see Chapter 9) that have a shape function representation but for which a coordinate series representation is not easily obtainable. In such cases the shape function approach is preferable.

Exercise 2.3   For the Poisson equation and linear triangular element, show that the two-dimensional version of Eq. (2.48), with respect to global coordinates,

$$k_{\alpha\beta}^e = \int\int_{D_e} \left(\frac{\partial N_\alpha}{\partial x} \frac{\partial N_\beta}{\partial x} + \frac{\partial N_\alpha}{\partial y} \frac{\partial N_\beta}{\partial y} \right) dx\, dy, \qquad (2.105)$$

becomes

$$k_{\alpha\beta}^e = \int\int_{D_e} \left(\frac{\partial N_\alpha}{\partial \xi} \frac{\partial N_\beta}{\partial \xi} + \frac{\partial N_\alpha}{\partial \eta} \frac{\partial N_\beta}{\partial \eta} \right) d\xi\, d\eta, \qquad (2.106)$$

when local coordinates are used.

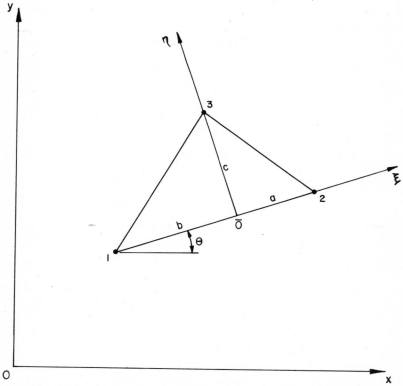

Fig. 2.8   Local coordinate system for the three-node triangular element.

Exercise 2.4   For a linear trial function and the three-node triangular element in the local coordinate system $\bar{O}\xi\eta$ shown in Fig. 2.8, develop a finite element formulation using the polynomial series procedure outlined in Section 2.1.2.

Exercise 2.5   Repeat Exercise 2.4, using the shape function approach as outlined in Note 1.

REFERENCES

1. D. H. Norrie and G. de Vries, "The Finite Element Method—Fundamentals and Applications." Academic Press, New York, 1973.
2. J. F. Lee and F. W. Sears, "Thermodynamics—An Introductory Text for Engineering Students." Addison-Wesley, Reading, Massachusetts, 1963.
3. E. H. Thompson, "Algebra of Matrices." Adam Hilger, London, 1969.
4. E. Kreyzig, "Advanced Engineering Mathematics." Wiley, New York, 1962.
5. W. Kaplan, "Advanced Calculus," 2nd ed. Addison-Wesley, Reading, Massachusetts, 1973.

3

COMPUTER PROGRAMMING OF
THE FINITE ELEMENT METHOD

3.1 COMPUTATIONAL DEVELOPMENT AND THE FINITE ELEMENT METHOD

The widespread adoption of the finite element method for increasingly diverse problems has been facilitated by:

(a)   the inherent generality of the method,

(b)   its natural formulation in matrix language,

(c)   the availability of efficient procedures to solve very large sets of equations, and

(d)   the capability of modern computers.

The first two of these factors will become evident as the reader progresses through this book and will not be commented on further. In relation to the third factor, it is worth noting that *efficient* procedures for very large sets of equations have *been available only since* 1950, as the following quotation from Birkhoff [1] shows:

> Over fifty years ago, mathematicians were familiar with the system of linear equations and knew that the solution could be computed by the algorithm of Gauss elimination . . . . Thus, the solution of the Poisson

equation to any desired accuracy could, in principle, be achieved by *discrete* methods .... However, to actually compute a reasonably accurate solution ... in 1945, using desk machines, would probably have cost at least $10,000. At that time, the most effective method of solving systems of linear equations did not use elimination at all. It used ... "relaxation" ... but each new problem took man-months of dedicated effort by a perceptive expert. I suggested to David Young in 1948 the desirability of trying to automate the relaxation methods ... so that they could be programmed and efficiently executed on a computer. In principle, the proposal was to solve ... by an automated iterative (relaxation) technique. Actually, Gauss and Jacobi had already used iterative methods ... in the nineteenth century, but their algorithms converged too slowly. Young's aim was to find a more rapidly convergent algorithm. After two years of hard work, Young succeeded in doing this .... (His) *successive over-relaxation* (SOR) algorithm reduced by a factor of 10 or more the number of iterations required .... Since then, the efficiency of computers has increased enormously. Today, variants of Gauss elimination (e.g. Cholesky decomposition) are available ... which solve problems of the kind described having 500 unknowns for less than a dollar per case on most large computers. However, in many important engineering problems ... as many as 50,000 unknowns are required to achieve adequate detail and accuracy. For these, even on the largest computers, variants of Gauss elimination are still impractical, and we must use variants of the SOR methods such as the *cyclic Chebyshev* algorithm.

As the foregoing makes clear, the fourth factor, large-scale computer capability, is essential for solving the large sets of equations arising from real engineering problems.

In a later chapter, the procedures available for solving sets of linear equations will be considered at some length. To complement the formulations given previously, a finite element program is presented in the following section. Although the development of *efficient* computer programs for finite element applications is of considerable importance, only the basic concepts of finite element programming will be presented at this stage.

3.2 A FINITE ELEMENT PROGRAM FOR THE LAPLACE PROBLEM

In this section, a FORTRAN IV computer program is developed to solve the Laplace equation using the global coordinate formulation of Section 2.1.1. The particular problem considered is the two-dimensional heat flow through the square bar shown in Fig. 3.1. On two sides of the bar, linear

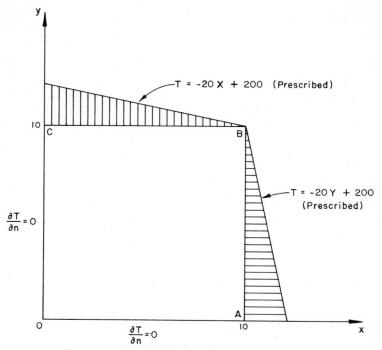

Fig. 3.1   Two-dimensional heat flow in a square bar.

temperature distributions are prescribed. On the remaining two sides, there is infinite resistance to heat flow, that is, perfect insulation. It is required to determine the lines of constant temperature or *isotherms* within the bar using the finite element method.

The region $OABC$ is subdivided into 50 three-node triangular elements with the total number of nodes being 36, as shown in Fig. 3.2. The basic input data for the program comprises:

(a)   the x and y coordinates of all the nodes, and
(b)   the relationship between the node identifiers and node numbers for all the elements.

These two sets of data are shown in Tables 3.1 and 3.2, respectively.

The programming strategy is outlined in the next section in a flow chart, which should be examined concurrently with the documented program following. Subsequent sections present the input data and the results from the program.

3.2.1   Flow Chart for the Computer Program

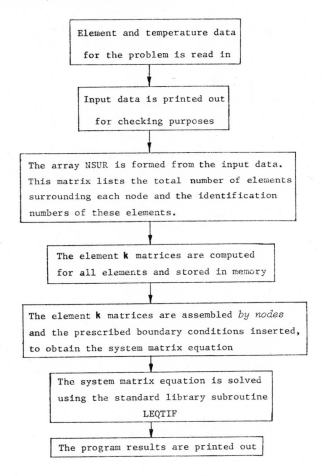

Element and temperature data
for the problem is read in

Input data is printed out
for checking purposes

The array NSUR is formed from the input data.
This matrix lists the total number of elements
surrounding each node and the identification
numbers of these elements.

The element **k** matrices are computed
for all elements and stored in memory

The element **k** matrices are assembled *by nodes*
and the prescribed boundary conditions inserted,
to obtain the system matrix equation

The system matrix equation is solved
using the standard library subroutine
LEQTIF

The program results are printed out

This program is a hybrid. Its basic structure is that of an *assembly by nodes* program, but to allow its assembly procedure to be easily replaced by *assembly by elements* (see Exercise 3.3), the element **k** matrices are computed and retained in storage for subsequent use.

Since the assembly of the element matrix equations in this program is by nodes, those elements that surround each node need to be ascertained. This information is compiled in the array NSUR(I, J). The variable I specifies not only the matrix row but also which node is under consideration. Column $J = 1$ lists the total number of elements surrounding node I, and columns

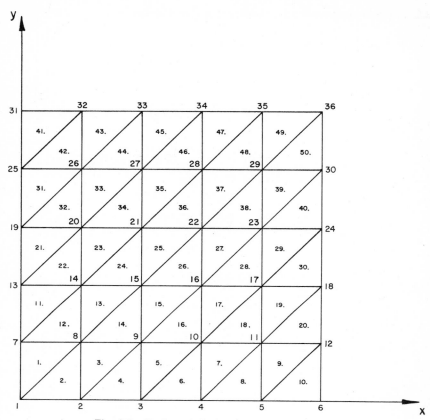

Fig. 3.2 Region subdivided into 50 finite elements.

TABLE 3.1

Nodal Coordinates

Node	Coordinate x	y	Node	Coordinate x	y	Node	Coordinate x	y
1	0.0	0.0	13	0.0	4.0	25	0.0	8.0
2	2.0	0.0	14	2.0	4.0	26	2.0	8.0
3	4.0	0.0	15	4.0	4.0	27	4.0	8.0
4	6.0	0.0	16	6.0	4.0	28	6.0	8.0
5	8.0	0.0	17	8.0	4.0	29	8.0	8.0
6	10.0	0.0	18	10.0	4.0	30	10.0	8.0
7	0.0	2.0	19	0.0	6.0	31	0.0	10.0
8	2.0	2.0	20	2.0	6.0	32	2.0	10.0
9	4.0	2.0	21	4.0	6.0	33	4.0	10.0
10	6.0	2.0	22	6.0	6.0	34	6.0	10.0
11	8.0	2.0	23	8.0	6.0	35	8.0	10.0
12	10.0	2.0	24	10.0	6.0	36	10.0	10.0

TABLE 3.2

Relationship between System Node Numbers
and Element Node Identifiers

	Node number				Node number				Node number		
Element	i	j	m	Element	i	j	m	Element	i	j	m
1	1	8	7	18	10	11	17	35	21	28	27
2	1	2	8	19	11	18	17	36	21	22	28
3	2	9	8	20	11	12	18	37	22	29	28
4	2	3	9	21	13	20	19	38	22	23	29
5	3	10	9	22	13	14	20	39	23	30	29
6	3	4	10	23	14	21	20	40	23	24	30
7	4	11	10	24	14	15	21	41	25	32	31
8	4	5	11	25	15	22	21	42	25	26	32
9	5	12	11	26	15	16	22	43	26	33	32
10	5	6	12	27	16	23	22	44	26	27	33
11	7	14	13	28	16	17	23	45	27	34	33
12	7	8	14	29	17	24	23	46	27	28	34
13	8	15	14	30	17	18	24	47	28	35	34
14	8	9	15	31	19	26	25	48	28	29	35
15	9	16	15	32	19	20	26	49	29	36	35
16	9	10	16	33	20	27	26	50	29	30	36
17	10	17	16	34	20	21	27				

$J = 2, 3, \ldots$, record the identification numbers of these surrounding elements. The strategy for constructing the array is the following. Initially, column 1 is set to zero for all nodes, as shown in Table 3.3. For simplicity, this table (and those subsequent) contains data for only the first ten nodes. The first element to be processed is element 1, which, as can be seen from Table 3.2, has nodes numbered 1, 8, and 7. In the array, the total number of surrounding elements listed for each of these nodes is incremented by one. The identification number of the element, namely 1, is recorded for each node in the next available column $J = 2$, as shown in Table 3.4. Next, element 2, with nodes 1, 2, and 8, is processed in the same way. The total number of surrounding elements listed for these nodes is incremented by one and the identification number of the element, namely 2, recorded in the next available column, as Table 3.5 illustrates. Continuing with the above procedure it may be verified that after the first 18 elements have been processed the array NSUR will have the entries shown in Table 3.6. The array NSUR is fully formed when all elements have been processed in turn.

TABLE 3.3

Array NSUR after Initialization

Node I	Total number of surrounding elements J = 1	Identification numbers of surrounding elements					
		J = 2	J = 3	J = 4	J = 5	J = 6	J = 7
1	0						
2	0						
3	0						
4	0						
5	0						
6	0						
7	0						
8	0						
9	0						
10	0						

TABLE 3.4

Array NSUR after Processing of Element 1

Node I	Total number of surrounding elements J = 1	Identification numbers of surrounding elements					
		J = 2	J = 3	J = 4	J = 5	J = 6	J = 7
1	1	1					
2	0						
3	0						
4	0						
5	0						
6	0						
7	1	1					
8	1	1					
9	0						
10	0						

TABLE 3.5

Array NSUR after Processing of Elements 1 and 2

Node I	Total number of surrounding elements J = 1	Identification numbers of surrounding elements					
		J = 2	J = 3	J = 4	J = 5	J = 6	J = 7
1	2	1	2				
2	1	2					
3	0						
4	0						
5	0						
6	0						
7	1	1					
8	2	1	2				
9	0						
10	0						

(NSUR(I, J))

TABLE 3.6

Array NSUR after Processing of Elements 1–18

Node I	Total number of surrounding elements J = 1	Identification numbers of surrounding elements					
		J = 2	J = 3	J = 4	J = 5	J = 6	J = 7
1	2	1	2				
2	3	2	3	4			
3	3	4	5	6			
4	3	6	7	8			
5	3	8	9	10			
6	1	10					
7	3	1	11	12			
8	6	1	2	3	12	13	14
9	6	3	4	5	14	15	16
10	6	5	6	7	16	17	18

(NSUR(I, J))

3.2.2   Computer Program

```
C       .....FINITE ELEMENT METHOD,PROGRAM 2.....
C       PORTION OF A HEAT SINK,CONSISTING OF A
C       TWO-DIMENSIONAL SQUARE BAR.
C       THE RESULTING SYSTEM MATRIX EQUATION IS SOLVED
C       USING THE STANDARD LIBRARY SUBROUTINE LEQT1F.
C
C       .....THE FOLLOWING IS A LIST OF SYMBOLS USED.....
C       NPOIN      -      TOTAL NUMBER OF NODES
C       NELEM      -      TOTAL NUMBER OF ELEMENTS
C       NPRES      -      TOTAL NUMBER OF NODES WHERE THE
C                         FUNCTION IS PRESCRIBED
C       NPT(I)     -      NODE NUMBER OF NODE WHERE THE FUNCTION
C                         IS PRESCRIBED,WHERE I=1,2,....,NPRES
C       VAL(I)     -      THE PRESCRIBED VALUE OF THE FUNCTION AT
C                         NODE NPT(I),WHERE I=1,2,....,NPRES
C       X(I),Y(I)  -      X,Y COORDINATES,RESPECTIVELY,OF NODE I
C       NOD(I,J)   -      THE THREE NODES,CORRESPONDING TO THE THREE
C                         NODE IDENTIFIERS J=1,2,3,OF ELEMENT I
C       XX(I),YY(I)-      X,Y COORDINATES,RESPECTIVELY,OF THE THREE
C                         NODES,CORRESPONDING TO THE THREE NODE
C                         IDENTIFIERS I=1,2,3,OF ANY ELEMENT
C       A(I),B(I),
C       C(I)       -      THE PARAMETERS DEFINED IN EQS.(2.11)
C                         OF ANY ELEMENT,WHERE I=1,2,3
C       DELTA      -      AREA OF ANY TRIANGLE
C       STE(IE,M,N)-      THE M-N TH ENTRY OF THE ELEMENT K MATRIX
C                         OF ELEMENT IE,WHERE M=1,2,3 AND N=1,2,3 ARE
C                         THE NODE IDENTIFIERS
C       ST(I,J)    -      THE I-J TH ELEMENT OF THE SYSTEM K MATRIX
C       RHS(I,1)   -      RIGHT-HAND SIDE MATRIX IN THE SYSTEM MATRIX
C                         EQUATION-DOUBLY SUBSCRIPTED TO SATISFY THE
C                         REQUIREMENTS OF THE SUBROUTINE LEQT1F
C       NSUR(I,J)  -      AN ARRAY CONTAINING THE TOTAL NUMBER OF
C                         ELEMENTS SURROUNDING NODE I IN COLUMN J=1,
C                         AND THE ELEMENT IDENTIFICATION NUMBERS
C                         IN COLUMNS J=2,3,ETC.
        PROGRAM PRGM2(INPUT,OUTPUT,TAPE5,TAPE6)
        DIMENSION NOD(50,3),X(36),Y(36),NPT(11),VAL(11),NSUR(36,7)
        DIMENSION XX(3),YY(3),A(3),B(3),C(3),STE(50,3,3),ST(36,36)
        DIMENSION RHS(36,1),WKAREA(36)
C
C       .....THE TOTAL NUMBER OF NODES,THE TOTAL NUMBER OF
C             ELEMENTS,AND THE TOTAL NUMBER OF NODES WHERE THE
C             FUNCTION IS PRESCRIBED ARE READ IN.....
        READ(5,10) NPOIN,NELEM,NPRES
10      FORMAT(3I3)
C
C       .....THE THREE NODES,CORRESPONDING TO THE THREE NODE
C             IDENTIFIERS J=1,2,AND 3,ARE READ IN FOR ALL ELEMENTS.....
        READ(5,20) (I,(NOD(I,J),J=1,3),II=1,NELEM)
20      FORMAT(24I3)
C
C       .....THE X AND Y COORDINATES ARE READ IN FOR ALL NODES.....
        READ(5,30) (I,X(I),Y(I),J=1,NPOIN)
30      FORMAT(5(I3,2F5.1))
C
C       .....THE NODES WHERE THE FUNCTION IS PRESCRIBED AND
C             THEIR PRESCRIBED VALUES ARE READ IN.....
        READ(5,40) (NPT(I),VAL(I),I=1,NPRES)
40      FORMAT(7(I3,F7.2))
```

```
C
C         .....THE TOTAL NUMBER OF NODES IS PRINTED OUT.....
          WRITE(6,50) NPOIN
50        FORMAT(/////,1X,22H TOTAL NUMBER OF NODES,I3)
C
C         .....THE TOTAL NUMBER OF ELEMENTS IS PRINTED OUT.....
          WRITE(6,60) NELEM
60        FORMAT(1X,25H TOTAL NUMBER OF ELEMENTS,I3)
C
C         .....THE TOTAL NUMBER OF NODES WHERE THE FUNCTION IS
C             PRESCRIBED IS PRINTED OUT.....
          WRITE(6,70) NPRES
70        FORMAT(1X,37H TOTAL NUMBER OF PRESCRIBED VARIABLES,I3)
C
C         .....THE X AND Y COORDINATES ARE PRINTED OUT
C             FOR ALL NODES.....
          WRITE(6,80)
80        FORMAT(//,1X,40H THE NODES AND THEIR X AND Y COORDINATES,/)
          WRITE(6,90)
90        FORMAT(1X,17H NODE    X       Y,2(21H     NODE    X       Y))
          WRITE(6,100) (I,X(I),Y(I),I=1,NPOIN)
100       FORMAT((1X,I3,F8.2,F7.2,2(I6,F8.2,F7.2)))
C
C         .....THE ELEMENTS AND THEIR THREE NODES,CORRESPONDING TO THE
C             THREE NODE IDENTIFIERS J=1,2,AND 3,ARE PRINTED OUT.....
          WRITE(6,110)
110       FORMAT(//,1X,29H THE ELEMENTS AND THEIR NODES,/)
          WRITE(6,120)
120       FORMAT(1X,13HELEM  I  J  M,3(16H   ELEM  I  J  M))
          WRITE(6,130) (I,(NOD(I,J),J=1,3),I=1,NELEM)
130       FORMAT((1X,I3,I4,2I3,3(I6,I4,2I3)))
C
C         .....THE NODES WHERE THE FUNCTION IS PRESCRIBED
C             AND THEIR PRESCRIBED VALUES ARE PRINTED OUT.....
          WRITE(6,140)
140       FORMAT(//,1X,38H NODES WITH PRESCRIBED FUNCTION VALUES,/)
          WRITE(6,150)
150       FORMAT(1X,12H NODE  VALUE,3(15H    NODE   VALUE))
          WRITE(6,160) (NPT(I),VAL(I),I=1,NPRES)
160       FORMAT((1X,I3,F9.3,3(I6,F9.3)))
C
C         .....THE ARRAY NSUR(P,Q),CONTAINING THE TOTAL NUMBER OF
C             ELEMENTS SURROUNDING NODE P IN COLUMN Q=1,AND
C             THE IDENTIFICATION NUMBERS OF THE SURROUNDING
C             ELEMENTS IN COLUMNS Q=2,3,ETC.,IS DETERMINED.....
          DO 170 I=1,NPOIN
170       NSUR(I,1)=0
          DO 190 I=1,NELEM
          DO 180 J=1,3
          LK=NOD(I,J)
          NSUR(LK,1)=NSUR(LK,1)+1
          LL=NSUR(LK,1)+1
180       NSUR(LK,LL)=I
190       CONTINUE
C
C         .....THE ELEMENT K MATRICES,BASED ON THE NODE IDENTIFIERS
C             1,2,AND 3,ARE OBTAINED FOR ALL ELEMENTS AND STORED
C             IN MEMORY.....
          DO 270 I=1,NELEM
          DO 200 J=1,3
          LK=NOD(I,J)
          XX(J)=X(LK)
```

```
200     YY(J)=Y(LK)
        DO 240 J=1,3
        LK=J+1
        LL=J+2
        IF(LK-3)230,220,210
210     LK=1
        LL=2
        GO TO 230
220     LL=1
230     A(J)=XX(LK)*YY(LL)-XX(LL)*YY(LK)
        B(J)=YY(LK)-YY(LL)
240     C(J)=XX(LL)-XX(LK)
        DELTA=(C(3)*B(2)-C(2)*B(3))/2.0
        DO 260 IR=1,3
        DO 250 IC=1,3
250     STE(I,IR,IC)=(B(IR)*B(IC)+C(IR)*C(IC))/(4.0*DELTA)
260     CONTINUE
270     CONTINUE
C
C       .....THE ELEMENT K MATRICES OF ALL ELEMENTS ARE ASSEMBLED
C             BY NODES,THE PRESCRIBED BOUNDARY CONDITIONS ARE
C             INSERTED,AND THE FINAL SYSTEM MATRIX EQUATION IS
C             OBTAINED.....
        DO 290 I=1,NPOIN
        DO 280 J=1,NPOIN
280     ST(I,J)=0.0
290     RHS(I,1)=0.0
        DO 370 NODE=1,NPOIN
        DO 310 I=1,NPRES
        IF(NODE-NPT(I))310,300,310
300     ST(NODE,NODE)=1.0
        RHS(NODE,1)=VAL(I)
        GO TO 370
310     CONTINUE
        IE=NSUR(NODE,1)
        IEL=IE+1
        DO 360 ITEL=2,IEL
        LEL=NSUR(NODE,ITEL)
        DO 320 I=1,3
        IR=I
        IF(NOD(LEL,I)-NODE)320,340,320
320     CONTINUE
        WRITE(6,330)
330     FORMAT(//,1X,32H ERROR IN ELEMENT NODE NUMBERING)
        GO TO 410
340     DO 350 IC=1,3
        ICO=NOD(LEL,IC)
350     ST(NODE,ICO)=ST(NODE,ICO)+STE(LEL,IR,IC)
360     CONTINUE
370     CONTINUE
C
C       .....THE SYSTEM MATRIX EQUATION IS SOLVED USING
C             THE STANDARD LIBRARY SUBROUTINE LEQT1F.....
        MM=1
        IDGT=0
        CALL LEQT1F(ST,1,NPOIN,NPOIN,RHS,0,WKAREA,IER)
C
C       .....THE SOLUTION HAS BEEN OBTAINED AND IS PRINTED OUT.....
        WRITE(6,380)
380     FORMAT(//,1X,31H THE SOLUTION HAS BEEN OBTAINED)
        WRITE(6,390)
390     FORMAT(/,1X,12H NODE   TEMP,3(15H   NODE    TEMP))
        WRITE(6,400) (I,RHS(I,1),I=1,NPOIN)
400     FORMAT((1X,I3,F9.3,3(I6,F9.3)))
410     STOP
        END
```

3.2.3 Input Data

```
36 50 11
 1  1  8  7  2  1  2  8  3  2  9  8  4  2  3  9  5  3 10  9  6  3  4 10
 7  4 11 10  8  4  5 11  9  5 12 11 10  5  6 12 11  7 14 13 12  7  8 14
13  8 15 14 14  8  9 15 15  9 16 15 16  9 10 16 17 10 17 16 18 10 11 17
19 11 18 17 20 11 12 18 21 13 20 19 22 13 14 20 23 14 21 20 24 14 15 21
25 15 22 21 26 15 16 22 27 16 23 22 28 16 17 23 29 17 24 23 30 17 18 24
31 19 26 25 32 19 20 26 33 20 27 26 34 20 21 27 35 21 28 27 36 21 22 28
37 22 29 28 38 22 23 29 39 23 30 29 40 23 24 30 41 25 32 31 42 25 26 32
43 26 33 32 44 26 27 33 45 27 34 33 46 27 28 34 47 28 35 34 48 28 29 35
49 29 36 35 50 29 30 36
 1  0.0  0.0  2  2.0  0.0  3  4.0  0.0  4  6.0  0.0  5  8.0  0.0
 6 10.0  0.0  7  0.0  2.0  8  2.0  2.0  9  4.0  2.0 10  6.0  2.0
11  8.0  2.0 12 10.0  2.0 13  0.0  4.0 14  2.0  4.0 15  4.0  4.0
16  6.0  4.0 17  8.0  4.0 18 10.0  4.0 19  0.0  6.0 20  2.0  6.0
21  4.0  6.0 22  6.0  6.0 23  8.0  6.0 24 10.0  6.0 25  0.0  8.0
26  2.0  8.0 27  4.0  8.0 28  6.0  8.0 29  8.0  8.0 30 10.0  8.0
31  0.0 10.0 32  2.0 10.0 33  4.0 10.0 34  6.0 10.0 35  8.0 10.0
36 10.0 10.0
 6 200.00 12 160.00 18 120.00 24  80.00 30  40.00 31 200.00 32 160.00
33 120.00 34  80.00 35  40.00 36   0.00
```

3.2.4 Results

```
TOTAL NUMBER OF NODES 36
TOTAL NUMBER OF ELEMENTS 50
TOTAL NUMBER OF PRESCRIBED VARIABLES 11

THE NODES AND THEIR X AND Y COORDINATES
```

NODE	X	Y	NODE	X	Y	NODE	X	Y
1	0.00	0.00	2	2.00	0.00	3	4.00	0.00
4	6.00	0.00	5	8.00	0.00	6	10.00	0.00
7	0.00	2.00	8	2.00	2.00	9	4.00	2.00
10	6.00	2.00	11	8.00	2.00	12	10.00	2.00
13	0.00	4.00	14	2.00	4.00	15	4.00	4.00
16	6.00	4.00	17	8.00	4.00	18	10.00	4.00
19	0.00	6.00	20	2.00	6.00	21	4.00	6.00
22	6.00	6.00	23	8.00	6.00	24	10.00	6.00
25	0.00	8.00	26	2.00	8.00	27	4.00	8.00
28	6.00	8.00	29	8.00	8.00	30	10.00	8.00
31	0.00	10.00	32	2.00	10.00	33	4.00	10.00
34	6.00	10.00	35	8.00	10.00	36	10.00	10.00

```
THE ELEMENTS AND THEIR NODES
```

ELEM	I	J	M	ELEM	I	J	M	ELEM	I	J	M	ELEM	I	J	M
1	1	8	7	2	1	2	8	3	2	9	8	4	2	3	9
5	3	10	9	6	3	4	10	7	4	11	10	8	4	5	11
9	5	12	11	10	5	6	12	11	7	14	13	12	7	8	14
13	8	15	14	14	8	9	15	15	9	16	15	16	9	10	16
17	10	17	16	18	10	11	17	19	11	18	17	20	11	12	18
21	13	20	19	22	13	14	20	23	14	21	20	24	14	15	21
25	15	22	21	26	15	16	22	27	16	23	22	28	16	17	23
29	17	24	23	30	17	18	24	31	19	26	25	32	19	20	26
33	20	27	26	34	20	21	27	35	21	28	27	36	21	22	28
37	22	29	28	38	22	23	29	39	23	30	29	40	23	24	30
41	25	32	31	42	25	26	32	43	26	33	32	44	26	27	33
45	27	34	33	46	27	28	34	47	28	35	34	48	28	29	35
49	29	36	35	50	29	30	36								

```
NODES WITH PRESCRIBED FUNCTION VALUES

NODE   VALUE      NODE   VALUE      NODE   VALUE      NODE   VALUE
  6   200.000      12   160.000      18   120.000      24    80.000
 30    40.000      31   200.000      32   160.000      33   120.000
 34    80.000      35    40.000      36     0.000

THE SOLUTION HAS BEEN OBTAINED

NODE    TEMP      NODE    TEMP      NODE    TEMP      NODE    TEMP
  1   132.826      2   132.826      3   133.925      4   139.417
  5   156.020      6   200.000      7   132.826      8   132.277
  9   131.728     10   133.861     11   142.331     12   160.000
 13   133.925     14   131.728     15   126.849     16   121.969
 17   119.444     18   120.000     19   139.417     20   133.861
 21   121.969     22   107.723     23    93.476     24    80.000
 25   156.020     26   142.331     27   119.444     28    93.476
 29    66.738     30    40.000     31   200.000     32   160.000
 33   120.000     34    80.000     35    40.000     36     0.000
```

These results are plotted in the form of isotherms, that is, lines of constant temperature, in Fig. 3.3.

Exercise 3.1   Modify the program given in Section 3.2.2, where appropriate, to solve the heat conduction problem of Section 2.1.1 using the 16 triangular elements of Fig. 2.4. Verify the results previously obtained.

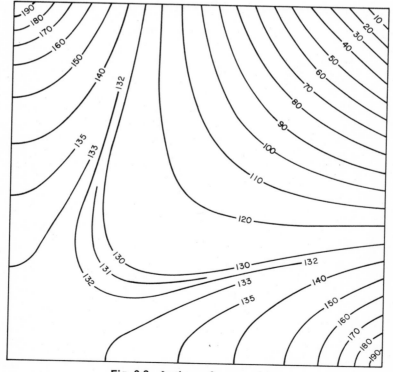

Fig. 3.3   Isotherms for square bar.

Exercise 3.2   Modify the program of Section 3.2.2 according to the *assembly by nodes* algorithm described following Eq. (2.40). The element matrices are not to be computed, and the program is to be structured throughout on an assembly by nodes basis. Verify the results previously obtained, using the same input data.

Exercise 3.3   Rewrite the program given in Section 3.2.2 so that it is structured throughout on an *assembly by elements* basis. Thus, each element should be taken in turn and its element matrix calculated and assembled into the system matrix. The array NSUR is not required. Verify the results previously obtained, using the same input data.

Exercise 3.4   Develop computer programs using the formulations of Exercises 2.4 and 2.5 to solve the conduction problem of Fig. 2.1, with the finite element mesh shown in Fig. 2.7. Compare the results with those for the quadratic trial function given in Eqs. (2.99).

Exercise 3.5   Use one of the computer programs developed in Exercise 3.4 to obtain the isotherms for the two-dimensional heat conduction problems shown in Figs. 3.4 and 3.5. In both cases, subdivide the region into the finite element mesh shown in Fig. 5.2.

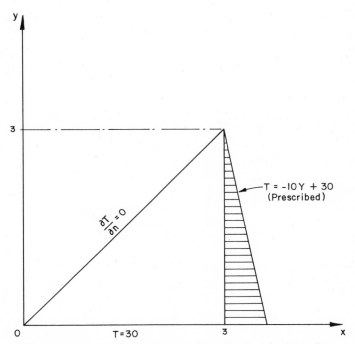

Fig. 3.4   Heat conduction in a two-dimensional triangular region.

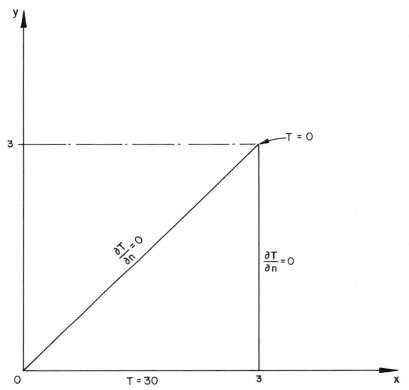

Fig. 3.5 Heat conduction in a two-dimensional triangular region.

3.3 PROGRAM MODIFICATIONS

The finite element program given in Section 3.2.2 is quite straightforward and follows the formulation presented in Section 2.1. It is not, however, a particularly efficient program. Modifications to improve such a program are therefore described in this section. The first four of these are quite simple and worth including in any finite element program whether large or small. The remaining modifications are more sophisticated and are especially valuable for larger, more complex problems.

3.3.1 Symmetry of System Matrix

As noted previously, the system **K** matrix is commonly symmetric *before* the boundary conditions are introduced. In fact, it can be shown that the variational finite element method always yields a linear system matrix equation, with a symmetric system **K** matrix if the functional is quadratic or quadratic-linear [2].

Whenever the system **K** matrix is symmetric, computation and storage of the matrix can be restricted to either the upper or lower triangular portions with the remaining elements generated from symmetry as needed.

3.3.2 Subprograms

Segments of a main computer program can often with advantage be replaced by subprograms. Liberal use of subprograms not only gives a simpler main program, but makes the whole program easier to follow, document and lay out, which are features especially important if modifications are subsequently made. Replacement of one element type by another, for example, can be accomplished by the substitution of a different subprogram. The widely adopted *modular concept*, by which finite element programs for various problems and elements are assembled from prewritten segments would be difficult to implement without subprograms. Another advantage is that each subprogram can be run and tested as a separate program, which greatly eases debugging.

3.3.3 Improved Simultaneous Equation Solvers

The standard library subroutine LEQTIF used to solve the system matrix equation in Section 3.2.2 is not particularly suited for finite element use since it does not take advantage of either the symmetry or banded nature of the system matrix. In Chapters 6 and 10, procedures that make use of such matrix properties will be discussed, and it will be shown how the problem characteristics can be used to determine the most suitable method.

An equation solver designed for a symmetric **K** matrix is more efficient and uses less storage than the corresponding procedure for an unsymmetric matrix. If a symmetric solver is to be used, it is essential that the original symmetry of the system **K** matrix is retained when correcting for Dirichlet boundary conditions. Two methods that allow such boundary conditions to be inserted without disturbing the symmetry are therefore discussed in the next section.

3.3.4 Retaining Symmetry when Inserting Boundary Conditions

Method 1 This procedure [3] can most simply be presented through an example. Consider the system matrix equation

$$
\begin{bmatrix}
K_{11} & K_{12} & K_{13} & K_{14} & K_{15} \\
K_{21} & K_{22} & K_{23} & K_{24} & K_{25} \\
K_{31} & K_{32} & K_{33} & K_{34} & K_{35} \\
K_{41} & K_{42} & K_{43} & K_{44} & K_{45} \\
K_{51} & K_{52} & K_{53} & K_{54} & K_{55}
\end{bmatrix}
\begin{bmatrix}
\bar{\phi}_1 \\
\bar{\phi}_2 \\
\bar{\phi}_3 \\
\bar{\phi}_4 \\
\bar{\phi}_5
\end{bmatrix}
=
\begin{bmatrix}
R_1 \\
R_2 \\
R_3 \\
R_4 \\
R_5
\end{bmatrix}.
\tag{3.1}
$$

Suppose the boundary conditions to be inserted are

$$\bar{\phi}_1 = c_1, \qquad \bar{\phi}_4 = c_4. \tag{3.2}$$

Since $\bar{\phi}_1$ and $\bar{\phi}_4$ are explicitly specified, the previously obtained equations for $\bar{\phi}_1$ and $\bar{\phi}_4$ in Eq. (3.1) are inapplicable and must be replaced by Eqs. (3.2). The substitution disturbs the symmetry of the **K** matrix, but this can be restored by inserting $\bar{\phi}_1 = c_1$ and $\bar{\phi}_4 = c_4$ in each of the other component equations and modifying the right-hand sides accordingly. For example, the second component equation in Eq. (3.1) is

$$K_{21}\bar{\phi}_1 + K_{22}\bar{\phi}_2 + K_{23}\bar{\phi}_3 + K_{24}\bar{\phi}_4 + K_{25}\bar{\phi}_5 = R_2, \tag{3.3}$$

which, after substitution for $\bar{\phi}_1$ and $\bar{\phi}_4$ from Eqs. (3.2), reduces to

$$K_{21}c_1 + K_{22}\bar{\phi}_2 + K_{23}\bar{\phi}_3 + K_{24}c_4 + K_{25}\bar{\phi}_5 = R_2 \tag{3.4}$$

or

$$0 + K_{22}\bar{\phi}_2 + K_{23}\bar{\phi}_3 + 0 + K_{25}\bar{\phi}_5 = (R_2 - K_{21}c_1 - K_{24}c_4). \tag{3.5}$$

Using this procedure for the component equations, Eq. (3.1) becomes

$$
\begin{bmatrix}
1 & 0 & 0 & 0 & 0 \\
0 & K_{22} & K_{23} & 0 & K_{25} \\
0 & K_{32} & K_{33} & 0 & K_{35} \\
0 & 0 & 0 & 1 & 0 \\
0 & K_{52} & K_{53} & 0 & K_{55}
\end{bmatrix}
\begin{bmatrix}
\bar{\phi}_1 \\
\bar{\phi}_2 \\
\bar{\phi}_3 \\
\bar{\phi}_4 \\
\bar{\phi}_5
\end{bmatrix}
=
\begin{bmatrix}
c_1 \\
R_2 - K_{21}c_1 - K_{24}c_4 \\
R_3 - K_{31}c_1 - K_{34}c_4 \\
c_4 \\
R_5 - K_{51}c_1 - K_{54}c_4
\end{bmatrix}, \tag{3.6}
$$

which is seen to be symmetric. The method can be summarized in the following rule:

If a nodal variable $\bar{\phi}_i$ is prescribed as $\bar{\phi}_i = c_i$, replace R_i by the prescribed value and K_{ii} by unity, overwriting the remainder of the ith row and the ith column of **K** with zeros. Then, for each nonprescribed R_j, subtract $K_{ji}c_i$ from R_j.

It is worth pointing out that for a large system of equations with many prescribed Dirichlet conditions of the type (3.2), the above procedure can be used to reduce the size of the final system matrix. For example, Eq. (3.6) can be written as

$$
\begin{bmatrix}
K_{22} & K_{23} & K_{25} \\
K_{32} & K_{33} & K_{35} \\
K_{52} & K_{53} & K_{55}
\end{bmatrix}
\begin{bmatrix}
\bar{\phi}_2 \\
\bar{\phi}_3 \\
\bar{\phi}_5
\end{bmatrix}
=
\begin{bmatrix}
R_2 - K_{21}c_1 - K_{24}c_4 \\
R_3 - K_{31}c_1 - K_{34}c_4 \\
R_5 - K_{51}c_1 - K_{54}c_4
\end{bmatrix} \tag{3.7}
$$

since $\bar{\phi}_1$ and $\bar{\phi}_4$ are already known from Eq. (3.2), which reduces the number of unknowns from 5 to 3.

Method 2   The Payne–Irons rule, already mentioned in Chapter 1, *approximates* the prescribed boundary conditions to a high degree of accuracy. It is summarized again here in the following way:

If a nodal variable $\bar{\phi}_i$ is prescribed as $\bar{\phi}_i = c_i$, replace K_{ii} by $K_{ii}B$ and R_i by $K_{ii}c_iB$, where B is a large number, say, 1.0×10^{12}. Repeat for the remaining prescribed variables.

The first method increases the sparseness of the system matrix, which may sometimes be an advantage but the second method is somewhat easier to program. An advantage of retaining symmetry with either method is that only half the final system matrix needs to be stored.

3.3.5   Macroprogramming and Overlays

When a finite element program is used to solve a number of problems of the same type but of different size, the changing of the DIMENSION statements to suit each problem soon becomes tedious. In large programs, to economize on central memory, EQUIVALENCE and COMMON statements[†] may be used but any subsequent change to DIMENSION statements generally requires considerable familiarity with the program, particularly where storage of an array overlaps another array by means of an EQUIVALENCE statement. Both of these problems can be overcome by *macrocoding*.

A *macro* is defined as a sequence of source statements that are saved separately and subsequently assembled into the program through a macro call whenever required. Simply stated, a macro is a named program segment that can be called up whenever that particular coding is required in the main program or in any of its subprograms. It is usually handled by a processor separate from that for the main body of the program and in this way differs from a SUBROUTINE. For FORTRAN, a macro processor that is often used is that developed by Day and Shaw at University College, London [4].

In general, the size of the arrays in a finite element program relate to characteristic parameters of the problem, such as number of nodes and order of interpolation. For each subprogram, the array dimensions can be defined by a macro in terms of these parameters. The dimensions for the main program are obtained by calling these subprogram macros plus one or more macros specific to the main program. For a different problem, all dimension changes are accomplished by simply changing those few macro statements incorporating characteristic parameters [4].

[†] EQUIVALENCE assigns two or more variables in the same program or subprogram to the same storage location, as distinct from COMMON which assigns a variable from a subprogram to the same storage location as a variable from the main program (or another subprogram).

With very large programs, particularly when core storage is limited, it is often advantageous to divide the program into independent parts, called *overlays*, which are loaded and executed as required. The manner in which the overlays are interconnected depends upon the computer architecture and its operating system.

3.3.6   Automatic Mesh Generation

For problems with large numbers of elements, the preparation of input data can be tedious and time-consuming. Automatic mesh generation is one way of speeding this task. When provided with certain basic information, such as the location of corner and boundary nodes, element type, and density of elements, a typical mesh generation program will compute the desired mesh and provide a listing of the nodes and their coordinates. While such a program can be incorporated within a finite element solution program, it is usual to run it separately so that the output can be checked. Combining the automatic mesh generation program with computer graphics is particularly valuable since the output can be monitored visually, and even presented in isometric or perspective views. For further information on mesh generation, see [5-9].

3.3.7   Node Renumbering

As will be seen in Chapter 6, the bandwidth of the system **K** matrix depends upon the manner in which the nodes are numbered. Unless the equation solver is such that zeros in the band are excluded from computation, the time and cost of solution will be dependent upon the bandwidth. In such cases, it is desirable to number the nodes in such a way that the bandwidth is minimized. Although this is easy to do in simple problems, this may not be the case for complex geometries. Fortunately, there are available programs that will take the input data for an already-numbered region and renumber the nodes so that the bandwidth is minimized. Used as a preprogram before the main finite element program, with a reverse-numbering program as a post-program, the *renumbering algorithm* can be contained internally within the processing between the original input data and the final output data. The original numbering is thus carried through to the output data for the problem. For further information on node renumbering programs, see [10-13].

3.3.8   Meshing of Assembly and Solution

As will be seen in Chapter 10, the complete system **K** matrix need not be retained in core storage during the solution of the system matrix equation. Typically, for the direct solution methods, only an active triangle of matrix elements is required, whereas for the indirect methods often only several rows are needed at any one time.

With large finite element programs, core storage on the available computer may be insufficient to retain the full system **K** matrix in memory. In this case the necessary active area of the matrix can successively be read in from auxiliary storage and the system matrix equation solved progressively. Alternatively, that active portion of the matrix required at any one time can be assembled as it is needed. Such a *meshing* of assembly and matrix solving can very successfully economize on core storage but needs careful program design if the costs of the additional computing and auxiliary storage are not to outweigh the savings from core memory. When an elimination (reduction) direct solution method is used for the assembled portion of the system matrix, the meshing procedure is known as *simultaneous assembly and reduction*. The *frontal solution method* explored and propagated by Irons [14] and the *reordering method* of King [10] both illustrate the technique. Assembly and reduction, together with other approaches to reduce core storage, are outlined in the recent text by Bathe and Wilson [15].

REFERENCES

1. G. Birkhoff, Mathematics and computer science, *Amer. Sci.* **63**, No. 1, 83–91 (January–February 1975).
2. D. H. Norrie and G. de Vries, "The Finite Element Method." Academic Press, New York, 1973.
3. C. A. Felippa and R. W. Clough, The finite element method in solid mechanics, *in* "Numerical Solution of Field Problems in Continuum Physics" (*SIAM–AMS Proc.*), Vol. 2, pp. 210–252. Amer. Math. Soc., Providence, Rhode Island, 1970.
4. A. C. Day, A Macro-Processor for FORTRAN. Tech. Rep. No. 2, Computer Centre, University College, London, England (February 1971).
5. W. R. Buell and B. A. Bush, Mesh generation—a survey, *J. E. Ind.* **95**, No. 1, 332–338 (1973).
6. J. Suhara and J. Fukuda, Automatic mesh generation for finite element analysis, *Proc. U. S.–Japan Seminar Comput. Methods Struct. Mech. Design, 2nd, Berkeley, California, August 1972* (J. T. Oden, R. W. Clough, and Y. Yamamoto, eds.), pp. 607–624. UAH Press, Huntsville, Alabama, 1972.
7. H. A. Kamel and H. K. Eisenstein, Automatic mesh generation in two- and three-dimensional interconnected domains, *Proc. Symp. Internat. Un. Theor. Appl. Mech. High Speed Comput. Elastic Struct., Univ. Liege, Belgium, August 1970* (F. de Veubeke, ed.), pp. 455–475. Univ. of Liege Press, Belgium, 1971.
8. C. A. Felippa, An alpha numeric finite element mesh plotter, *Internat. J. Numer. Methods Engrg.* **5**, No. 2, 217–236 (1972).
9. O. C. Zienkiewicz and D. V. Phillips, An automatic mesh generation scheme for plane and curved surfaces by isoparametric coordinates, *Internat. J. Numer. Methods Engrg.* **3**, No. 4, 519–528 (1971).
10. I. P. King, An automatic re-ordering scheme for simultaneous equations derived from network systems, *Internat. J. Numer. Methods Engrg.* **2**, No. 4, 523–533 (1970).

11.   R. J. Collins, Bandwidth reduction by automatic renumbering, *Internat. J. Numer. Methods Engrg.* **6**, No. 3, 345–356 (1973).
12.   H. R. Grooms, Algorithm for matrix bandwidth reduction, *Proc. ASCE, J. Struct. Div.* **98**, St. 1, 203–214 (1972).
13.   G. Akras and G. Dhatt, An automatic relabelling algorithm for bandwidth minimization, *Proc. Canad. Congr. Appl. Mech., 5th, Fredericton. 26–30 May, 1975*, pp. 691–692.
14.   B. M. Irons, A frontal solution program for finite element analysis, *Internat. J. Numer. Methods Engrg.* **2**, 5–32 (1970).
15.   K-J. Bathe and E. J. Wilson, "Numerical Methods in Finite Element Analysis." Prentice-Hall, Englewood Cliffs, New Jersey, 1976.

4

BOUNDARY CONDITIONS

In the previous sections, the only boundary conditions considered were of the Dirichlet and Neumann types. Other boundary conditions exist, some of which are quite complex and difficult to incorporate in the finite element formulation. The present chapter defines the most commonly encountered types of boundary condition and then indicates how the functional can be modified so that boundary conditions of various kinds can be satisfied. In the closing section, other approaches to boundary conditions are briefly referenced.

4.1 CLASSIFICATION OF BOUNDARY CONDITIONS

The boundary conditions most frequently encountered in scientific and engineering problems are of the Dirichlet, Neumann, and Cauchy types, sometimes described as being of the first, second, and third kinds, respectively. Where a boundary has two or more sections, on each of which a different condition is specified, a mixed boundary condition is said to exist.

For the Dirichlet boundary condition, the value of the dependent variable is specified on the boundary S. Thus, in a two-dimensional problem where

ϕ is the dependent variable, the relation

$$\phi = g(x, y) \qquad \text{on } S, \tag{4.1}$$

prescribes a Dirichlet condition where $g(x, y)$ is explicitly known. The specification of voltage on a boundary in the case of an electrical conduction
field or of temperature in a thermal conduction field are examples of Dirichlet
boundary conditions.

For the Neumann boundary condition, the normal derivative of the
dependent variable is specified on the boundary. In a two-dimensional
problem, such a condition could be written as

$$\frac{\partial \phi}{\partial n} + p = 0 \qquad \text{on } S, \tag{4.2}$$

where p is explicitly known as a function of position and n is in the direction
of the normal to S. The kinematic boundary condition in a fluid field, for
which the normal component of fluid velocity at the boundary equals the
normal component of the boundary velocity, is a specific example of the
Neumann condition.

The Cauchy condition occurs when both the dependent variable and its
derivative are specified in relation to each other on the boundary, as in

$$\frac{\partial \phi}{\partial n} + p + q\phi = 0 \qquad \text{on } S, \tag{4.3}$$

where p and q are known functions of position along the boundary S. This
condition occurs, for example, when there is a resistance layer on the
boundary.

The above boundary conditions involve only the dependent variable
and/or its first derivative. More complex boundary conditions encountered
in practice can involve higher derivatives as well. The method described in
the next section, for incorporating boundary conditions in a finite element
formulation, is illustrated using low-order boundary conditions for simplicity. It can, however, be extended to more difficult cases.

4.2   BOUNDARY CONDITIONS VIA LINE OR SURFACE INTEGRALS

It was noted earlier that the boundary conditions in variational procedures
can be classified as either *principal* or *natural* boundary conditions. From
the variational calculus it is known that any trial function, in addition to
being admissible, must satisfy the principal boundary conditions, whereas
the natural boundary conditions are satisfied as a natural consequence of
the variational procedure.

In Chapter 7, the conditions under which the functional

$$\chi = \int_D \left[\left(\frac{\partial T}{\partial x} \right)^2 + \left(\frac{\partial T}{\partial y} \right)^2 \right] dx\,dy + \int_S pT\,dS \qquad (4.4)$$

becomes stationary will be examined in detail. In summary, it will be shown that *that particular function* $T(x, y)$ causing the functional to be stationary also satisfies the field equation

$$\nabla^2 T = 0 \qquad \text{in } D \qquad (4.5)$$

subject to the Dirichlet condition

$$T = g \qquad \text{on } S_1 \qquad (4.6a)$$

and the Neumann condition

$$\frac{\partial T}{\partial n} + p = 0 \qquad \text{on } S_2 \qquad (4.6b)$$

where the boundary $S = S_1 + S_2$ encloses D. In Eqs. (4.6) the variables g and p are given functions of position along S_1 and S_2, respectively.

It will be seen from Eq. (4.6b) that if p is chosen as zero on S_2, the problem defined by Eqs. (4.5) and (4.6) becomes identical to the heat conduction problem considered in Chapter 2. The second term in Eq. (4.4) vanishes and the functional reverts to that presented earlier. If these last few steps are reversed, a situation exists in which the functional for a problem with a Cauchy boundary condition is obtained from the functional appropriate to a Neumann boundary condition, simply by adding an extra integral. This raises the question of whether the boundary conditions for any problem can be modified as desired, by the inclusion of a suitable additional integral in the functional. In many cases this is so, although the determination of this *proper* additional integral is not always simple. For a two-dimensional problem, where this approach is possible, the extra integral will be a line integral. In the three-dimensional case, the additional term will be a surface integral. Further information on the relationship between such integrals and the boundary conditions of the problem will be found in standard treatments of the variational calculus [1, 2]. The remainder of this section will therefore focus on the question of how to deal with the line (surface) integral numerically, once the functional and its associated principal and natural boundary conditions have been obtained.

The incorporation of a line or surface integral into the finite element formulation will be illustrated through an example. Consider the quasi-harmonic equation in three dimensions

$$\frac{\partial}{\partial x}\left(k_x \frac{\partial T}{\partial x} \right) + \frac{\partial}{\partial y}\left(k_y \frac{\partial T}{\partial y} \right) + \frac{\partial}{\partial z}\left(k_z \frac{\partial T}{\partial z} \right) = R(x, y, z), \qquad (4.7)$$

where k_x, k_y, k_z, and R are functions of x, y, and z. This equation governs a number of physical phenomenon where the medium is *nonisotropic*, that is, its properties vary with direction. In this case, the form of the equation corresponds to the principal axes for the medium property k coinciding with the x, y, and z axes. For example, for thermal conductivity, T would be the temperature, R the internal heat generation, and k_x, k_y, and k_z the thermal conductivities in the x, y, and z directions, respectively. Other phenomena for which Eq. (4.7) applies can be found listed in various sources [3, 4 etc.]. If the medium is *isotropic*, $k_x = k_y = k_z$, and Eq. (4.7) reduces to Poisson's equation. If the source term $R(x, y, z)$ vanishes also, Eq. (4.7) becomes Laplace's equation.

Consider now the two-dimensional quasi-harmonic equation, which from Eq. (4.7) is given as

$$\frac{\partial}{\partial x}\left(k_x \frac{\partial T}{\partial x}\right) + \frac{\partial}{\partial y}\left(k_y \frac{\partial T}{\partial y}\right) = R(x, y). \tag{4.8}$$

Let the boundary conditions be a Dirichlet specification on part of the boundary

$$T = g(x, y) \qquad \text{on } S_1, \tag{4.9a}$$

and a Cauchy condition on the remaining part

$$k_x \frac{\partial T}{\partial x} n_x + k_y \frac{\partial T}{\partial y} n_y + p + qT = 0 \qquad \text{on } S_2, \tag{4.9b}$$

where n_x, n_y are the x, y components of the unit outward normal to S, and where g, p, and q are given functions of position along S_1 and S_2, respectively.

It can be shown using the variational calculus that the functional for this problem is

$$\chi = \frac{1}{2} \int\int_D \left[k_x \left(\frac{\partial T}{\partial x}\right)^2 + k_y \left(\frac{\partial T}{\partial y}\right)^2 + 2RT \right] dx\,dy + \int_{S_2} (pT + \tfrac{1}{2}qT^2)\,dS_2.$$
$$\tag{4.10}$$

Equation (4.10) can conveniently be written as

$$\chi = \chi_D + \chi_{S_2}, \tag{4.11}$$

where the right-hand side terms comprise the domain and surface integrals, respectively. The finite element formulation for a domain integral such as χ_D is similar to that presented in earlier chapters. The presence of the surface integral in the functional leads to additional terms in the element matrix equation, whose evaluation will now be considered.

It is recalled that the trial function $\hat{T}^e$, in polynomial form, can be written in terms of global coordinates x and y as

$$\hat{T}^e = \alpha_1 + \alpha_2 x + \alpha_3 y + \alpha_4 x^2 + \cdots, \qquad (4.12)$$

where the order of the series and the values of the coefficients are appropriate to the element being used. Applying Eq. (4.12), in turn, at the nodes of the element yields the set of equations

$$\mathbf{A}\boldsymbol{\alpha} = \mathbf{T}, \qquad (4.13)$$

where $\mathbf{A} = [a_{ij}]$ is a coefficient matrix involving nodal coordinates, $\boldsymbol{\alpha} = [\alpha_i]$ is the column matrix with entries $\alpha_1, \alpha_2, \ldots$, and $\mathbf{T} = [T_i]$ is the element nodal vector.[†] Inverting the matrix $\mathbf{A}$, denoting the inverse by $\mathbf{B} = \mathbf{A}^{-1}$ and premultiplying Eq. (4.13) by $\mathbf{B}$, yields $\boldsymbol{\alpha}$ as

$$\boldsymbol{\alpha} = \mathbf{B}\mathbf{T}. \qquad (4.14)$$

If Eq. (4.12) is written as

$$\hat{T}^e = \begin{bmatrix} 1 & x & y & x^2 & \cdots \end{bmatrix} \begin{bmatrix} \alpha_1 \\ \alpha_2 \\ \vdots \end{bmatrix}, \qquad (4.15)$$

the notation

$$\mathbf{X} = \begin{bmatrix} 1 & x & y & x^2 & \cdots \end{bmatrix}, \qquad (4.16)$$

allows the trial function $\hat{T}^e$ to be obtained in the form

$$\hat{T}^e = \mathbf{X}\boldsymbol{\alpha}. \qquad (4.17)$$

Consider now the line integral in Eq. (4.10), which can be written as a summation over those element boundaries that *coincide*[‡] with S_2. For an element that has a side along the boundary S_2, the element boundary contribution can, from Eq. (4.10), be written as

$$\chi^e_{S_2} = \int_0^{L^e} \left[p\underline{\hat{T}}^e + \tfrac{1}{2}q(\underline{\hat{T}}^e)^2 \right] dS, \qquad (4.18)$$

where L^e is the length of the side along S_2, and $\underline{\hat{T}}^e$ is the trial function representation *along this side*. More generally, the line integral of Eq. (4.10) is given in terms of element contributions as

$$\int_{S_2} (pT + \tfrac{1}{2}qT^2)dS_2 = \sum_{e=1}^{l} \chi^e_{S_2}, \qquad (4.19)$$

[†] The nodal parameters T_i will include derivatives of T in addition to the nodal values of T itself, if the element is of the derivative type (e.g., Hermitian).

[‡] Within the sense of the approximation.

where l is the total number of elements in D. The functional for the problem, given by Eq. (4.10), can thus be written in the form

$$\chi = \chi_D + \chi_{S_2} = \sum_{e=1}^{l} [\chi_D{}^e + \chi_{S_2}^e]. \tag{4.20}$$

It needs noting that although the summation in Eq. (4.20) is prescribed over all elements, only those elements that have a side along S_2 will actually contribute to χ_{S_2}.

If the equation of the side is given by

$$y = ax + b, \tag{4.21}$$

substitution into Eq. (4.16) allows the interpolation matrix of Eq. (4.15) to be written as

$$\underline{\mathbf{X}} = \begin{bmatrix} 1 & x & (ax + b) & x^2 & \cdots \end{bmatrix}. \tag{4.22}$$

A bar has been used in Eq. (4.22) to indicate that the matrix $\mathbf{X}$ is evaluated along the side. The trial function representation along the side, denoted by $\hat{T}^e$, can be written from Eqs. (4.15) and (4.22) as

$$\hat{T}^e = \underline{\mathbf{X}}\boldsymbol{\alpha}. \tag{4.23}$$

The element boundary contribution in Eq. (4.18), after substitution from Eq. (4.23), becomes

$$\chi_{S_2}^e = \int_0^{L^e} [p\underline{\mathbf{X}}\boldsymbol{\alpha} + \tfrac{1}{2}q(\underline{\mathbf{X}}\boldsymbol{\alpha})^2] \, dS. \tag{4.24}$$

Use of Eq. (4.14) allows Eq. (4.24) to be also written as

$$\chi_{S_2}^e = \int_0^{L^e} [p\underline{\mathbf{X}}\mathbf{B}\mathbf{T} + \tfrac{1}{2}q(\underline{\mathbf{X}}\mathbf{B}\mathbf{T})^2] \, dS. \tag{4.25}$$

From previous chapters it is clear that the element matrix equation, based on a functional containing the domain integral only, has the form

$$\partial \chi_D{}^e / \partial \mathbf{T} = \mathbf{k}^e \mathbf{T}, \tag{4.26}$$

For the functional of Eq. (4.10), the contribution $\partial \chi_{S_2}^e / \partial \mathbf{T}$ must be added to Eq. (4.26) to account for the line integral. From Eq. (4.25) this extra term is obtained by differentiation (Appendix B) as

$$\frac{\partial \chi_{S_2}^e}{\partial \mathbf{T}} = \int_0^{L^e} [p\mathbf{B}^{\mathrm{T}}\underline{\mathbf{X}}^{\mathrm{T}} + q\mathbf{B}^{\mathrm{T}}\underline{\mathbf{X}}^{\mathrm{T}}\underline{\mathbf{X}}\mathbf{B}\mathbf{T}] \, dS. \tag{4.27}$$

Since the matrix $\mathbf{B}$ is a function of nodal coordinates only, Eq. (4.27) can be written as

$$\frac{\partial \chi_{S_2}^e}{\partial \mathbf{T}} = \mathbf{B}^{\mathrm{T}} \left[\int_0^{L^e} p\underline{\mathbf{X}}^{\mathrm{T}} \, dS \right] + \mathbf{B}^{\mathrm{T}} \left[\int_0^{L^e} q\underline{\mathbf{X}}^{\mathrm{T}}\underline{\mathbf{X}} \, dS \right] \mathbf{B}\mathbf{T}. \tag{4.28}$$

Using the notation

$$\mathbf{P} = \int_0^{L^e} p\underline{\mathbf{X}}^\mathrm{T}\, dS, \qquad \mathbf{Q} = \int_0^{L^e} q\underline{\mathbf{X}}^\mathrm{T}\underline{\mathbf{X}}\, dS \tag{4.29}$$

allows Eq. (4.28) to be written as

$$\frac{\partial \chi_{S_2}^e}{\partial \mathbf{T}} = \mathbf{B}^\mathrm{T}\mathbf{P} + \mathbf{B}^\mathrm{T}\mathbf{Q}\mathbf{B}\mathbf{T}. \tag{4.30}$$

Adding Eqs. (4.26) and (4.30), finally yields the element matrix equation for an element that has a side coinciding with S_2 as

$$\frac{\partial \chi^e}{\partial \mathbf{T}} = \frac{\partial \chi_D^{\,e}}{\partial \mathbf{T}} + \frac{\partial \chi_{S_2}^e}{\partial \mathbf{T}} = \overline{\mathbf{k}}^e\mathbf{T} + \mathbf{f}^e, \tag{4.31}$$

where

$$\overline{\mathbf{k}}^e = \mathbf{k} + \mathbf{B}^\mathrm{T}\mathbf{Q}\mathbf{B}, \tag{4.32a}$$

$$\mathbf{f}^e = \mathbf{B}^\mathrm{T}\mathbf{P}. \tag{4.32b}$$

The additional terms $\mathbf{B}^\mathrm{T}\mathbf{Q}\mathbf{B}$ and $\mathbf{B}^\mathrm{T}\mathbf{P}$ in the element matrix equation have been evaluated in the foregoing with respect to a global frame of reference. Appropriate modification will allow local coordinates to be used instead, with a considerable simplification in the procedure.

4.3 ALTERNATIVE FORMULATIONS FOR BOUNDARY CONDITIONS

In addition to the procedure described in Section 4.2, there are other approaches to satisfying boundary conditions in the finite element method. It is shown in Chapter 7, for example, that constraint equations can be incorporated in a variational formulation through the use of *Lagrange multipliers*. Since boundary conditions can be regarded as constraint equations, the value of such a formulation is evident. In the Lagrange multiplier approach, the boundary conditions are introduced directly into the system matrix equation. Although this has the virtue of simplicity, it also has the disadvantage that the *augmented* system matrix equation must now be solved for additional unknowns, the Lagrange multipliers themselves. The details of this method are beyond the scope of this text and the reader is referred to the literature [5–7].

In preceding chapters, the elements have had only values of the function as nodal parameters. Elements that include derivatives of the function among the nodal parameters are also possible and are introduced in the next chapter. Such *derivative* or *Hermitian* elements have the advantage that boundary conditions involving function derivatives can be directly inserted into the element matrix as *equivalent Dirichlet conditions*. This approach to boundary conditions is also taken up in the next chapter.

REFERENCES

1. P. N. Berg, Calculus of variation, *in* "Handbook of Engineering Mechanics" (W. Flügge, ed.), Chapter 16. McGraw-Hill, New York, 1962.
2. R. S. Schecter, "The Variational Method in Engineering." McGraw-Hill, New York, 1967.
3. K. H. Huebner, "The Finite Element Method for Engineers." Wiley, New York, 1975.
4. D. H. Norrie and G. de Vries, "Finite Element Bibliography." Plenum, New York, 1976.
5. O. C. Zienkiewicz, "The Finite Element Method in Engineering Science." McGraw-Hill, New York, 1971.
6. R. H. Gallagher, "Finite Element Analysis." Prentice-Hall, Englewood Cliffs, New Jersey, 1975.
7. B. E. Green, R. E. Jones, R. W. McLay, and D. W. Strome, Generalized variational principles in the finite element method, *AIAA J.* **7**, No. 7, 1254–1269 (1969).

5

HERMITIAN ELEMENTS, CONDENSATION, AND COUPLED BOUNDARY CONDITIONS

In previous chapters the nodal parameters of the elements consisted of the values of the function (variable) at the nodes. Such elements, commonly known as *Lagrangian*, are frequently employed in practice. In many cases, however, it is advantageous to use elements that use the values of both the function and its derivative(s) at the nodes. These so-called *Hermitian* elements, although discussed subsequently, are illustrated in the present chapter, using as an example the four-node cubic triangle.

When Hermitian or *derivative* elements are employed in a local coordinate formulation, it is necessary to transform the derivatives from one coordinate system to another using the *transformation matrix*; this procedure is also presented in this chapter.

An advantage of Hermitian elements is that derivative boundary conditions (Neumann or Cauchy) can often be handled as *equivalent Dirichlet conditions*. Where the derivative conditions involve $\partial/\partial n$, however, *coupled equivalent Dirichlet* conditions may result and the insertion of these requires some care. Such coupled conditions are also considered in this chapter.

5.1 PROBLEM STATEMENT AND SELECTION OF ELEMENT

The Laplace problem has been used extensively in previous chapters to illustrate basic concepts of the variational finite element method. It is again considered in this chapter as a convenient vehicle for demonstration. It should be noted, however, that the principles presented are also applicable to phenomena in other areas. Consider, then, two-dimensional heat flow through the triangular region OAB shown in Fig. 5.1. The point B is maintained at $0°C$, the lower surface OA at $30°C$, and the side walls AB and BO are perfectly insulated. It is required to find the isotherms within the triangular block.

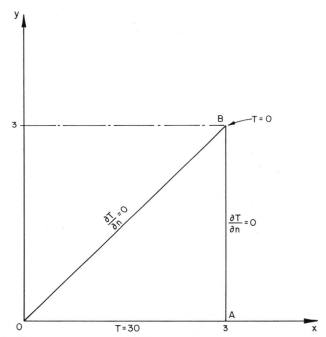

Fig. 5.1 Heat conduction in a two-dimensional triangular region.

Let the region be subdivided into nine three-node triangular elements (Fig. 5.2) with a total of 10 nodes, that is, $n = 10$ and $l = 9$. The problem is formulated in local coordinates with the frame of reference $\bar{O}\xi\eta$ chosen as shown in Fig. 5.3, where the node identifiers 1, 2, and 3 are assigned in a counterclockwise direction. The x and y coordinates for each node are listed in Table 5.1. For each element, the node identifiers and their relationship to the system node numbers, as well as the characteristic dimensions a, b, and c, are presented in Table 5.2.

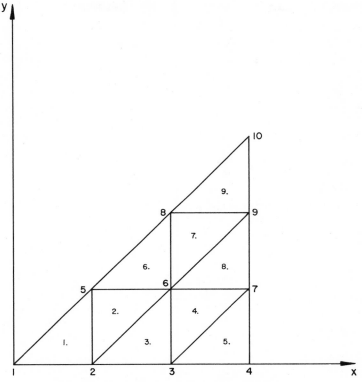

Fig. 5.2 Region subdivided into nine finite elements.

The trial function is chosen to be the complete cubic polynomial in ξ and η; hence for element e^i,

$$\hat{T}_L^e = \alpha_1 + \alpha_2\xi + \alpha_3\eta + \alpha_4\xi^2 + \alpha_5\xi\eta + \alpha_6\eta^2 + \alpha_7\xi^3$$
$$+ \alpha_8\xi^2\eta + \alpha_9\xi\eta^2 + \alpha_{10}\eta^3, \tag{5.1}$$

where the subscript L, indicating the local coordinate system, prevents confusion with the interpolation in global coordinates. To evaluate the 10 constants $\alpha_1, \alpha_2, \ldots, \alpha_{10}$, 10 nodal parameters need to be assigned to the element; these are chosen to be the three values of the function and its first derivatives at each node, together with the value of the function at the centroid,[†] as shown in Fig. 5.4.

[†] This choice makes the trial function continuous, with piecewise-continuous first derivatives, across the region.

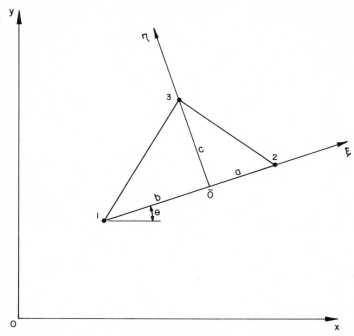

Fig. 5.3 Local coordinate system for the three-node triangular element.

TABLE 5.1

Nodal Coordinates

Node	Coordinate x	y	Node	Coordinate x	y
1	0	0	6	2	1
2	1	0	7	3	1
3	2	0	8	2	2
4	3	0	9	3	2
5	1	1	10	3	3

The coordinates of the centroid C, denoted $(\bar{\xi}, \bar{\eta})$, are given by

$$\bar{\xi} = \frac{a - b}{3}, \qquad \bar{\eta} = \frac{c}{3}. \tag{5.2}$$

The temperature at node 3 is denoted by $\bar{T}_3$, and its ξ and η derivatives by $\bar{T}_{\xi 3}$ and $\bar{T}_{\eta 3}$, respectively, with a similar convention applying at nodes 1 and 2. $\bar{T}_c$ is the temperature at the centroid.

TABLE 5.2

Node Numbers Corresponding to
Element Node Identifiers

Element	Node numbers			Parameters		
	1	2	3	a	b	c
1	2	5	1	1	0	1
2	5	2	6	1	0	1
3	3	6	2	1	0	1
4	6	3	7	1	0	1
5	4	7	3	1	0	1
6	6	8	5	1	0	1
7	8	6	9	1	0	1
8	7	9	6	1	0	1
9	9	10	8	1	0	1

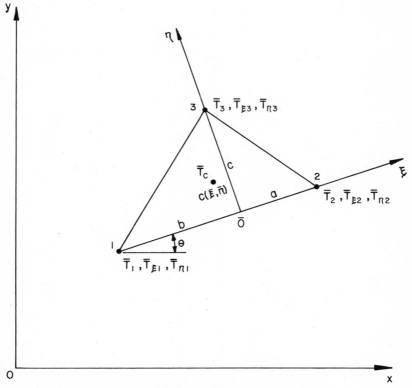

Fig. 5.4   Nodal parameters for the triangular element.

Applying the function of Eq. (5.1), its ξ derivative, and its η derivative, in that order, in turn to the nodes 1, 2, and 3, and finally to the centroid C, gives

$$
\begin{bmatrix}
1 & -b & 0 & b^2 & 0 & 0 & -b^3 & 0 & 0 & 0 \\
0 & 1 & 0 & -2b & 0 & 0 & 3b^2 & 0 & 0 & 0 \\
0 & 0 & 1 & 0 & -b & 0 & 0 & b^2 & 0 & 0 \\
1 & a & 0 & a^2 & 0 & 0 & a^3 & 0 & 0 & 0 \\
0 & 1 & 0 & 2a & 0 & 0 & 3a^2 & 0 & 0 & 0 \\
0 & 0 & 1 & 0 & a & 0 & 0 & a^2 & 0 & 0 \\
1 & 0 & c & 0 & 0 & c^2 & 0 & 0 & 0 & c^3 \\
0 & 1 & 0 & 0 & c & 0 & 0 & 0 & c^2 & 0 \\
0 & 0 & 1 & 0 & 0 & 2c & 0 & 0 & 0 & 3c^2 \\
1 & (a-b)/3 & c/3 & (a-b)^2/9 & c(a-b)/9 & c^2/9 & (a-b)^3/27 & c(a-b)^2/27 & c^2(a-b)/27 & c^3/27
\end{bmatrix}
\begin{bmatrix}
\alpha_1 \\ \alpha_2 \\ \alpha_3 \\ \alpha_4 \\ \alpha_5 \\ \alpha_6 \\ \alpha_7 \\ \alpha_8 \\ \alpha_9 \\ \alpha_{10}
\end{bmatrix}
=
\begin{bmatrix}
\bar{T}_1 \\ \bar{T}_{\xi 1} \\ \bar{T}_{\eta 1} \\ \bar{T}_2 \\ \bar{T}_{\xi 2} \\ \bar{T}_{\eta 2} \\ \bar{T}_3 \\ \bar{T}_{\xi 3} \\ \bar{T}_{\eta 3} \\ \bar{T}_c
\end{bmatrix}
\tag{5.3}
$$

The system of equations (5.3) can be written more simply in matrix form as

$$\mathbf{A\alpha} = \mathbf{T}_L. \tag{5.4}$$

In Eq. (5.4), the matrix $\mathbf{A}$ with elements a_{ij}, that is,

$$\mathbf{A} = [a_{ij}], \qquad i = 1, 2, \ldots, 10, \quad j = 1, 2, \ldots, 10, \tag{5.5}$$

is recognized as the coefficient matrix of Eq. (5.3), $\mathbf{\alpha}$ is the column matrix

$$\mathbf{\alpha} = [\alpha_i] = \begin{bmatrix} \alpha_1 \\ \alpha_2 \\ \vdots \\ \alpha_{10} \end{bmatrix}, \tag{5.6}$$

and $\mathbf{T}_L$ is the element nodal vector with respect to the local coordinate system, namely

$$\mathbf{T}_L = [T_j]_L = \begin{bmatrix} \bar{T}_1 \\ \bar{T}_{\xi 1} \\ \bar{T}_{\eta 1} \\ \bar{T}_2 \\ \bar{T}_{\xi 2} \\ \bar{T}_{\eta 2} \\ \bar{T}_3 \\ \bar{T}_{\xi 3} \\ \bar{T}_{\eta 3} \\ \bar{T}_c \end{bmatrix}. \tag{5.7}$$

Equation (5.4) can be solved uniquely for $\mathbf{\alpha}$ if and only if the matrix $\mathbf{A}$ is nonsingular, that is, if the determinant of $\mathbf{A}$ does not vanish. Evaluating the determinant yields

$$\det \mathbf{A} = |\mathbf{A}| = -c^7(a + b)^7/27. \tag{5.8}$$

Since the area of the triangle equals $\frac{1}{2}c(a + b)$ and never equals zero, it is seen from Eq. (5.8) that the determinant of $\mathbf{A}$ can never vanish and hence $\mathbf{A}$ is nonsingular and invertible. Consequently, premultiplying Eq. (5.4) by the inverse of $\mathbf{A}$, denoted by $\mathbf{A}^{-1}$, gives

$$\boldsymbol{\alpha} = \mathbf{A}^{-1}\mathbf{T}_L. \tag{5.9}$$

Introducing the notation

$$\mathbf{B} = \mathbf{A}^{-1} = [b_{ij}], \quad i = 1, 2, \ldots, 10, \quad j = 1, 2, \ldots, 10, \tag{5.10}$$

allows Eq. (5.9) to be written as

$$\boldsymbol{\alpha} = \mathbf{B}\mathbf{T}_L, \tag{5.11}$$

or, alternatively, using the summation convention, as

$$\alpha_i = b_{ij}T_j, \quad i = 1, 2, \ldots, 10, \quad j = 1, 2, \ldots, 10. \tag{5.12}$$

5.2   DETERMINATION OF THE ELEMENT MATRIX EQUATIONS

The trial function [Eq. (5.1)] can be written in the form

$$\hat{T}_L^e = \sum_{i=1}^{10} \alpha_i \xi^{m_i}\eta^{n_i}, \tag{5.13}$$

where the indices m_i and n_i are given in Table 5.3. Differentation of Eq. (5.13) with respect to ξ and η and substitution of these derivatives into the element contribution [Eq. (2.69)]

$$\chi^e = \frac{1}{2} \int \int_e \left[\left(\frac{\partial \hat{T}^e}{\partial \xi} \right)^2 + \left(\frac{\partial \hat{T}^e}{\partial \eta} \right)^2 \right] d\xi \, d\eta, \tag{5.14}$$

results in the same expression as obtained previously in Eq. (2.74), except

TABLE 5.3

Indices for the Polynomial Series

	Indices			Indices	
i	m_i	n_i	i	m_i	n_i
1	0	0	6	0	2
2	1	0	7	3	0
3	0	1	8	2	1
4	2	0	9	1	2
5	1	1	10	0	3

that the upper summation index 6 now becomes 10, that is,

$$\chi^e = \frac{1}{2} \sum_{i=1}^{10} \sum_{j=1}^{10} \alpha_i g_{ij} \alpha_j. \tag{5.15}$$

The g_{ij} are given by the same relationship as in Eq. (2.90), except for the range of the indices i, j; hence

$$g_{ij} = m_i m_j h(m_i + m_j - 2, n_i + n_j) + n_i n_j h(m_i + m_j, n_i + n_j - 2),$$
$$i = 1, 2, \ldots, 10, \quad j = 1, 2, \ldots, 10, \tag{5.16}$$

where from Eq. (2.89)

$$h(m, n) = \frac{c^{n+1}[a^{m+1} - (-b)^{m+1}]m!n!}{(m + n + 2)!}. \tag{5.17}$$

In matrix form, Eq. (5.15) can be written as

$$\chi^e = \tfrac{1}{2}\boldsymbol{\alpha}^T \mathbf{G} \boldsymbol{\alpha}, \tag{5.18}$$

which on account of Eq. (5.11) becomes

$$\chi^e = \tfrac{1}{2}\mathbf{T}_L{}^T(\mathbf{B}^T\mathbf{G}\mathbf{B})\mathbf{T}_L. \tag{5.19}$$

From the definition of $\mathbf{T}_L$ in Eq. (5.7), it is seen that this matrix lists the nodal parameters with respect to the local coordinate system $\bar{O}\xi\eta$. It is now necessary to make a transformation to the global system Oxy. The relationship between the first derivatives in the two systems was given in Chapter 2 as

$$\begin{bmatrix} \partial T/\partial \xi \\ \partial T/\partial \eta \end{bmatrix} = \begin{bmatrix} \cos\theta & \sin\theta \\ -\sin\theta & \cos\theta \end{bmatrix} \begin{bmatrix} \partial T/\partial x \\ \partial T/\partial y \end{bmatrix}. \tag{2.65b}$$

where θ is the angle between the two frames of reference. Consequently, the nodal parameters in the $\bar{O}\xi\eta$ system are related to those in the Oxy system through the matrix equation

$$\begin{bmatrix} \bar{T}_1 \\ \bar{T}_{\xi 1} \\ \bar{T}_{\eta 1} \\ \bar{T}_2 \\ \bar{T}_{\xi 2} \\ \bar{T}_{\eta 2} \\ \bar{T}_3 \\ \bar{T}_{\xi 3} \\ \bar{T}_{\eta 3} \\ \bar{T}_c \end{bmatrix} = \begin{bmatrix} 1 & 0 & 0 & 0 & 0 & 0 & 0 & 0 & 0 & 0 \\ 0 & \cos\theta & \sin\theta & 0 & 0 & 0 & 0 & 0 & 0 & 0 \\ 0 & -\sin\theta & \cos\theta & 0 & 0 & 0 & 0 & 0 & 0 & 0 \\ 0 & 0 & 0 & 1 & 0 & 0 & 0 & 0 & 0 & 0 \\ 0 & 0 & 0 & 0 & \cos\theta & \sin\theta & 0 & 0 & 0 & 0 \\ 0 & 0 & 0 & 0 & -\sin\theta & \cos\theta & 0 & 0 & 0 & 0 \\ 0 & 0 & 0 & 0 & 0 & 0 & 1 & 0 & 0 & 0 \\ 0 & 0 & 0 & 0 & 0 & 0 & 0 & \cos\theta & \sin\theta & 0 \\ 0 & 0 & 0 & 0 & 0 & 0 & 0 & -\sin\theta & \cos\theta & 0 \\ 0 & 0 & 0 & 0 & 0 & 0 & 0 & 0 & 0 & 1 \end{bmatrix} \begin{bmatrix} \bar{T}_1 \\ \bar{T}_{x1} \\ \bar{T}_{y1} \\ \bar{T}_2 \\ \bar{T}_{x2} \\ \bar{T}_{y2} \\ \bar{T}_3 \\ \bar{T}_{x3} \\ \bar{T}_{y3} \\ \bar{T}_c \end{bmatrix}, \tag{5.20}$$

which can be written as

$$\mathbf{T}_L = \mathbf{RT}, \tag{5.21}$$

where **T** is the element nodal vector with respect to the global system, that is,

$$\mathbf{T} = [T_j] = \begin{bmatrix} \bar{T}_1 \\ \bar{T}_{x1} \\ \bar{T}_{y1} \\ \bar{T}_2 \\ \bar{T}_{x2} \\ \bar{T}_{y2} \\ \bar{T}_3 \\ \bar{T}_{x3} \\ \bar{T}_{y3} \\ \bar{T}_c \end{bmatrix}, \tag{5.22}$$

and $\bar{T}_L$ is defined in Eq. (5.7). The transformation matrix **R** is defined as the coefficient matrix in Eq. (5.20).

Substituting Eq. (5.21) into Eq. (5.19) results in

$$\chi^e = \tfrac{1}{2}\mathbf{T}^T\mathbf{R}^T\mathbf{B}^T\mathbf{GBRT}, \tag{5.23}$$

or

$$\chi^e = \tfrac{1}{2}\mathbf{T}^T\mathbf{kT}, \tag{5.24}$$

where

$$\mathbf{k} = \mathbf{R}^T\mathbf{B}^T\mathbf{GBR}, \tag{5.25}$$

is the element **k** matrix. Differentiating Eq. (5.24) finally yields the element matrix equation for any element e as

$$\partial\chi^e/\partial\mathbf{T} = \mathbf{k}^e\mathbf{T}^e, \tag{5.26}$$

where the superscripts e have been added to distinguish between different elements. To illustrate the above procedure for the problem posed in Fig. 5.1, a, b, and c are substituted from Table 5.1 into the coefficient matrix **A** of Eq. (5.3) to give

$$\mathbf{A} = \begin{bmatrix} 1 & 0 & 0 & 0 & 0 & 0 & 0 & 0 & 0 & 0 \\ 0 & 1 & 0 & 0 & 0 & 0 & 0 & 0 & 0 & 0 \\ 0 & 0 & 1 & 0 & 0 & 0 & 0 & 0 & 0 & 0 \\ 1 & 1 & 0 & 1 & 0 & 0 & 1 & 0 & 0 & 0 \\ 0 & 1 & 0 & 2 & 0 & 0 & 3 & 0 & 0 & 0 \\ 0 & 0 & 1 & 0 & 1 & 0 & 0 & 1 & 0 & 0 \\ 1 & 0 & 1 & 0 & 0 & 1 & 0 & 0 & 0 & 1 \\ 0 & 1 & 0 & 0 & 1 & 0 & 0 & 0 & 1 & 0 \\ 0 & 0 & 1 & 0 & 0 & 2 & 0 & 0 & 0 & 3 \\ 1 & \frac{1}{3} & \frac{1}{3} & \frac{1}{9} & \frac{1}{9} & \frac{1}{9} & \frac{1}{27} & \frac{1}{27} & \frac{1}{27} & \frac{1}{27} \end{bmatrix}. \tag{5.27}$$

The inverse of this matrix will be found to be

$$\mathbf{B} = \mathbf{A}^{-1} = \begin{bmatrix} 1 & 0 & 0 & 0 & 0 & 0 & 0 & 0 & 0 & 0 \\ 0 & 1 & 0 & 0 & 0 & 0 & 0 & 0 & 0 & 0 \\ 0 & 0 & 1 & 0 & 0 & 0 & 0 & 0 & 0 & 0 \\ -3 & -2 & 0 & 3 & -1 & 0 & 0 & 0 & 0 & 0 \\ -13 & -3 & -3 & -7 & 2 & -1 & -7 & -1 & 2 & 27 \\ -3 & 0 & -2 & 0 & 0 & 0 & 3 & 0 & -1 & 0 \\ 2 & 1 & 0 & -2 & 1 & 0 & 0 & 0 & 0 & 0 \\ 13 & 3 & 2 & 7 & -2 & 2 & 7 & 1 & -2 & -27 \\ 13 & 2 & 3 & 7 & -2 & 1 & 7 & 2 & -2 & -27 \\ 2 & 0 & 1 & 0 & 0 & 0 & -2 & 0 & 1 & 0 \end{bmatrix}. \tag{5.28}$$

From Eq. (5.16) and Table 5.3, the matrix $\mathbf{G}$ is then given as Eq. (5.29). Evaluating Eq. (5.29) with the aid of Eq. (5.17) yields $\mathbf{G}$ as

$$\mathbf{G} = \frac{1}{180} \begin{bmatrix} 0 & 0 & 0 & 0 & 0 & 0 & 0 & 0 & 0 & 0 \\ 0 & 90 & 0 & 60 & 30 & 0 & 45 & 15 & 15 & 0 \\ 0 & 0 & 90 & 0 & 30 & 60 & 0 & 15 & 15 & 45 \\ 0 & 60 & 0 & 60 & 15 & 0 & 54 & 12 & 6 & 0 \\ 0 & 30 & 30 & 15 & 30 & 15 & 9 & 15 & 15 & 9 \\ 0 & 0 & 60 & 0 & 15 & 60 & 0 & 6 & 12 & 54 \\ 0 & 45 & 0 & 54 & 9 & 0 & 54 & 9 & 3 & 0 \\ 0 & 15 & 15 & 12 & 15 & 6 & 9 & 10 & 6 & 3 \\ 0 & 15 & 15 & 6 & 15 & 12 & 3 & 6 & 10 & 9 \\ 0 & 0 & 45 & 0 & 9 & 54 & 0 & 3 & 9 & 54 \end{bmatrix}. \tag{5.30}$$

The product $\mathbf{B}^{\mathrm{T}}\mathbf{G}\mathbf{B}$ in Eq. (5.25), after substitution from Eqs. (5.28) and (5.30), can be evaluated as

$$\mathbf{B}^{\mathrm{T}}\mathbf{G}\mathbf{B} = \frac{1}{180} \begin{bmatrix} 398 & 38 & 38 & 71 & -22 & 3 & 71 & 3 & -22 & -540 \\ 38 & 10 & 0 & 11 & -4 & -1 & 5 & 1 & -1 & -54 \\ 38 & 0 & 10 & 5 & -1 & 1 & 11 & -1 & -4 & -54 \\ 71 & 11 & 5 & 248 & -52 & 42 & 140 & 36 & -34 & -459 \\ -22 & -4 & -1 & -52 & 14 & -9 & -34 & -9 & 8 & 108 \\ 3 & -1 & 1 & 42 & -9 & 14 & 36 & 10 & -9 & -81 \\ 71 & 5 & 11 & 140 & -34 & 36 & 248 & 42 & -52 & -459 \\ 3 & 1 & -1 & 36 & -9 & 10 & 42 & 14 & -9 & -81 \\ -22 & -1 & -4 & -34 & 8 & -9 & -52 & -9 & 14 & 108 \\ -540 & -54 & -54 & -459 & 108 & -81 & -459 & -81 & 108 & 1458 \end{bmatrix}. \tag{5.31}$$

$$
\mathbf{G} =
\begin{bmatrix}
0 & 0 & 0 & 0 & 0 & 0 & 0 & 0 & 0 & 0 \\
0 & h(0,0) & 0 & 2h(1,0) & h(0,1) & 0 & 3h(2,0) & 2h(1,1) & h(0,2) & 0 \\
0 & 0 & h(0,0) & 0 & h(1,0) & 2h(0,1) & 0 & h(2,0) & 2h(1,1) & 3h(0,2) \\
0 & 2h(1,0) & 0 & 4h(2,0) & 2h(1,1) & 0 & 6h(3,0) & 4h(2,1) & 2h(1,2) & 0 \\
0 & h(0,1) & h(1,0) & 2h(1,1) & h(0,2)+h(2,0) & 2h(1,1) & 3h(2,1) & 2h(1,2)+h(3,0) & h(0,3)+2h(2,1) & 3h(1,2) \\
0 & 0 & 2h(0,1) & 0 & 2h(1,1) & 4h(0,2) & 0 & 2h(2,1) & 4h(1,2) & 6h(0,3) \\
0 & 3h(2,0) & 0 & 6h(3,0) & 3h(2,1) & 0 & 9h(4,0) & 6h(3,1) & 3h(2,2) & 0 \\
0 & 2h(1,1) & h(2,0) & 4h(2,1) & 2h(1,2)+h(3,0) & 2h(2,1) & 6h(3,1) & 4h(2,2)+h(4,0) & 2h(1,3)+2h(3,1) & 3h(2,2) \\
0 & h(0,2) & 2h(1,1) & 2h(1,2) & h(0,3)+2h(2,1) & 4h(1,2) & 3h(2,2) & 2h(1,3)+2h(3,1) & h(0,4)+4h(2,2) & 6h(1,3) \\
0 & 0 & 3h(0,2) & 0 & 3h(1,2) & 6h(0,3) & 0 & 3h(2,2) & 6h(1,3) & 9h(0,4)
\end{bmatrix}
\tag{5.29}
$$

The rotation matrix **R** for each element can be determined from Eq. (5.20). It will be seen from Fig. 5.2 and Table 5.1 that

$$\theta = 90°, \qquad \cos\theta = 0, \quad \sin\theta = 1 \qquad \text{for} \quad e = 1, 3, 5, 6, 8, 9, \qquad (5.32a)$$

$$\theta = -90°, \quad \cos\theta = 0, \quad \sin\theta = -1 \quad \text{for} \quad e = 2, 4, 7, \qquad (5.32b)$$

and that there are therefore only two different rotation matrices to be evaluated. Substitution of Eqs. (5.32) into Eq. (5.20) gives these rotation matrices as

$$\mathbf{R} = \begin{bmatrix} 1 & 0 & 0 & 0 & 0 & 0 & 0 & 0 & 0 & 0 \\ 0 & 0 & 1 & 0 & 0 & 0 & 0 & 0 & 0 & 0 \\ 0 & -1 & 0 & 0 & 0 & 0 & 0 & 0 & 0 & 0 \\ 0 & 0 & 0 & 1 & 0 & 0 & 0 & 0 & 0 & 0 \\ 0 & 0 & 0 & 0 & 0 & 1 & 0 & 0 & 0 & 0 \\ 0 & 0 & 0 & 0 & -1 & 0 & 0 & 0 & 0 & 0 \\ 0 & 0 & 0 & 0 & 0 & 0 & 1 & 0 & 0 & 0 \\ 0 & 0 & 0 & 0 & 0 & 0 & 0 & 0 & 1 & 0 \\ 0 & 0 & 0 & 0 & 0 & 0 & 0 & -1 & 0 & 0 \\ 0 & 0 & 0 & 0 & 0 & 0 & 0 & 0 & 0 & 1 \end{bmatrix} \qquad (5.33)$$

for elements 1, 3, 5, 6, 8, 9, and

$$\mathbf{R} = \begin{bmatrix} 1 & 0 & 0 & 0 & 0 & 0 & 0 & 0 & 0 & 0 \\ 0 & 0 & -1 & 0 & 0 & 0 & 0 & 0 & 0 & 0 \\ 0 & 1 & 0 & 0 & 0 & 0 & 0 & 0 & 0 & 0 \\ 0 & 0 & 0 & 1 & 0 & 0 & 0 & 0 & 0 & 0 \\ 0 & 0 & 0 & 0 & 0 & -1 & 0 & 0 & 0 & 0 \\ 0 & 0 & 0 & 0 & 1 & 0 & 0 & 0 & 0 & 0 \\ 0 & 0 & 0 & 0 & 0 & 0 & 1 & 0 & 0 & 0 \\ 0 & 0 & 0 & 0 & 0 & 0 & 0 & -1 & 0 \\ 0 & 0 & 0 & 0 & 0 & 0 & 0 & 1 & 0 & 0 \\ 0 & 0 & 0 & 0 & 0 & 0 & 0 & 0 & 0 & 1 \end{bmatrix} \qquad (5.34)$$

for elements 2, 4, 7.

Evaluating the products $\mathbf{R}^T\mathbf{B}^T\mathbf{GBR}$ using Eqs. (5.31), (5.33), and (5.34) now yields the element **k** matrices as

$$
\mathbf{k} = \mathbf{R}^\mathrm{T}\mathbf{B}^\mathrm{T}\mathbf{GBR} = \frac{1}{180}
\begin{bmatrix}
398 & -38 & 38 & 71 & -3 & -22 & 71 & 22 & 3 & -540 \\
-38 & 10 & 0 & -5 & 1 & 1 & -11 & -4 & 1 & 54 \\
38 & 0 & 10 & 11 & 1 & -4 & 5 & 1 & 1 & -54 \\
71 & -5 & 11 & 248 & -42 & -52 & 140 & 34 & 36 & -459 \\
-3 & 1 & 1 & -42 & 14 & 9 & -36 & -9 & -10 & 81 \\
-22 & 1 & -4 & -52 & 9 & 14 & -34 & -8 & -9 & 108 \\
71 & -11 & 5 & 140 & -36 & -34 & 248 & 52 & 42 & -459 \\
22 & -4 & 1 & 34 & -9 & -8 & 52 & 14 & 9 & -108 \\
3 & 1 & 1 & 36 & -10 & -9 & 42 & 9 & 14 & -81 \\
-540 & 54 & -54 & -459 & 81 & 108 & -459 & -108 & -81 & 1458
\end{bmatrix}
$$

$$(5.35)$$

for elements 1, 3, 5, 6, 8, 9, and

$$
\mathbf{k} = \mathbf{R}^\mathrm{T}\mathbf{B}^\mathrm{T}\mathbf{GBR} = \frac{1}{180}
\begin{bmatrix}
398 & 38 & -38 & 71 & 3 & 22 & 71 & -22 & -3 & -540 \\
38 & 10 & 0 & 5 & 1 & 1 & 11 & -4 & 1 & -54 \\
-38 & 0 & 10 & -11 & 1 & -4 & -5 & 1 & 1 & 54 \\
71 & 5 & -11 & 248 & 42 & 52 & 140 & -34 & -36 & -459 \\
3 & 1 & 1 & 42 & 14 & 9 & 36 & -9 & -10 & -81 \\
22 & 1 & -4 & 52 & 9 & 14 & 34 & -8 & -9 & -108 \\
71 & 11 & -5 & 140 & 36 & 34 & 248 & -52 & -42 & -459 \\
-22 & -4 & 1 & -34 & -9 & -8 & -52 & 14 & 9 & 108 \\
-3 & 1 & 1 & -36 & -10 & -9 & -42 & 9 & 14 & 81 \\
-540 & -54 & 54 & -459 & -81 & -108 & -459 & 108 & 81 & 1458
\end{bmatrix}
$$

$$(5.36)$$

for elements 2, 4, 7.

5.3 CONDENSATION

For the centroidal node, the only contributing element is the element containing that node; hence from the standard minimizing conditions there is obtained

$$
\frac{\partial \chi}{\partial \overline{T}_c} = \sum_{e=1}^{l} \frac{\partial \chi^e}{\partial \overline{T}_c} = \frac{\partial \chi^{e_c}}{\partial \overline{T}_c} = 0, \tag{5.37}
$$

where C refers to the centroid of element e_c. Using Eq. (5.37), it is possible to eliminate the centroidal nodal parameter from each element matrix equation. The procedure is called *condensation* and can be used, more generally, to eliminate any internal nodal parameters. Using Eq. (5.26) with Eq. (5.37), yields for the centroidal nodal parameter $\overline{T}_c$ of any element e

$$
\frac{\partial \chi}{\partial \overline{T}_c} = k_{10,j}T_j = 0, \qquad j = 1, 2, \ldots, 10, \tag{5.38}
$$

or

$$\frac{\partial \chi}{\partial \bar{T}_c} = k_{10,j} T_j + k_{10,10} \bar{T}_c = 0, \qquad j = 1, 2, \ldots, 9, \qquad (5.39)$$

where the T_j are the elements of the matrix $\mathbf{T}$ [see Eq. (5.22)] and where the superscripts on the k's has been dropped for convenience. Equation (5.39) may be solved for the centroidal nodal parameter to give

$$\bar{T}_c = -\frac{k_{10,j}}{k_{10,10}} T_j, \qquad j = 1, 2, \ldots, 9. \qquad (5.40)$$

Substitution of Eq. (5.40) into the remaining nine expressions for

$$\frac{\partial \chi}{\partial \bar{T}_i} = k_{ij} T_j, \qquad i = 1, 2, \ldots, 9, \quad j = 1, 2, \ldots, 10, \qquad (5.41)$$

gives the condensed form of the element matrix equation as

$$\frac{\partial \chi}{\partial \bar{T}_i} = k_{ij} T_j - \frac{k_{i,10} k_{10,j}}{k_{10,10}} T_j, \qquad i = 1, 2, \ldots, 9, \quad j = 1, 2, \ldots, 9. \qquad (5.42)$$

From Eq. (5.42) the condensed element $\mathbf{k}$ matrix, denoted by $\bar{\mathbf{k}}$ and referred to as the *element $\bar{\mathbf{k}}$ matrix*, is seen to have as its elements

$$\bar{k}_{ij} = k_{ij} - \frac{k_{i,10} k_{10,j}}{k_{10,10}}, \qquad i = 1, 2, \ldots, 9, \quad j = 1, 2, \ldots, 9. \qquad (5.43)$$

Using the element $\mathbf{k}$ matrix given in Eq. (5.35) for elements 1, 3, 5, 6, 8, 9, it is noted from Eq. (5.43) that the entries of the element $\bar{\mathbf{k}}$ matrix for these elements become

$$\bar{k}_{ij} = k_{ij} - \frac{k_{i,10} k_{10,j}}{1458}. \qquad (5.44)$$

Evaluating Eq. (5.44), using the entries k_{ij} from Eq. (5.35), gives

$$\bar{\mathbf{k}} = \frac{1}{180} \begin{bmatrix} 198 & -18 & 18 & -99 & 27 & 18 & -99 & -18 & -27 \\ -18 & 8 & 2 & 12 & -2 & -3 & 6 & 0 & 4 \\ 18 & 2 & 8 & -6 & 4 & 0 & -12 & -3 & -2 \\ -99 & 12 & -6 & 103.5 & -16.5 & -18 & -4.5 & 0 & 10.5 \\ 27 & -2 & 4 & -16.5 & 9.5 & 3 & -10.5 & -3 & -5.5 \\ 18 & -3 & 0 & -18 & 3 & 6 & 0 & 0 & -3 \\ -99 & 6 & -12 & -4.5 & -10.5 & 0 & 103.5 & 18 & 16.5 \\ -18 & 0 & -3 & 0 & -3 & 0 & 18 & 6 & 3 \\ -27 & 4 & -2 & 10.5 & -5.5 & -3 & 16.5 & 3 & 0.5 \end{bmatrix} \qquad (5.45)$$

for elements 1, 3, 5, 6, 8, 9. In similar fashion there is obtained

$$\bar{k} = \frac{1}{180} \begin{bmatrix} 198 & 18 & -18 & -99 & -27 & -18 & -99 & 18 & 27 \\ 18 & 8 & 2 & -12 & -2 & -3 & -6 & 0 & 4 \\ -18 & 2 & 8 & 6 & 4 & 0 & 12 & -3 & -2 \\ -99 & -12 & 6 & 103.5 & 16.5 & 18 & -4.5 & 0 & -10.5 \\ -27 & -2 & 4 & 16.5 & 9.5 & 3 & 10.5 & -3 & -5.5 \\ -18 & -3 & 0 & 18 & 3 & 6 & 0 & 0 & -3 \\ -99 & -6 & 12 & -4.5 & 10.5 & 0 & 103.5 & -18 & -16.5 \\ 18 & 0 & -3 & 0 & -3 & 0 & -18 & 6 & 3 \\ 27 & 4 & -2 & -10.5 & -5.5 & -3 & -16.5 & 3 & 9.5 \end{bmatrix} \quad (5.46)$$

for elements 2, 4, 7.

5.4 ASSEMBLY AND INSERTION OF BOUNDARY CONDITIONS

Assembly into the system **K** matrix can now take place, either by nodes or by elements. After the system matrix equation has been obtained, it must be corrected for the Dirichlet boundary conditions. From Figs. 5.2 and 5.3, it is seen that the *principal* boundary conditions are

$$\bar{T}_1 = 30, \qquad \bar{T}_2 = 30, \qquad \bar{T}_3 = 30,$$
$$\bar{T}_4 = 30, \qquad \text{and} \qquad \bar{T}_{10} = 0, \qquad (5.47)$$

which may be inserted in the usual way.

The Neumann conditions

$$\partial T / \partial n = 0 \qquad \text{on } AB, \qquad (5.48a)$$

$$\partial T / \partial n = 0 \qquad \text{on } BO, \qquad (5.48b)$$

are *natural* boundary conditions and are satisfied automatically, at least in the same approximate sense as the rest of the solution. An alternative to leaving these conditions [Eqs. (5.48)] to be satisfied *naturally* and *approximately* is their exact prescription through the nodal values of the appropriate derivatives. The Neumann boundary conditions then become what will here be referred to as *equivalent Dirichlet conditions*.

The Neumann condition [Eq. (5.48a)] can be written as

$$\frac{\partial T}{\partial n}\bigg|_{\text{along } AB} = \frac{\partial T}{\partial x}\bigg|_{\text{along } AB} = 0, \qquad (5.49)$$

and the appropriate nodal values along AB can therefore be prescribed as

$$\bar{T}_{x4} = 0, \qquad \bar{T}_{x7} = 0, \qquad \bar{T}_{x9} = 0, \qquad \bar{T}_{x10} = 0. \qquad (5.50)$$

Along BO, the Neumann boundary condition [Eq. (5.48b)] can be written as

$$\frac{\partial T}{\partial n} = \frac{\partial T}{\partial x} n_x + \frac{\partial T}{\partial y} n_y = 0 \qquad \text{on } BO, \qquad (5.51)$$

where n_x and n_y are the x and y components of the unit outward normal $\mathbf{n}$ to BO, respectively, given by $-1/\sqrt{2}$ and $1/\sqrt{2}$. Consequently Eq. (5.51) reduces to

$$-\frac{\partial T}{\partial x} + \frac{\partial T}{\partial y} = 0 \qquad \text{on } BO. \qquad (5.52)$$

In terms of the appropriate nodal values on BO, the equivalent Dirichlet conditions become

$$-\bar{T}_{x1} + \bar{T}_{y1} = 0, \qquad (5.53a)$$
$$-\bar{T}_{x5} + \bar{T}_{y5} = 0, \qquad (5.53b)$$
$$-\bar{T}_{x8} + \bar{T}_{y8} = 0, \qquad (5.53c)$$
$$-\bar{T}_{x10} + \bar{T}_{y10} = 0. \qquad (5.53d)$$

From Eqs. (5.50), $\bar{T}_{x10} = 0$, and hence Eq. (5.53d) reduces to

$$\bar{T}_{y10} = 0. \qquad (5.54)$$

The remaining conditions, Eqs. (5.53a), (5.53b), and (5.53c), are not *independent* equivalent Dirichlet conditions, but are *coupled* equivalent Dirichlet conditions. There are two ways in which these can be incorporated in the formulation:

(i)   by defining the trial functions in those elements that have a node with a prescribed coupled Dirichlet condition in such a way that this coupling is built into the trial function; or

(ii)   by modifying the element $\bar{\mathbf{k}}$ matrix so that the coupled condition is satisfied.

The first method is considered in a subsequent exercise and the second method is outlined below. Consider the functional χ, written as

$$\chi = \chi(\bar{T}_1, \bar{T}_{x1}, \bar{T}_{y1}, \ldots, \bar{T}_5, \bar{T}_{x5}, \bar{T}_{y5}, \ldots, \bar{T}_{10}, \bar{T}_{x10}, \bar{T}_{y10}). \qquad (5.55a)$$

The minimization conditions and the element matrix equations have been derived on the assumption that each of the nodal variables on the right-hand side of Eq. (5.55) is independent. It is now proposed however to incorporate the constraint

$$\bar{T}_{y5} = \bar{T}_{x5}, \tag{5.56}$$

deriving from Eq. (5.53b) in the formulation. This means that $\bar{T}_{y5}$ in Eq. (5.55a) is not independent as previously assumed but is a function of $\bar{T}_{x5}$, and this equation therefore becomes

$$\chi = \chi(\bar{T}_1, \bar{T}_{x1}, \bar{T}_{y1}, \ldots, \bar{T}_5, \bar{T}_{x5}, \bar{T}_{y5}(\bar{T}_{x5}), \ldots, \bar{T}_{10}, \bar{T}_{x10}, \bar{T}_{y10}). \tag{5.55b}$$

Consequently, the previous minimization conditions

$$\partial \chi / \partial \bar{T}_{x5} = 0, \qquad \partial \chi / \partial \bar{T}_{y5} = 0 \tag{5.57a, b}$$

based on Eq. (5.55a), must be replaced by the single minimization condition

$$(\partial \chi / \partial \bar{T}_{x5})_m = 0 \tag{5.57c}$$

based on Eq. (5.55b), where the derivative on the left-hand side of Eq. (5.57c) is subscripted to indicate that it is not identical with the derivative on the left-hand side of Eq. (5.57a). From Eq. (5.55b) the chain rule of calculus [1] allows the *modified* minimization condition of Eq. (5.57c) to be written as

$$\left(\frac{\partial \chi}{\partial \bar{T}_{x5}}\right)_m = \frac{\partial \chi}{\partial \bar{T}_{x5}} + \frac{\partial \chi}{\partial \bar{T}_{y5}}\left(\frac{\partial \bar{T}_{y5}}{\partial \bar{T}_{x5}}\right) = 0, \tag{5.58a}$$

which by virtue of Eq. (5.56) becomes

$$\left(\frac{\partial \chi}{\partial \bar{T}_{x5}}\right)_m = \frac{\partial \chi}{\partial \bar{T}_{x5}} + \frac{\partial \chi}{\partial \bar{T}_{y5}} \cdot 1 = 0. \tag{5.58b}$$

The right-hand side derivatives in Eq. (5.58b) can be identified with those given in Eqs. (5.57a,b). Thus, Eq. (5.58b) shows how the two previously obtained nodal equations for $\bar{T}_{x5}$ and $\bar{T}_{y5}$ respectively, as listed in the original system matrix equation, are to be combined to obtain a single minimization condition when the constraint of Eq. (5.56) is imposed. In the present case, the previous nodal equation for $\bar{T}_{y5}$ is multiplied by 1 and added to the previous nodal equation for $\bar{T}_{x5}$ to obtain the single replacement nodal equation.

Instead of the required modifications being made to the system matrix equation, they can be made to the element $\bar{\mathbf{k}}$ matrices of each of the elements that surround node 5 (in this case e_1, e_2, and e_6). This follows from the fact that the $\bar{T}_{x5}$ and $\bar{T}_{y5}$ rows and columns in the system $\mathbf{K}$ matrix are assembled from the $\bar{T}_{x5}$ and $\bar{T}_{y5}$ rows and columns of the element $\bar{\mathbf{k}}$ matrices of these elements. This approach is presented below.

The changes in the element $\bar{\mathbf{k}}$ matrices necessary to impose the boundary condition given in Eq. (5.56) are the following. First, for those elements that have the node under consideration in common, the row corresponding to $\bar{T}_{y5}$ in the element $\bar{\mathbf{k}}$ matrix is multiplied by 1 and added to the row corresponding to $\bar{T}_{x5}$, in accordance with Eq. (5.58b). Secondly, to effect the substitution of $\bar{T}_{y5}$ by $\bar{T}_{x5}$ in the system equation, according to Eq. (5.56), the column corresponding to $\bar{T}_{y5}$ is multiplied by 1 and added to the column corresponding to $\bar{T}_{x5}$. Finally, since the dependency of $\bar{T}_{y5}$ is now incorporated in the nodal equation of $\bar{T}_{x5}$, the row and column corresponding to $\bar{T}_{y5}$ must be replaced by zeros to complete the elimination of the original nodal equation for $\bar{T}_{y5}$ from the formulation.

At this point, the following difficulty is noted and must be overcome. The zeros in the row corresponding to $\bar{T}_{y5}$ in the *system matrix* cause the matrix to be singular and prevent the subsequent solution of the system matrix equation. This can easily be circumvented by removing the equation for this variable from the system matrix equation (by deleting the appropriate row and column) and thereby *reducing* the size of the system matrix equation. Another possibility is to retain the equation for this variable, and to modify each contributing element $\bar{\mathbf{k}}$ matrix in either one of the following ways:

(1)   In the row corresponding to $\bar{T}_{y5}$, which was previously set to zero, enter a 1 on its diagonal position. After assembly of these element $\bar{\mathbf{k}}$ matrices, since the zero in the right-hand side matrix of the system matrix equation has not been modified, this procedure will produce the erroneous result

$$\bar{T}_{y5} = 0. \tag{5.59}$$

The result [Eq. (5.59)] is then discarded since it is known from Eq. (5.56) and the modification procedure outlined above, that $\bar{T}_{y5}$ in fact equals $\bar{T}_{x5}$. This approach, although it may not look appealing, is useful in practice since the inherent symmetry of the assembled system matrix is retained.

(2)   Insert the condition $-\bar{T}_{x5} + \bar{T}_{y5} = 0$ [see Eqs. (5.53b) and (5.56)] directly into the row corresponding to $\bar{T}_{y5}$ in the element $\bar{\mathbf{k}}$ matrix by entering -1 and 1 in the columns corresponding to $\bar{T}_{x5}$ and $\bar{T}_{y5}$, respectively. After assembly of the element $\bar{\mathbf{k}}$ matrices, since the original zero on the right-hand side of the system matrix equation has not been modified, this procedure incorporates Eq. (5.53b) as one of the system equations. This condition has already been enforced by the modifications to the element matrix equations

and there is no necessity for its restatement. It is noted, however, that this method makes the system $\mathbf{K}$ matrix nonsymmetric, which is often a disadvantage. Its virtue is that the correct value of $\bar{T}_{y5}$ appears as part of the solution.

Using the second of the above approaches, together with the preceding modifications, to impose the boundary conditions [Eqs. (5.53)] yields the corrected element $\bar{\mathbf{k}}$ matrix, denoted by $\bar{\mathbf{k}}_c$, for each of the elements 1, 6, 9 as

$$
\bar{\mathbf{k}}_c =
\begin{bmatrix}
198 & -18 & 18 & -99 & 45 & 0 & -99 & -45 & 0 \\
-18 & 8 & 2 & 12 & -5 & 0 & 6 & 4 & 0 \\
18 & 2 & 8 & -6 & 4 & 0 & -12 & -5 & 0 \\
-99 & 12 & -6 & 103.5 & -34.5 & 0 & -4.5 & 10.5 & 0 \\
45 & -5 & 4 & -34.5 & 21.5 & 0 & -10.5 & -11.5 & 0 \\
0 & 0 & 0 & 0 & -1 & 1 & 0 & 0 & 0 \\
-99 & 6 & -12 & -4.5 & -10.5 & 0 & 103.5 & 34.5 & 0 \\
-45 & 4 & -5 & 10.5 & -11.5 & 0 & 34.5 & 21.5 & 0 \\
0 & 0 & 0 & 0 & 0 & 0 & 0 & -1 & 1
\end{bmatrix}.
\quad (5.60)
$$

Similarly, for elements 2 and 7 the element $\bar{\mathbf{k}}$ matrix in its corrected form becomes

$$
\bar{\mathbf{k}}_c =
\begin{bmatrix}
198 & 0 & 0 & -99 & -27 & -18 & -99 & 18 & 27 \\
0 & 20 & 0 & -6 & 2 & -3 & 6 & -3 & 2 \\
0 & -1 & 1 & 0 & 0 & 0 & 0 & 0 & 0 \\
-99 & -6 & 0 & 103.5 & 16.5 & 18 & -4.5 & 0 & -10.5 \\
-27 & 2 & 0 & 16.5 & 9.5 & 3 & 10.5 & -3 & -5.5 \\
-18 & -3 & 0 & 18 & 3 & 6 & 0 & 0 & -3 \\
-99 & 6 & 0 & -4.5 & 10.5 & 0 & 103.5 & -18 & -16.5 \\
18 & -3 & 0 & 0 & -3 & 0 & -18 & 6 & 3 \\
27 & 2 & 0 & -10.5 & -5.5 & -3 & -16.5 & 3 & 9.5
\end{bmatrix}.
\quad (5.61)
$$

The element $\bar{\mathbf{k}}$ (or $\bar{\mathbf{k}}_c$) matrices can now be assembled into the system $\mathbf{K}$ matrix: for the elements 3, 5, 8, from Eq. (5.45); for 4, from Eq. (5.46); for 1, 6, 9, from Eq. (5.60); and for 2 and 7, from Eq. (5.61). Inserting the appropriate Dirichlet boundary conditions in the right-hand side matrix, and correcting the system $\mathbf{K}$ matrix accordingly, finally results in the system matrix equation (5.62). Only the nonzero elements are shown in the coefficient matrix of this equation.

$$
\begin{bmatrix}
1 \\
& 1 \\
& -1 & 1 \\
& & & 1 \\
& & & & 1 \\
-12 & -1 & & 52\tfrac{1}{2} & 8 & 23\tfrac{1}{2} & -27 & 4 & -2 & & & & -24 & 1 & & 10\tfrac{1}{2} & -5\tfrac{1}{2} \\
& & & & 1 \\
& & & & & 1 \\
& & -12 & -3 & -2 & 52\tfrac{1}{2} & 8 & 23\tfrac{1}{2} & -27 & 4 & -2 & & & & & -24 & 1 \\
& & & & & & 1 \\
& & & & & & & 1 \\
& & & -12 & -3 & -2 & 18 & 2 & 8 \\
-4\tfrac{1}{2} & 10\tfrac{1}{2} & -198 & -15 & -24 & & & & & & & & 405 & & & -198 & 24 \\
-10\tfrac{1}{2} & -11\tfrac{1}{2} & 39 & -3 & 1 & & & & & & & & & 63 & & -39 & 1 \\
& & & & & & & & & & & & & -3 & 3 \\
& & -9 & 10\tfrac{1}{2} & 10\tfrac{1}{2} & -198 & -15 & -24 & & & & & -198 & -39 & & 810 \\
& & -10\tfrac{1}{2} & -6 & -5\tfrac{1}{2} & 15 & -4 & 1 & & & & & 24 & 1 & & & 47 \\
& & -10\tfrac{1}{2} & -5\tfrac{1}{2} & -6 & 24 & 1 & & & & & & 15 & -3 & & & 16 \\
& & & & & -9 & 10\tfrac{1}{2} & 10\tfrac{1}{2} & -99 & 12 & -6 & & & & & -198 & -24 \\
& & & & & -10\tfrac{1}{2} & -5\tfrac{1}{2} & -6 & 18 & -3 & & & & & & 15 & 1 \\
& & & & & & & & & & & & -4\tfrac{1}{2} & 10\tfrac{1}{2} & & -198 & -15 \\
& & & & & & & & & & & & -10\tfrac{1}{2} & -11\tfrac{1}{2} & & 39 & -3 \\
& & & & & & & & & & & & & & & -9 & 10\tfrac{1}{2} \\
& & & & & & & & & & & & & & & -10\tfrac{1}{2} & -5\tfrac{1}{2}
\end{bmatrix}
$$

It is noted from Fig. 5.1 that since T is prescribed as a constant along OA, its derivative in the x direction must also vanish there, that is,

$$\partial T/\partial x = 0 \qquad \text{along } OA. \tag{5.63}$$

Consequently, the additional equivalent Dirichlet conditions

$$\bar{T}_{x1} = 0, \qquad \bar{T}_{x2} = 0, \qquad \bar{T}_{x3} = 0, \qquad \bar{T}_{x4} = 0, \tag{5.64}$$

were also imposed on the system matrix; see Eq. (5.62). Solution of Eq. (5.62) by any standard procedure yields the nodal parameters $\bar{T}_1, \bar{T}_{x_1}, \bar{T}_{y_1}, \ldots, \bar{T}_{y_{10}}$.

Exercise 5.1   In Section 5.4, it was shown that a coupled equivalent Dirichlet condition (specified with respect to the global coordinates) could be incorporated in the formulation by suitably modifying the element $\bar{\mathbf{k}}$ matrix after it is obtained in global form. This exercise shows that the same conditions (but written with respect to local coordinates) can be incorporated before the element $\mathbf{k}$ matrix is transformed from the local to the global

$$
\begin{bmatrix}
-6 & & & & & & & & \\
10\tfrac{1}{2} & -5\tfrac{1}{2} & -6 & & & & & & \\
-6 & 4 & & & & & & & \\
15 & & & & -4\tfrac{1}{2} & -10\tfrac{1}{2} & & & \\
-3 & & & & 10\tfrac{1}{2} & -11\tfrac{1}{2} & & & \\
-198 & 24 & 15 & -198 & 39 & & -9 & -10\tfrac{1}{2} & -10\tfrac{1}{2} \\
16 & -24 & 1 & -15 & -3 & & 10\tfrac{1}{2} & -6 & -5\tfrac{1}{2} \\
47 & -15 & 1 & -4 & -24 & 1 & 10\tfrac{1}{2} & -5\tfrac{1}{2} & -6 \\
-15 & 405 & -52\tfrac{1}{2} & -16\tfrac{1}{2} & & & -99 & 27 & 18 \\
 & 1 & & & & & & & \\
-4 & -16\tfrac{1}{2} & 8 & 23\tfrac{1}{2} & & & -6 & 4 & \\
-24 & & & 405 & & & -198 & 24 & 15 & -4\tfrac{1}{2} & -10\tfrac{1}{4} \\
1 & & & & 63 & & -39 & 1 & -3 & 10\tfrac{1}{2} & -11\tfrac{1}{2} \\
 & & & & -3 & 3 & & & & & \\
10\tfrac{1}{2} & -99 & 12 & -6 & -198 & -39 & 405 & -52\tfrac{1}{2} & -16\tfrac{1}{2} & -99 & 45 \\
 & & & & & & & 1 & & & \\
-6 & 18 & -3 & & 15 & -3 & -16\tfrac{1}{2} & 8 & 23\tfrac{1}{2} & -6 & 4 \\
 & & & & & & & & & 1 & \\
 & & & & & & & & & 1 & \\
 & & & & & & & & & -1 & 1
\end{bmatrix}
\begin{bmatrix}
T_1 \\ T_{x1} \\ T_{y1} \\ T_2 \\ T_{x2} \\ T_{y2} \\ T_3 \\ T_{x3} \\ T_{y3} \\ T_4 \\ T_{x4} \\ T_{y4} \\ T_5 \\ T_{x5} \\ T_{y5} \\ T_6 \\ T_{x6} \\ T_{y6} \\ T_7 \\ T_{x7} \\ T_{y7} \\ T_8 \\ T_{x8} \\ T_{y8} \\ T_9 \\ T_{x9} \\ T_{y9} \\ T_{10} \\ T_{x10} \\ T_{y10}
\end{bmatrix}
=
\begin{bmatrix}
30 \\ 0 \\ 0 \\ 30 \\ 0 \\ 0 \\ 30 \\ 0 \\ 0 \\ 30 \\ 0 \\ 0 \\ 0 \\ 0 \\ 0 \\ 0 \\ 0 \\ 0 \\ 0 \\ 0 \\ 0 \\ 0 \\ 0 \\ 0 \\ 0 \\ 0 \\ 0 \\ 0 \\ 0 \\ 0
\end{bmatrix}. \tag{5.62}
$$

system. This approach builds such coupled conditions into the element trial function. In Fig. 5.5, the global coordinate system Oxy and the local coordinate system $\bar{O}\xi\eta$ (see Fig. 5.3 and Table 5.2) is shown for element 1 of the subdivided region illustrated in Fig. 5.2.

(a) Using the transformation

$$
\begin{bmatrix} \xi \\ \eta \end{bmatrix} = \begin{bmatrix} \cos\theta & \sin\theta \\ -\sin\theta & \cos\theta \end{bmatrix} \begin{bmatrix} x \\ y \end{bmatrix}, \tag{5.65}
$$

and noting that

$$
\frac{\partial T}{\partial n} = \frac{\partial T}{\partial x} n_x + \frac{\partial T}{\partial y} n_y, \tag{5.66}
$$

where

$$
n_x = \frac{\partial x}{\partial n}, \qquad n_y = \frac{\partial y}{\partial n}, \tag{5.67}
$$

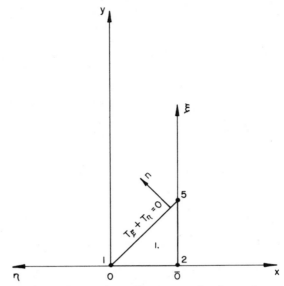

Fig. 5.5   Local coordinate system for element 1.

are the x, y components respectively of the unit outward normal $\mathbf{n}$ to side 5-1, (see Fig. 5.5), show that in the local system

$$\frac{\partial T}{\partial n} = \frac{\partial T}{\partial \xi} n_\xi + \frac{\partial T}{\partial \eta} n_\eta, \qquad (5.68)$$

where n_ξ and n_η are the ξ and η components of $\mathbf{n}$, respectively. Since $n_\xi = 1/\sqrt{2}$ and $n_\eta = 1/\sqrt{2}$, for both nodes 5 and 1, it follows that the previous coupled equivalent Dirichlet conditions [Eqs. (5.53a) and (5.53b)] can be expressed in local coordinates in the form

$$T_\xi + T_\eta = 0 \qquad \text{at nodes 5 and 1.} \qquad (5.69)$$

(b)   Differentiate the trial function [Eq. (5.1)] to obtain expressions for T_ξ and T_η, and substitute these relations together with the coordinates of nodes 5 and 1 into Eq. (5.69) to obtain the coupled boundary conditions at each of the nodes 5 and 1.

(c)   Form the coefficient matrix $\mathbf{A}$ as before [see Eq. (5.3)] and replace the equations corresponding to $\overline{T}_{\xi 5}$ and $\overline{T}_{\xi 1}$, in the fifth and eighth rows, respectively, by the coupled boundary conditions derived from Eq. (5.69).

Hence, show that

$$
\begin{bmatrix}
1 & 0 & 0 & 0 & 0 & 0 & 0 & 0 & 0 & 0 \\
0 & 1 & 0 & 0 & 0 & 0 & 0 & 0 & 0 & 0 \\
0 & 0 & 1 & 0 & 0 & 0 & 0 & 0 & 0 & 0 \\
1 & 1 & 0 & 1 & 0 & 0 & 1 & 0 & 0 & 0 \\
0 & 1 & 1 & 2 & 1 & 0 & 3 & 1 & 0 & 0 \\
0 & 0 & 1 & 0 & 1 & 0 & 0 & 1 & 0 & 0 \\
1 & 0 & 1 & 0 & 0 & 1 & 0 & 0 & 0 & 1 \\
0 & 1 & 1 & 0 & 1 & 2 & 0 & 0 & 1 & 3 \\
0 & 0 & 1 & 0 & 0 & 2 & 0 & 0 & 0 & 3 \\
1 & \frac{1}{3} & \frac{1}{3} & \frac{1}{9} & \frac{1}{9} & \frac{1}{9} & \frac{1}{27} & \frac{1}{27} & \frac{1}{27} & \frac{1}{27}
\end{bmatrix}
\begin{bmatrix}
\alpha_1 \\ \alpha_2 \\ \alpha_3 \\ \alpha_4 \\ \alpha_5 \\ \alpha_6 \\ \alpha_7 \\ \alpha_8 \\ \alpha_9 \\ \alpha_{10}
\end{bmatrix}
=
\begin{bmatrix}
\bar{T}_2 \\ \bar{T}_{\xi 2} \\ \bar{T}_{\eta 2} \\ \bar{T}_5 \\ 0 \\ \bar{T}_{\eta 5} \\ \bar{T}_1 \\ 0 \\ \bar{T}_{\eta 1} \\ \bar{T}_c
\end{bmatrix}.
\tag{5.70}
$$

(d) Invert the coefficient matrix $\mathbf{A}$ of Eq. (5.70) and verify that

$$
\mathbf{B} = \mathbf{A}^{-1} =
\begin{bmatrix}
1 & 0 & 0 & 0 & 0 & 0 & 0 & 0 & 0 & 0 \\
0 & 1 & 0 & 0 & 0 & 0 & 0 & 0 & 0 & 0 \\
0 & 0 & 1 & 0 & 0 & 0 & 0 & 0 & 0 & 0 \\
-3 & -2 & 0 & 3 & 0 & 1 & 0 & 0 & 0 & 0 \\
-13 & -3 & -3 & -7 & 0 & -3 & -7 & 0 & 3 & 27 \\
-3 & 0 & -2 & 0 & 0 & 0 & 3 & 0 & -1 & 0 \\
2 & 1 & 0 & -2 & 0 & -1 & 0 & 0 & 0 & 0 \\
13 & 3 & 2 & 7 & 0 & 4 & 7 & 0 & -3 & -27 \\
13 & 2 & 3 & 7 & 0 & 3 & 7 & 0 & -4 & -27 \\
2 & 0 & 1 & 0 & 0 & 0 & -2 & 0 & 1 & 0
\end{bmatrix}.
\tag{5.71}
$$

(e) Premultiply Eq. (5.70) by the inverse of $\mathbf{A}$ from Eq. (5.71), and show that

$$
\begin{bmatrix}
\alpha_1 \\ \alpha_2 \\ \alpha_3 \\ \alpha_4 \\ \alpha_5 \\ \alpha_6 \\ \alpha_7 \\ \alpha_8 \\ \alpha_9 \\ \alpha_{10}
\end{bmatrix}
=
\begin{bmatrix}
1 & 0 & 0 & 0 & 0 & 0 & 0 & 0 & 0 & 0 \\
0 & 1 & 0 & 0 & 0 & 0 & 0 & 0 & 0 & 0 \\
0 & 0 & 1 & 0 & 0 & 0 & 0 & 0 & 0 & 0 \\
-3 & -2 & 0 & 3 & 0 & 1 & 0 & 0 & 0 & 0 \\
-13 & -3 & -3 & -7 & 0 & -3 & -7 & 0 & 3 & 27 \\
-3 & 0 & -2 & 0 & 0 & 0 & 3 & 0 & -1 & 0 \\
2 & 1 & 0 & -2 & 0 & -1 & 0 & 0 & 0 & 0 \\
13 & 3 & 2 & 7 & 0 & 4 & 7 & 0 & -3 & -27 \\
13 & 2 & 3 & 7 & 0 & 3 & 7 & 0 & -4 & -27 \\
2 & 0 & 1 & 0 & 0 & 0 & -2 & 0 & 1 & 0
\end{bmatrix}
\begin{bmatrix}
\bar{T}_2 \\ \bar{T}_{\xi 2} \\ \bar{T}_{\eta 2} \\ \bar{T}_5 \\ 0 \\ \bar{T}_{\eta 5} \\ \bar{T}_1 \\ 0 \\ \bar{T}_{\eta 1} \\ \bar{T}_c
\end{bmatrix}.
\tag{5.72a}
$$

Also, verify that Eq. (5.72a) can be written in the following equivalent form

$$
\begin{bmatrix} \alpha_1 \\ \alpha_2 \\ \alpha_3 \\ \alpha_4 \\ \alpha_5 \\ \alpha_6 \\ \alpha_7 \\ \alpha_8 \\ \alpha_9 \\ \alpha_{10} \end{bmatrix}
=
\begin{bmatrix}
1 & 0 & 0 & 0 & 0 & 0 & 0 & 0 & 0 & 0 \\
0 & 1 & 0 & 0 & 0 & 0 & 0 & 0 & 0 & 0 \\
0 & 0 & 1 & 0 & 0 & 0 & 0 & 0 & 0 & 0 \\
-3 & -2 & 0 & 3 & 0 & 1 & 0 & 0 & 0 & 0 \\
-13 & -3 & -3 & -7 & 0 & -3 & -7 & 0 & 3 & 27 \\
-3 & 0 & -2 & 0 & 0 & 0 & 3 & 0 & -1 & 0 \\
2 & 1 & 0 & -2 & 0 & -1 & 0 & 0 & 0 & 0 \\
13 & 3 & 2 & 7 & 0 & 4 & 7 & 0 & -3 & -27 \\
13 & 2 & 3 & 7 & 0 & 3 & 7 & 0 & -4 & -27 \\
2 & 0 & 1 & 0 & 0 & 0 & -2 & 0 & 1 & 0
\end{bmatrix}
\begin{bmatrix} \bar{T}_2 \\ \bar{T}_{\xi 2} \\ \bar{T}_{\eta 2} \\ \bar{T}_5 \\ \bar{T}_{\xi 5} \\ \bar{T}_{\eta 5} \\ \bar{T}_1 \\ \bar{T}_{\xi 1} \\ \bar{T}_{\eta 1} \\ \bar{T}_c \end{bmatrix},
\qquad (5.72\text{b})
$$

where the coefficient matrix above will subsequently be referred to as the *modified* matrix **B**, and denoted as $\bar{\mathbf{B}}$.

(f)   Using **G** from Eq. (5.30) and the modified matrix $\bar{\mathbf{B}}$ from Eq. (5.72b), show that the product $\bar{\mathbf{B}}^{\mathsf{T}}\mathbf{G}\bar{\mathbf{B}}$ becomes

$$
= \frac{1}{180}
\begin{bmatrix}
398 & 38 & 38 & 71 & 0 & 25 & 71 & 0 & -25 & -540 \\
38 & 10 & 0 & 11 & 0 & 3 & 5 & 0 & -2 & -54 \\
38 & 0 & 10 & 5 & 0 & 2 & 11 & 0 & -3 & -54 \\
71 & 11 & 5 & 248 & 0 & 94 & 140 & 0 & -70 & -459 \\
0 & 0 & 0 & 0 & 0 & 0 & 0 & 0 & 0 & 0 \\
25 & 3 & 2 & 94 & 0 & 46 & 70 & 0 & -36 & -189 \\
71 & 5 & 11 & 140 & 0 & 70 & 248 & 0 & -94 & -459 \\
0 & 0 & 0 & 0 & 0 & 0 & 0 & 0 & 0 & 0 \\
-25 & -2 & -3 & -70 & 0 & -36 & -94 & 0 & 46 & 189 \\
-540 & -54 & -54 & -459 & 0 & -189 & -459 & 0 & 189 & 1458
\end{bmatrix}.
$$

$$(5.73)$$

(g)   Verify that the resulting rows and columns of zeros in the above matrix will assemble with similar rows and columns from the other elements surrounding the nodes 5 and 1 to yield corresponding rows and columns of zeros in the system matrix. The system **K** matrix will thus be singular. To avoid this and to ensure that values for $\bar{T}_{\xi 5}$ and $T_{\xi 1}$ are obtained as part of the final solution, the condition $T_{\xi} + T_{\eta} = 0$ at nodes 5 and 1 are to be inserted in the element **k** matrix. This can be accomplished by inserting the coefficients of the equations $\bar{T}_{\xi 5} + \bar{T}_{\eta 5} = 0$ and $T_{\xi 1} + \bar{T}_{\eta 1} = 0$ into the matrix given in Eq. (5.73), to yield the corrected form of $\bar{\mathbf{B}}^{\mathsf{T}}\mathbf{G}\bar{\mathbf{B}}$, denoted by $\overline{\bar{\mathbf{B}}^{\mathsf{T}}\mathbf{G}\bar{\mathbf{B}}}$, as

$$= \frac{1}{180}
\begin{bmatrix}
398 & 38 & 38 & 71 & 0 & 25 & 71 & 0 & -25 & -540 \\
38 & 10 & 0 & 11 & 0 & 3 & 5 & 0 & -2 & -54 \\
38 & 0 & 10 & 5 & 0 & 2 & 11 & 0 & -3 & -54 \\
71 & 11 & 5 & 248 & 0 & 94 & 140 & 0 & -70 & -459 \\
0 & 0 & 0 & 0 & 1 & 1 & 0 & 0 & 0 & 0 \\
25 & 3 & 2 & 94 & 0 & 46 & 70 & 0 & -36 & -189 \\
71 & 5 & 11 & 140 & 0 & 70 & 248 & 0 & -94 & -459 \\
0 & 0 & 0 & 0 & 0 & 0 & 0 & 1 & 1 & 0 \\
-25 & -2 & -3 & -70 & 0 & -36 & -94 & 0 & 46 & 189 \\
-540 & -54 & -54 & -459 & 0 & -189 & -459 & 0 & 189 & 1458
\end{bmatrix}.$$

$$(5.74)$$

Verify that the above procedure yields this result.

(h) Using Eqs. (5.33) and (5.74) show that the corrected element **k** matrix, in *global form*, becomes

$$\mathbf{k}_c = \mathbf{R}^\mathsf{T}\overline{\mathbf{B}}^\mathsf{T}\mathbf{G}\overline{\mathbf{B}}\mathbf{R} = \frac{1}{180}
\begin{bmatrix}
398 & -38 & 38 & 71 & -25 & 0 & 71 & 25 & 0 & -540 \\
-38 & 10 & 0 & -5 & 2 & 0 & -11 & -3 & 0 & 54 \\
38 & 0 & 10 & 11 & -3 & 0 & 5 & 2 & 0 & -54 \\
71 & -5 & 11 & 248 & -94 & 0 & 140 & 70 & 0 & -459 \\
-25 & 2 & -3 & -94 & 46 & 0 & -70 & -36 & 0 & 189 \\
0 & 0 & 0 & 0 & -1 & 1 & 0 & 0 & 0 & 0 \\
71 & -11 & 5 & 140 & -70 & 0 & 248 & 94 & 0 & -459 \\
25 & -3 & 2 & 70 & -36 & 0 & 94 & 46 & 0 & -189 \\
0 & 0 & 0 & 0 & 0 & 0 & 0 & -1 & 1 & 0 \\
-540 & 54 & -54 & -459 & 189 & 0 & -459 & -189 & 0 & 1458
\end{bmatrix}.$$

$$(5.75)$$

(i) Using the condensation procedure presented in Section 5.3, show that the condensed, corrected element **k** matrix in global form becomes

$$\overline{\mathbf{k}}_c = \frac{1}{180}
\begin{bmatrix}
198 & -18 & 18 & -99 & 45 & 0 & -99 & -45 & 0 \\
-18 & 8 & 2 & 12 & -5 & 0 & 6 & 4 & 0 \\
18 & 2 & 8 & -6 & 4 & 0 & -12 & -5 & 0 \\
-99 & 12 & -6 & 103.5 & -34.5 & 0 & -4.5 & 10.5 & 0 \\
45 & -5 & 4 & -34.5 & 21.5 & 0 & -10.5 & -11.5 & 0 \\
0 & 0 & 0 & 0 & -1 & 1 & 0 & 0 & 0 \\
-99 & 6 & -12 & -4.5 & -10.5 & 0 & 103.5 & 34.5 & 0 \\
-45 & 4 & -5 & 10.5 & -11.5 & 0 & 34.5 & 21.5 & 0 \\
0 & 0 & 0 & 0 & 0 & 0 & 0 & -1 & 1
\end{bmatrix}.$$

$$(5.76)$$

Comment: Equation (5.76) is the same as Eq. (5.60), showing the equivalence of the two procedures for incorporating coupled boundary conditions. The first method is to be preferred since it is more direct and requires less computation.

Exercise 5.2 Rework the previous exercise using the local coordinate system for element 1 shown in Fig. 5.6. The matrix manipulations can be carried out using a suitably modified version of the computer program presented in Section 5.5. Why is symmetry retained in the matrix $\bar{\mathbf{k}}_c$ in this exercise but not in the previous one?

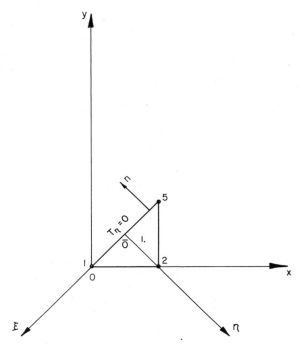

Fig. 5.6 Local coordinate system for element 1.

Hints: The sequence of node numbers for the element is 5, 1, and 2, and the characteristic lengths are $a = \sqrt{2}/2$, $b = \sqrt{2}/2$, $c = \sqrt{2}/2$. The condition $\partial T/\partial n = 0$ at points 5 and 1 now becomes $\partial T/\partial \eta = 0$. Replace the equations corresponding to $T_{\eta 5}$ and $T_{\eta 1}$ to ensure that the Neumann condition is satisfied at nodes 5 and 1.

Answer:

$$
\bar{K}_c = \frac{1}{180}
\begin{bmatrix}
103.5 & -17.25 & -17.25 & -4.5 & 5.25 & 5.25 & -99 & 12 & -6 \\
-17.25 & 5.87 & 4.87 & -5.25 & -2.87 & -2.87 & 22.5 & -2.5 & 2 \\
-17.25 & 4.87 & 5.87 & -5.25 & -2.87 & -2.87 & 22.5 & -2.5 & 2 \\
-4.5 & -5.25 & -5.25 & 103.5 & 17.25 & 17.25 & -99 & 6 & -12 \\
5.25 & -2.87 & -2.87 & 17.25 & 5.87 & 4.87 & -22.5 & 2 & -2.5 \\
5.25 & -2.87 & -2.87 & 17.25 & 4.87 & 5.87 & -22.5 & 2 & -2.5 \\
-99 & 22.5 & 22.5 & -99 & -22.5 & -22.5 & 198 & -18 & 18 \\
12 & -2.5 & -2.5 & 6 & 2 & 2 & -18 & 8 & 2 \\
-6 & 2 & 2 & -12 & -2.5 & -2.5 & 18 & 2 & 8
\end{bmatrix}.
\quad (5.77)
$$

Exercise 5.3   For element 4, taking the nodes in the sequence 6, 3, 7, as in Table 5.2, the element $\bar{k}$ matrix in global form is given from Eq. (5.46) as

$$
\bar{k}^4 = \frac{1}{180}
\begin{bmatrix}
198 & 18 & -18 & -99 & -27 & -18 & -99 & 18 & 27 \\
18 & 8 & 2 & -12 & -2 & -3 & -6 & 0 & 4 \\
-18 & 2 & 8 & 6 & 4 & 0 & 12 & -3 & -2 \\
-99 & -12 & 6 & 103.5 & 16.5 & 18 & -4.5 & 0 & -10.5 \\
-27 & -2 & 4 & 16.5 & 9.5 & 3 & 10.5 & -3 & -5.5 \\
-18 & -3 & 0 & 18 & 3 & 6 & 0 & 0 & -3 \\
-99 & -6 & 12 & -4.5 & 10.5 & 0 & 103.5 & -18 & -16.5 \\
18 & 0 & -3 & 0 & -3 & 0 & -18 & 6 & 3 \\
27 & 4 & -2 & -10.5 & -5.5 & -3 & -16.5 & 3 & 9.5
\end{bmatrix}.
$$

$$(5.78)$$

In partitioned form, this matrix can be written as

$$
\bar{k}^4 = \frac{1}{180}
\begin{bmatrix}
\bar{k}^4_{66} & \bar{k}^4_{63} & \bar{k}^4_{67} \\
\bar{k}^4_{36} & \bar{k}^4_{33} & \bar{k}^4_{37} \\
\bar{k}^4_{76} & \bar{k}^4_{73} & \bar{k}^4_{77}
\end{bmatrix}.
\quad (5.79)
$$

(a)   Using the computer program developed in Exercise 5.2, and taking the nodes of element 4 in the sequence 3, 7, 6 so that the local coordinate system is as shown in Fig. 5.7, obtain the element $\bar{k}$ matrix for element 4, and show that it can be written in terms of the submatrices of Eq. (5.79) as

$$
\bar{k}^4 = \frac{1}{180}
\begin{bmatrix}
\bar{k}^4_{33} & \bar{k}^4_{37} & \bar{k}^4_{36} \\
\bar{k}^4_{73} & \bar{k}^4_{77} & \bar{k}^4_{76} \\
\bar{k}^4_{63} & \bar{k}^4_{67} & \bar{k}^4_{66}
\end{bmatrix}.
\quad (5.80)
$$

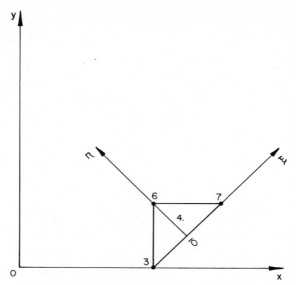

Fig. 5.7   Local coordinate system for element 4.

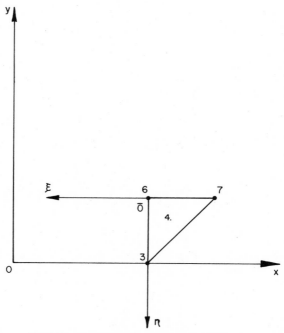

Fig. 5.8   Local coordinate system for element 4.

(b)   Repeat part (a) using the nodes of element 4 in the sequence 7, 6, 3 (see Fig. 5.8) and hence show that

$$\bar{k}^4 = \frac{1}{180}\begin{bmatrix} \bar{k}_{77}^4 & \bar{k}_{76}^4 & \bar{k}_{73}^4 \\ \bar{k}_{67}^4 & \bar{k}_{66}^4 & \bar{k}_{63}^4 \\ \bar{k}_{37}^4 & \bar{k}_{36}^4 & \bar{k}_{33}^4 \end{bmatrix}. \tag{5.81}$$

Comment: The above results [Eqs. (5.79)–(5.81)] verify that the entries of the element $\bar{k}$ matrix are independent of the choice of the local coordinate system and the node numbering.

5.5   COMPUTER PROGRAMMING

5.5.1   FORTRAN IV Computer Program

The formulation given in Sections 5.1–5.4 has been used in the following computer program to solve the Laplace problem of Fig. 5.1. The approach outlined in Section 5.4 for coupled equivalent Dirichlet conditions has been incorporated. Assembly of the element matrix equations is by elements. The program is presented without a flow chart or further documentation.

5.5.1.1   Main Computer Program

```
C       .....FINITE ELEMENT METHOD,PROGRAM 3.....
C       SOLUTION OF THE HEAT CONDUCTION PROBLEM SHOWN IN FIG.5.1.
C       THE RESULTING SYSTEM OF EQUATIONS IS SOLVED USING THE
C       STANDARD LIBRARY SUBROUTINE LEQT1F.
C       NPOIN      -    TOTAL NUMBER OF NODES
C       NELEM      -    TOTAL NUMBER OF ELEMENTS
C       NPR        -    TOTAL NUMBER OF PRESCRIBED VARIABLES,
C                       COMPRISING THE FUNCTION AND/OR ITS
C                       DERIVATIVE(S)
C       NPP(I,J)   -    NPP(I,2) IDENTIFIES THE DEGREE OF FREEDOM
C                       WHICH IS PRESCRIBED AT NODE NPP(I,1),
C                       WHERE I=1,2,....,NPR
C       VALP(I)    -    THE PRESCRIBED VALUE OF THE VARIABLE AT
C                       NODE NPP(I,1),WHERE I=1,2,....,NPR
C       NCP        -    TOTAL NUMBER OF NODES WHERE THE COUPLED
C                       CONDITION -TX+TY=0 IS PRESCRIBED
C       NPC(I)     -    NODES WHERE THE COUPLED CONDITION
C                       -TX+TY=0 IS PRESCRIBED,WHERE I=1,2,....,NCP
C       X(I),Y(I)  -    X,Y COORDINATES OF NODE I
C       NOD(I,J)   -    THE THREE NODES OF ELEMENT I,CORRESPONDING
C                       TO THE THREE NODE IDENTIFIERS J=1,2,3
        PROGRAM PRGRM3(INPUT,OUTPUT,TAPE5,TAPE6)
        DIMENSION NPP(12,2),VALP(12),NPC(4),X(10),Y(10),NOD(9,3)
        DIMENSION XX(3),YY(3),STE(10,10),STEINV(10,10),WKAREA(10)
        DIMENSION D(10,10),ROT(10,10),STNEW(9,9),N(3),IST(3)
        DIMENSION ISTNEW(3),ST(30,30),RHS(30,1),WKAR(30)
C       .....SUBROUTINE DATAIN IS CALLED TO READ IN THE REQUIRED
C            DATA.....
        CALL DATAIN(NPOIN,NELEM,NPR,NPP,VALP,NCP,NPC,X,Y,NOD)
```

```
C
C        .....SUBROUTINE DATAOUT IS CALLED TO PRINT OUT THE
C             INPUT DATA FOR CHECKING PURPOSES.....
         CALL DATAOUT(NPOIN,NELEM,NPR,NPP,VALP,NCP,NPC,X,Y,NOD)
         NPOIN3=3*NPOIN
C
C        .....THE SYSTEM K MATRIX,DENOTED BY ST,AND THE RIGHT-
C             HAND SIDE MATRIX,DENOTED BY RHS,ARE INITIALIZED
C             TO ZERO.....
         DO 5 I=1,NPOIN3
         DO 4 J=1,NPOIN3
4        ST(I,J)=0.0
5        RHS(I,1)=0.0
C
C        .....IN THE NEXT DO LOOP,FOR EACH ELEMENT IN TURN, THE
C             ELEMENT K MATRIX IS OBTAINED AND ASSEMBLED DIRECTLY
C             INTO THE SYSTEM K MATRIX.....
         DO 160 IE=1,NELEM
C
C        .....THE X AND Y COORDINATES OF THE NODES FOR THE
C             ELEMENT,IE,UNDER CONSIDERATION,ARE OBTAINED AND
C             DENOTED BY XX(I),YY(I),I=1,2,3.....
         DO 10 J=1,3
         LK=NOD(IE,J)
         XX(J)=X(LK)
10       YY(J)=Y(LK)
C
C        .....SUBROUTINE PARAM IS CALLED TO CALCULATE THE
C             PARAMETERS A,B,C,COS(THETA)=CN,SIN(THETA)=SN.....
         CALL PARAM(XX,YY,A,B,C,CN,SN,R)
C
C        .....SUBROUTINE COEFMAT IS CALLED TO CALCULATE THE
C             COEFFICIENT MATRIX A,HERE DENOTED BY STE,USING
C             THE PARAMETERS A,B,AND C OBTAINED ABOVE.....
         CALL COEFMAT(A,B,C,STE)
C
C        .....THE STANDARD LIBRARY SUBROUTINE LINV1F IS CALLED
C             TO INVERT THE MATRIX A(=STE) TO GIVE THE INVERSE
C             OF A,DENOTED IN THE TEXT AS B,AND IN THE PROGRAM
C             AS STEINV.....
         CALL LINV1F(STE,10,10,STEINV,0,WKAREA,IER)
C
C        .....SUBROUTINE INTEG IS CALLED TO CALCULATE THE
C             INTEGRATION MATRIX G,HERE DENOTED BY D.....
         CALL INTEG(A,B,C,D)
C
C        .....SUBROUTINE MULT IS CALLED TO CALCULATE THE MATRIX
C             PRODUCT (TRANSPOSE OF B)*G*B.THE RESULT OF THIS
C             PRODUCT IS RETURNED TO THE MAIN PROGRAM AS D.....
         CALL MULT(D,STEINV)
C
C        .....SUBROUTINE ROTAT IS CALLED TO CALCULATE THE
C             ROTATION MATRIX R,HERE DENOTED BY ROT.....
         CALL ROTAT(CN,SN,ROT)
C
C        .....SUBROUTINE MULT IS CALLED AGAIN TO CALCULATE
C             THE MATRIX PRODUCT (TRANSPOSE OF R)*(TRANSPOSE OF
C             B)*G*B*R.THE RESULT IS RETURNED TO THE MAIN PROGRAM
C             AS D.....
         CALL MULT(D,ROT)
C
C        .....THE CONDENSATION PROCEDURE,DISCUSSED IN SECTION 5.3,
C             IS NOW CARRIED OUT.....
```

```
          DO 30 I=1,9
          DO 20 J=1,9
20        STNEW(I,J)=D(I,J)-(D(I,10)*D(10,J))/D(10,10)
30        CONTINUE
C
C         .....THE VECTOR DENOTED BY N=(N(1),N(2),N(3)) IS SET TO
C              ZERO.THIS VECTOR IS USED TO INDICATE WHETHER ANY
C              NODE OF ELEMENT IE HAS THE CONDITION -TX+TY=0
C              PRESCRIBED.IF SO THE N(I) CORRESPONDING TO THAT
C              NODE IDENTIFIER I IS SET EQUAL TO ONE.....
          DO 65 I=1,3
65        N(I)=0
          DO 70 I=1,NCP
          LK=NPC(I)
          DO 60 J=1,3
          IF(NOD(IE,J)-LK)60,50,60
50        N(J)=1
60        CONTINUE
70        CONTINUE
C
C         .....SUBROUTINE CHANGE IS CALLED TO MODIFY THE ELEMENT
C              K MATRIX IF THE CONDITION -TX+TY=0 APPLIES AT ANY
C              OF THE NODES OF ELEMENT IE.....
          CALL CHANGE(STNEW,N)
C
C         .....THE CORRECT ROWS AND COLUMNS FOR THE SYSTEM K MATRIX
C              (ST) ARE DETERMINED AND DENOTED BY IR AND IC,
C              RESPECTIVELY.THE CORRESPONDING ROWS AND COLUMNS FOR
C              THE ELEMENT K MATRIX (STNEW) ARE OBTAINED AND DENOTED
C              BY I AND J.FINALLY THE ELEMENT K MATRIX IS ASSEMBLED
C              INTO THE SYSTEM K MATRIX.....
          DO 110 I=1,3
          LK=NOD(IE,I)
          IST(I)=(LK-1)*3
110       ISTNEW(I)=(I-1)*3
          DO 130 II=1,3
          DO 120 JJ=1,3
          DO 210 IRR=1,3
          IR=IST(II)+IRR
          I=ISTNEW(II)+IRR
          DO 220 ICC=1,3
          IC=IST(JJ)+ICC
          J=ISTNEW(JJ)+ICC
220       ST(IR,IC)=ST(IR,IC)+STNEW(I,J)
210       CONTINUE
120       CONTINUE
130       CONTINUE
C
C         .....IN THE FOLLOWING ALL NODES WITH PRESCRIBED VARIABLES
C              ARE CHECKED.FOR EACH PRESCRIBED VARIABLE,THE SYSTEM
C              K MATRIX IS CORRECTED.....
          DO 150 I=1,NPR
          LK=NPP(I,1)
          LOC=(LK-1)*3+NPP(I,2)
          DO 140 J=1,NPOIN3
140       ST(LOC,J)=0.0
          ST(LOC,LOC)=1.0
150       RHS(LOC,1)=VALP(I)
160       CONTINUE
C
C         .....THE FINAL SYSTEM MATRIX EQUATION HAS NOW BEEN OBTAINED
C              AND IS SOLVED BY CALLING THE STANDARD LIBRARY
C              SUBROUTINE LEQT1F.THE FINAL RESULTS ARE RETURNED
```

```
C              TO THE MAIN PROGRAM AS RHS.....
           CALL LEQT1F(ST,1,30,30,RHS,0,WKAR,IER)
C
C          .....THE FINAL RESULTS ARE PRINTED OUT.....
           WRITE(6,170)
170        FORMAT(//,31H THE SOLUTION HAS BEEN OBTAINED)
           WRITE(6,180)
180        FORMAT(/,1X,31H NODE        T        TX        TY )
           LK=0
           DO 190 I=1,28,3
           LK=LK+1
190        WRITE(6,200) LK,RHS(I,1),RHS(I+1,1),RHS(I+2,1)
200        FORMAT(2X,I3,4X,F6.2,3X,F6.2,3X,F6.2)
           STOP
           END
```

5.5.1.2   Subroutines

```
           SUBROUTINE DATAIN(NPOIN,NELEM,NPR,NPP,VALP,NCP,NPC,X,Y,NOD)
           DIMENSION NPP(12,2),VALP(12),NPC(4),X(10),Y(10),NOD(9,3)
C          .....THE TOTAL NUMBER OF NODES,THE TOTAL NUMBER OF
C              ELEMENTS,THE TOTAL NUMBER OF PRESCRIBED VARIABLES,
C              AND THE TOTAL NUMBER OF NODES WHERE THE COUPLED
C              CONDITION -TX+TY=0 IS PRESCRIBED ARE READ IN.....
           READ(5,10) NPOIN,NELEM,NPR,NCP
10         FORMAT(4I3)
C          .....THE THREE NODES,CORRESPONDING TO THE THREE NODE
C              IDENTIFIERS J=1,2,3, ARE READ IN FOR ALL ELEMENTS.....
           READ(5,20) (I,(NOD(I,J),J=1,3),JJ=1,NELEM)
20         FORMAT(24I3)
C
C          .....THE X AND Y COORDINATES ARE READ IN FOR ALL NODES.....
           READ(5,30) (I,X(I),Y(I),J=1,NPOIN)
30         FORMAT(5(I3,2F5.1))
C
C          .....THOSE NODES WHERE A VARIABLE IS PRESCRIBED,THE
C              DEGREE OF FREEDOM WHICH IS PRESCRIBED,AND THE
C              PRESCRIBED VALUE ARE READ IN.....
           READ(5,40) (NPP(I,1),NPP(I,2),VALP(I),I=1,NPR)
40         FORMAT(6(2I3,F6.2))
C
C          .....THOSE NODES WHERE THE COUPLED CONDITION -TX+TY=0
C              IS PRESCRIBED ARE READ IN.....
           READ(5,20) (NPC(I),I=1,NCP)
           RETURN
           END

           SUBROUTINE DATAOUT(NPOIN,NELEM,NPR,NPP,VALP,NCP,NPC,X,Y,NOD)
           DIMENSION NPP(12,2),VALP(12),NPC(4),X(10),Y(10),NOD(9,3)
C          .....THE TOTAL NUMBER OF NODES,THE TOTAL NUMBER OF
C              ELEMENTS,THE TOTAL NUMBER OF PRESCRIBED VARIABLES,
C              AND THE TOTAL NUMBER OF NODES WHERE THE COUPLED
C              CONDITION -TX+TY=0 IS PRESCRIBED ARE PRINTED OUT.....
           WRITE(6,10) NPOIN
10         FORMAT(//////,1X,22H TOTAL NUMBER OF NODES,I3)
           WRITE(6,20) NELEM
20         FORMAT(1X,25H TOTAL NUMBER OF ELEMENTS,I3)
           WRITE(6,30) NPR
30         FORMAT(1X,37H TOTAL NUMBER OF PRESCRIBED VARIABLES,I3)
           WRITE(6,40) NCP
40         FORMAT(1X,51H TOTAL NUMBER OF NODES WHERE -TX+TY=0 IS PRESCRIBED,I
          +3 )
C
```

```
C        .....THE X AND Y COORDINATES ARE PRINTED OUT FOR ALL NODES.....
         WRITE(6,50)
50       FORMAT(//,1X,40H THE NODES AND THEIR X AND Y COORDINATES,/)
         WRITE(6,60)
60       FORMAT(1X,15HNODE    X    Y,2(20H       NODE    X     Y))
         WRITE(6,70) (I,X(I),Y(I),I=1,NPOIN)
70       FORMAT((1X,I3,F7.1,F6.1,2(I7,F7.1,F6.1)))
C
C        .....THE ELEMENTS AND THEIR THREE NODES ARE PRINTED OUT.....
         WRITE(6,80)
80       FORMAT(//,1X,29H THE ELEMENTS AND THEIR NODES,/)
         WRITE(6,90)
90       FORMAT(1X,13HELEM  1  2  3,3(16H     ELEM  1  2  3))
         WRITE(6,100) (I,(NOD(I,J),J=1,3),I=1,NELEM)
100      FORMAT((1X,I3,I4,2I3,3(I6,I4,2I3)))
C
C        .....THOSE NODES WHERE THE COUPLED CONDITION -TX+TY=0
C             IS PRESCRIBED ARE PRINTED OUT.....
         WRITE(6,110)
         WRITE(6,120)
110      FORMAT(//,1X,38H THE NODES WHERE THE COUPLED CONDITION)
120      FORMAT(1X,23H -TX+TY=0 IS PRESCRIBED,/)
         WRITE(6,130) (NPC(I),I=1,NCP)
130      FORMAT(6I6)
C        .....THE NODES WHERE A VARIABLE IS PRESCRIBED,THE DEGREE
C             OF FREEDOM WHICH IS PRESCRIBED,AND THE PRESCRIBED
C             VALUE ARE PRINTED OUT
         WRITE(6,140)
140      FORMAT(//,1X,32H NODES WITH PRESCRIBED VARIABLES,/)
         WRITE(6,150)
150      FORMAT(1X,72H NODE   DEG.OF FREEDOM  PRES.VALUE      NODE    DEG.OF
        +FREEDOM  PRES.VALUE)
         WRITE(6,160) (NPP(I,1),NPP(I,2),VALP(I),I=1,NPR)
160      FORMAT(1X,I4,8X,I3,10X,F6.2,8X,I3,8X,I3,10X,F6.2)
         RETURN
         END

         SUBROUTINE PARAM(X,Y,A,B,C,CN,SN,R)
         DIMENSION X(3),Y(3)
         R=((X(2)-X(1))**2+(Y(2)-Y(1))**2)**0.5
         CN=(X(2)-X(1))/R
         SN=(Y(2)-Y(1))/R
         A=(X(2)-X(3))*CN-(Y(3)-Y(2))*SN
         B=(X(3)-X(1))*CN+(Y(3)-Y(1))*SN
         C=(Y(3)-Y(2))*CN+(X(2)-X(3))*SN
         RETURN
         END

         SUBROUTINE COEFMAT(A,B,C,STE)
         DIMENSION STE(10,10)
         A2=A**2 $ A3=A**3 $ B2=B**2 $ B3=B**3 $ C2=C**2 $ C3=C**3
C
C        .....THE MATRIX STE IS INITIALIZED TO ZERO.....
         DO 20 I=1,10
         DO 10 J=1,10
10       STE(I,J)=0.0
20       CONTINUE
C
C        .....THE NON-ZERO ENTRIES OF THE MATRIX STE ARE DETERMINED.....
         STE(1,1)=1.0 $ STE(1,2)=-B $ STE(1,4)=B2 $ STE(1,7)=-B3
         STE(2,2)=1.0 $ STE(2,4)=-2.0*B $ STE(2,7)=3.0*B2
         STE(3,3)=1.0 $ STE(3,5)=-B $ STE(3,8)=B2
         STE(4,1)=1.0 $ STE(4,2)=A $ STE(4,4)=A2 $ STE(4,7)=A3
         STE(5,2)=1.0 $ STE(5,4)=2.0*A $ STE(5,7)=3.0*A2
         STE(6,3)=1.0 $ STE(6,5)=A $ STE(6,8)=A2
         STE(7,1)=1.0 $ STE(7,3)=C $ STE(7,6)=C2 $ STE(7,10)=C3
```

```
      STE(8,2)=1.0 $ STE(8,5)=C $ STE(8,9)=C2
      STE(9,3)=1.0 $ STE(9,6)=2.0*C $ STE(9,10)=3.0*C2
      STE(10,1)=1.0 $ STE(10,2)=(A-B)/3.0 $ STE(10,3)=C/3.0
      STE(10,4)=((A-B)**2)/9.0 $ STE(10,5)=(C*(A-B))/9.0
      STE(10,6)=(C**2)/9.0 $ STE(10,7)=((A-B)**3)/27.0
      STE(10,8)=(((A-B)**2)*C)/27.0
      STE(10,9)=((A-B)*(C**2))/27.0
      STE(10,10)=(C**3)/27.0
      RETURN
      END

      SUBROUTINE INTEG(A,B,C,D)
      DIMENSION FAC(7),M(10),N(10),D(10,10)
C
C     .....IN THE FOLLOWING,FAC(I)=(I-1) FACTORIAL.....
      FAC(1)=1.0
      DO 10 I=1,6
      DET=I
10    FAC(I+1)=DET*FAC(I)
      M(1)=0 $ M(2)=1 $ M(3)=0 $ M(4)=2 $ M(5)=1 $ M(6)=0 $ M(7)=3
      M(8)=2 $ M(9)=1 $ M(10)=0
      N(1)=0 $ N(2)=0 $ N(3)=1 $ N(4)=0 $ N(5)=1 $ N(6)=2 $ N(7)=0
      N(8)=1 $ N(9)=2 $ N(10)=3
      DO 30 I=1,10
      DO 20 J=1,10
      M1=(M(I)+M(J)-2)+1
      M2=(M(I)+M(J))+1
      N1=(N(I)+N(J))+1
      N2=(N(I)+N(J)-2)+1
      COEF1=M(I)*M(J)
      COEF2=N(I)*N(J)
      IF(COEF1)40,50,60
40    WRITE(6,70)
70    FORMAT(*COEFFICIENT IN INTEG IS NEGATIVE*)
      RETURN
50    G1=1.0
      GO TO 80
60    G1=(C**N1)*((A**M1)-((-B)**M1))*FAC(M1)*FAC(N1)/FAC(M1+N1+1)
80    IF(COEF2)40,90,100
90    G2=1.0
      GO TO 20
100   G2=(C**N2)*((A**M2)-((-B)**M2))*FAC(M2)*FAC(N2)/FAC(M2+N2+1)
20    D(I,J)=G1*COEF1+G2*COEF2
30    CONTINUE
      RETURN
      END

      SUBROUTINE MULT(D,STE)
      DIMENSION STE(10,10),D(10,10),H(10,10)
C
C     .....THE PRODUCT D*STE IS CALCULATED.....
      DO 30 I=1,10
      DO 20 J=1,10
      H(I,J)=0.0
      DO 10 LK=1,10
10    H(I,J)=H(I,J)+D(I,LK)*STE(LK,J)
20    CONTINUE
30    CONTINUE
C
C     .....THE PRODUCT (STE TRANSPOSE*D*STE) IS CALCULATED.....
      DO 70 I=1,10
      DO 60 J=1,10
      D(I,J)=0.0
      DO 50 LK=1,10
50    D(I,J)=D(I,J)+STE(LK,I)*H(LK,J)
60    CONTINUE
70    CONTINUE
```

```
      RETURN
      END
      SUBROUTINE ROTAT(CN,SN,ROT)
      DIMENSION ROT(10,10)
C
C     .....THE MATRIX ROT IS INITIALIZED TO ZERO.....
      DO 20 I=1,10
      DO 10 J=1,10
10    ROT(I,J)=0.0
20    CONTINUE
C
C     .....THE NON-ZERO ENTRIES OF THE MATRIX ROT ARE DETERMINED.....
      ROT(1,1)=1.0 $ ROT(2,2)=CN $ ROT(2,3)=SN $ ROT(3,2)=-SN
      ROT(3,3)=CN $ ROT(4,4)=1.0 $ ROT(5,5)=CN $ ROT(5,6)=SN
      ROT(6,5)=-SN $ ROT(6,6)=CN $ ROT(7,7)=1.0 $ ROT(8,8)=CN
      ROT(8,9)=SN $ ROT(9,8)=-SN $ ROT(9,9)=CN $ ROT(10,10)=1.0
      RETURN
      END
      SUBROUTINE CHANGE(ST,N)
      DIMENSION N(3),ST(9,9)
      DO 40 I=1,3
      IF(N(I))40,40,10
10    IRTX=(I-1)*3+2
      IRTY=(I-1)*3+3
      DO 20 J=1,9
      ST(IRTX,J)=ST(IRTX,J)+ST(IRTY,J)
20    ST(IRTY,J)=0.0
      DO 30 J=1,9
      ST(J,IRTX)=ST(J,IRTX)+ST(J,IRTY)
30    ST(J,IRTY)=0.0
      ST(IRTY,IRTX)=-1.0
      ST(IRTY,IRTY)=1.0
40    CONTINUE
      RETURN
      END
```

5.5.1.3   Input Data

```
10   9  12   4
 1   2   5   1   2   5   2   6   3   3   6   2   4   6   3   7   5   4   7   3   6   6   8   5
 7   8   6   9   8   7   9   6   9   9  10   8
 1   0.0   0.0   2   1.0   0.0   3   2.0   0.0   4   3.0   0.0   5   1.0   1.0
 6   2.0   1.0   7   3.0   1.0   8   2.0   2.0   9   3.0   2.0  10   3.0   3.0
 1   1  30.0   1   2   0.0   2   1  30.0   2   2   0.0   3   1  30.0   3   2   0.0
 4   1  30.0   4   2   0.0   7   2   0.0   9   2   0.0  10   1   0.0  10   2   0.0
 1   5   8  10
```

5.5.1.4   Results

```
 TOTAL NUMBER OF NODES 10
 TOTAL NUMBER OF ELEMENTS   9
 TOTAL NUMBER OF PRESCRIBED VARIABLES 12
 TOTAL NUMBER OF NODES WHERE -TX+TY=0 IS PRESCRIBED   4


 THE NODES AND THEIR X AND Y COORDINATES
```

NODE	X	Y	NODE	X	Y	NODE	X	Y
1	0.0	0.0	2	1.0	0.0	3	2.0	0.0
4	3.0	0.0	5	1.0	1.0	6	2.0	1.0
7	3.0	1.0	8	2.0	2.0	9	3.0	2.0
10	3.0	3.0						

THE ELEMENTS AND THEIR NODES

ELEM	1	2	3	ELEM	1	2	3	ELEM	1	2	3	ELEM	1	2	3
1	2	5	1	2	5	2	6	3	3	6	2	4	6	3	7
5	4	7	3	6	6	8	5	7	8	6	9	8	7	9	6
9	9	10	8												

THE NODES WHERE THE COUPLED CONDITION
−TX+TY=0 IS PRESCRIBED

   1   5   8   10

NODES WITH PRESCRIBED VARIABLES

NODE	DEG.OF FREEDOM	PRES.VALUE	NODE	DEG.OF FREEDOM	PRES.VALUE
1	1	30.00	1	2	0.00
2	1	30.00	2	2	0.00
3	1	30.00	3	2	0.00
4	1	30.00	4	2	0.00
7	2	0.00	9	2	0.00
10	1	0.00	10	2	0.00

THE SOLUTION HAS BEEN OBTAINED

NODE	T	TX	TY
1	30.00	0.00	0.00
2	30.00	0.00	−1.99
3	30.00	0.00	−4.16
4	30.00	0.00	−5.16
5	27.85	−1.71	−1.71
6	25.65	−1.41	−4.41
7	24.51	0.00	−6.35
8	20.90	−2.05	−2.05
9	16.19	0.00	−11.01
10	0.00	0.00	0.00

5.6 EXTENSION TO HIGHER ORDER ELEMENTS AND MORE COMPLEX BOUNDARY CONDITIONS

The procedures outlined in this chapter for derivative elements and coupled boundary conditions are not limited to first-order derivatives and are applicable to higher order trial functions. More complex boundary conditions such as the Cauchy condition $\partial\phi/\partial n + q\phi + p = 0$, can be handled by modifying the methods presented earlier. Indeed, any boundary condition involving the function and/or its derivatives at a node can be so treated, provided these variables are nodal parameters of the element.

The program of Section 5.5 used assembly by elements, which is the most widely adopted approach. The whole of the final (nonsymmetric) system matrix was then retained in storage and used with a *full storage mode* equation solving subroutine. For large problems, the full storage mode is neither economic nor practical, and other approaches must be chosen. For example, it is shown in the next chapter that appropriate node numbering can minimize the bandwidth so that a band storage mode can be usefully employed. Another approach to reducing storage, also described in the next chapter, is partitioning and tridiagonalization. Other procedures for limiting storage requirements are considered in Chapter 10. For additional details, the literature should be consulted [2–4].

REFERENCES

1. W. Kaplan, "Advanced Calculus," 2nd ed. Addison-Wesley, 1973, p. 135.
2. D. H. Norrie and G. de Vries, "Finite Element Bibliography." Plenum, New York, 1976.
3. S. J. Fenves, N. Perrone, A. R. Robinson, and W. C. Schnobrich (eds.), "Numerical and Computer Methods in Structural Mechanics." Academic Press, New York, 1973.
4. K-J. Bathe and E. L. Wilson, "Numerical Methods in Finite Element Analysis." Prentice-Hall, Englewood Cliffs, New Jersey, 1976.

6

ECONOMIZATION OF CORE STORAGE, PARTITIONING, AND TRIDIAGONALIZATION

It is not uncommon for a scientist or engineer to have a computer program rejected because it is too large for the available facility. In such cases, the system is usually *storage-bound* rather than *computation-bound*; that is, the core storage required by the program exceeds that available. With finite element programs this situation is easy to get into, particularly with three-dimensional problems. Many procedures have been developed to reduce core storage requirements, even at the expense of additional computation; some of the simpler approaches are presented in this chapter. Other possibilities are considered in Chapter 10.

6.1 BANDWIDTH

The final system equation can be solved either by *direct* methods which yield the solution in one pass, or by *indirect* (*iterative*) methods which refine an initially approximate solution to improved levels of accuracy with successive passes. Direct methods can be coarsely grouped according to whether they have been designed for system K matrices which are *full*, *banded*, or *sparse*. In general, full and banded direct solution procedures are more efficient if **K** is symmetric; furthermore, for a given density,[†] the banded methods are more economic than full-matrix methods. Banded methods also become less costly than sparse methods as the band itself becomes more dense.

For a given matrix density, the efficiency of the banded direct methods typically increases as the bandwidth decreases. As the next section shows, storage can also be reduced as bandwidth is decreased. It is thus generally advantageous to minimize the bandwidth of the system **K** matrix when a banded solution procedure is to be adopted. In the following, the factors that influence the bandwidth are described.

Consider the portion of a two-dimensional finite element mesh shown in Fig. 6.1. Let the exact solution to the problem, here denoted by ϕ, be approximated by linear trial functions in each of the three-node triangular elements.

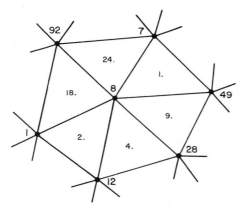

Fig. 6.1 Two-dimensional finite element mesh.

As shown previously, assembly by nodes results in the n nodal equations

$$\frac{\partial \chi}{\partial \overline{\phi}_p} = \sum_{e=1}^{l} \frac{\partial \chi^e}{\partial \overline{\phi}_p} = 0, \qquad p = 1, 2, \ldots, n, \tag{6.1}$$

where only the elements that surround the node p need to be included in the summation since the contributions of the other elements are zero.

As seen from Eq. (6.1) and Fig. 6.1, the equation for node 8 becomes

$$\frac{\partial \chi}{\partial \overline{\phi}_8} = \frac{\partial \chi^1}{\partial \overline{\phi}_8} + \frac{\partial \chi^2}{\partial \overline{\phi}_8} + \frac{\partial \chi^4}{\partial \overline{\phi}_8} + \frac{\partial \chi^9}{\partial \overline{\phi}_8} + \frac{\partial \chi^{18}}{\partial \overline{\phi}_8} + \frac{\partial \chi^{24}}{\partial \overline{\phi}_8} = 0. \tag{6.2}$$

[†] Density is defined as the ratio of nonzero elements to total elements in the matrix. For the estimation of density in advance, when straight-sided elements are used, see. [1].

When the appropriate element contributions χ^e are substituted into Eq. (6.2), this nodal equation can be reduced to the form

$$K_{8,1}\bar{\phi}_1 + K_{8,7}\bar{\phi}_7 + K_{8,8}\bar{\phi}_8 + K_{8,12}\bar{\phi}_{12}$$
$$+ K_{8,28}\bar{\phi}_{28} + K_{8,49}\bar{\phi}_{49} + K_{8,92}\bar{\phi}_{92} = 0, \quad (6.3)$$

It is noted again that only the nodes in the surrounding elements appear in Eq. (6.3). This is because the coefficients $K_{\gamma\delta}$ are zero for all other nodes.

Consider in Eq. (6.2), the contributing term from element 9 which can be written in the form

$$\frac{\partial \chi^9}{\partial \bar{\phi}_8} = k^9_{8,8}\bar{\phi}_8 + k^9_{8,28}\bar{\phi}_{28} + k^9_{8,49}\bar{\phi}_{49}. \quad (6.4)$$

Similar relationships are obtainable for the other terms of Eq. (6.2).

For each of the elements 1, 2, 4, 9, 18, and 24, the nodal equations analogous to Eq. (6.4) can be written in matrix form and expanded to system size by inserting zero terms where appropriate. It will then be seen that the square coefficient matrix of each of these expanded nodal equations has only s nonzero terms (where s is the total number of nodal parameters in the element, in this case 3) with one of these always being in the diagonal position. The width of this "band"[†] in such an expanded nodal equation is given by the difference between the column numbers of the rightmost and leftmost nonzero entries, plus one.

The same rule applies for any nodal equation of system size. For example, in the expanded matrix form[‡] of Eq. (6.3) the rightmost nonzero entry is in column 92 of the coefficient matrix, whereas the leftmost is in column 1. Consequently, the "bandwidth" for this nodal equation is equal to $b = 92 - 1 + 1 = 92$. Since the diagonal position is in column 8, it is seen that the band is not centered on the diagonal position. The width of the partial band (excluding the diagonal) to the left of the diagonal, denoted b_L, is the difference between the column numbers of the diagonal position and the leftmost nonzero entry. Similarly, the width of the partial band to the right, b_R, is the difference between the column numbers of the rightmost nonzero entry and the diagonal position. Thus, for Eq. (6.3), $b_L = 8 - 1 = 7$ and $b_R = 92 - 8 = 84$.

Consider now the whole assembled system $\mathbf{K}$ matrix and imagine two lines drawn parallel to the principal diagonal so that the resulting band between these lines includes all nonzero elements and leaves as many zero elements as possible in the remaining upper and lower triangular regions of

---

[†] The strip of elements from the leftmost nonzero element to the rightmost nonzero element.
[‡] From Eq. (6.2), this is equal to the sum of the expanded nodal equations for elements 1, 2, 4, 9, 18, 24.

the matrix. The width of the partial band to the left of the diagonal of the system $\mathbf{K}$ matrix, denoted B_L, is equal to largest of the b_L's of the system nodal equations. Similarly, the width of the partial band to the right of the diagonal of the system $\mathbf{K}$ matrix, denoted B_R, is equal to the largest of the b_R's of the system nodal equations. The bandwidth of the system $\mathbf{K}$ matrix (including the principal diagonal) is thus equal to

$$B = B_L + B_R + 1. \tag{6.5}$$

This bandwidth will be balanced equally with respect to the diagonal. To illustrate, the node and band data taken from Fig. 6.2 for the nodal equations of that system are listed in Table 6.1. From this table it is seen that

$$B_L = \text{Max}(b_L) = 14, \tag{6.6a}$$
$$B_R = \text{Max}(b_R) = 14, \tag{6.6b}$$

and from Eq. (6.5),

$$B = 29. \tag{6.7}$$

The semibandwidth B_S of the system $\mathbf{K}$ matrix, excluding the principal diagonal, is therefore 14, as can be verified from Fig. 6.2b. It is worth noting from the above that

$$B_L = B_R = B_S \tag{6.8}$$

and also that B_S is given by the maximum difference between the lowest and highest node numbers of any single element.

Where there is more than one parameter per node, as in Hermitian elements, the above procedure can be extended as follows. Suppose there

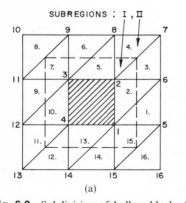
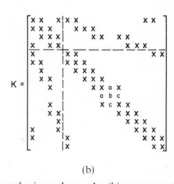

Fig. 6.2 Subdivision of hollow block: (a) numbering scheme A; (b) corresponding $\mathbf{K}$ matrix.

TABLE 6.1

Partial Bandwidths

Node	Largest adjacent node number	Bandwidth to the right b_R	Smallest adjacent node number	Bandwidth to the left b_L
1	15	14	1	0
2	8	6	1	1
3	12	9	2	1
4	14	10	1	3
5	16	11	1	4
6	7	1	1	5
7	8	1	2	5
8	9	1	2	6
9	12	3	3	6
10	11	1	9	1
11	12	1	3	8
12	13	1	3	9
13	14	1	4	9
14	15	1	1	13
15	16	1	1	14
16	15	-1	5	11

are q degrees of freedom (parameters) at each node. Then

$$B_S = (d + 1)q - 1, \tag{6.9}$$

where d is the maximum difference between the lowest and highest node numbers of any single element in the system.

To minimize the bandwidth of the system **K** matrix, therefore, the nodes of the system should be numbered in such a way as to keep this maximum difference as small as possible. Consider the two-dimensional hollow block, subdivided into 16 three-node triangular elements, shown in Figs. 6.2a and 6.3a. It will be noted that the nodes are numbered in different ways in the two figures. If there is only one parameter per node and if boundary conditions are neglected, the corresponding system **K** matrices become as shown schematically in Figs. 6.2b and 6.3b. The semibandwidth for the first numbering scheme is given by $B_S = 15 - 1 = 14$, whereas for the second scheme $B_S = 4$. The figures also show how partitioning of the region relates to partitioning of the corresponding system **K** matrix, which is discussed subsequently.

It is useful at this stage to consider the relationship of the physical interaction between the nodes and the entries of the element **K** matrix. To illus-

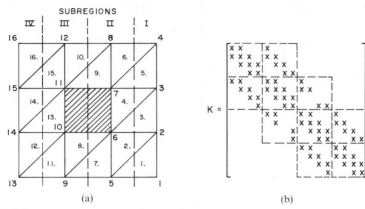

Fig. 6.3   Subdivision of hollow block: (a) numbering scheme B; (b) corresponding $\mathbf{K}$ matrix.

trate, consider the nodal equation for node 10 of element 8 (see Fig. 6.2a); this is given by

$$k^{e_8}_{10,9}\overline{\phi}_9 + k^{e_8}_{10,10}\overline{\phi}_{10} + k^{e_8}_{10,11}\overline{\phi}_{11} = 0, \tag{6.10}$$

where the node numbers corresponding to the three node identifiers 1, 2, and 3 were selected in the order 9, 10, and 11. By careful reexamination of the derivation of these terms, it is found that:

$k^{e_8}_{10,9}$ is the influence of node 9 on node 10;

$k^{e_8}_{10,10}$ is the influence of node 10 on itself;

$k^{e_8}_{10,11}$ is the influence of node 11 on node 10.

For self-adjoint problems, the influence of one node on another is reciprocal; that is, the influence of node m on node n is the same as the influence of node n on node m. The resulting system $\mathbf{K}$ matrices, at least in their uncorrected form, are therefore symmetric, as Figs. 6.2 and 6.3 show.

6.2   STORAGE MODES

Since the system $\mathbf{K}$ matrix is typically a sparse symmetric banded matrix, computer storage can be reduced by storing only the elements within the band. This can be done by storing the (principal) diagonal elements with either the codiagonal elements to the left or to the right; these are referred to as subdiagonal or superdiagonal storage modes, respectively. This procedure is illustrated in Table 6.2 for the case of subdiagonal storage of the symmetric system $\mathbf{K}$ matrix with semibandwidth $B_S = 2$. Only the first five rows of the matrix are shown. It is seen from Table 6.2 that in this

TABLE 6.2

Band Symmetric Storage Mode

Full storage mode (system **K** matrix with bandwidth $B_S = 2$)									Entries to be retained for "subdiagonal" storage			Band symmetric storage mode		
K_{11}	K_{12}	K_{13}	0	·	·	·	·	0	K_{11}			0	0	K_{11}
K_{21}	K_{22}	K_{23}	K_{24}	0	·	·	·	0	K_{21}	K_{22}		0	K_{21}	K_{22}
K_{31}	K_{32}	K_{33}	K_{34}	K_{35}	0	·	·	0	K_{31}	K_{32}	K_{33}	K_{31}	K_{32}	K_{33}
0	K_{42}	K_{43}	K_{44}	K_{45}	K_{46}	0	$0\cdots 0$		K_{42}	K_{43}	K_{44}	K_{42}	K_{43}	K_{44}
0	0	K_{53}	K_{54}	K_{55}	K_{56}	K_{57}	$0\cdots 0$		K_{53}	K_{54}	K_{55}	K_{53}	K_{54}	K_{55}

case the diagonal elements fall in the last column. If superdiagonal storage is used, the diagonal elements occur in the first column of the rectangular matrix.

For subdiagonal storage, an element K_{cd} of the system **K** matrix in the original full storage mode becomes an element K_{ce} in band symmetric storage mode, where

$$e = d + B_S - c + 1 \qquad (6.11)$$

and where B_S is the matrix semibandwidth. A similar equation applies for superdiagonal storage. It is often desirable to store a matrix or portion thereof as a *vector* of elements, that is, as a string of elements. The band symmetric storage mode shown in the right-hand side of Table 6.2 can be converted to vector storage mode either by storing successive rows or successive columns.

The banded system matrices obtained from finite element procedures seldom have profiles[†] that are straight lines parallel to the principal diagonal. It can therefore be advantageous to store successively the matrix column sections (between the profiles) as a vector, particularly when the solution algorithm is a column-oriented elimination [2]. Where the band is very sparse, one of the schemes described in Section 10.2.4 which exclude zero elements from storage may be preferable. However, unless carefully programmed, these methods can involve a large amount of data handling and become quite inefficient [3]. For sparse matrices, the hypermatrix storage scheme based on partitioning can be valuable, except when the solution algorithm is based on manipulating specific rows and columns [3].

[†] The upper profile is the demarcation line between nonzero elements and the upper right-hand-side region containing only zero elements. The lower profile is similarly defined. *Skyline* is used occasionally in place of the term *profile*.

6.3   RELATIVE MERITS OF ASSEMBLY BY NODES AND ELEMENTS

Conceptually, assembly of the system **K** matrix by nodes is somewhat simpler than assembly by elements and may be easier to program. A disadvantage of nodal assembly, however, is that the relevant element parameters once calculated, either need to be stored for their subsequent reuse with another node of the same element or allowed to be lost and recalculated when needed. In the former case, the penalty is additional storage, whilst in the latter more computation is required, compared with assembly by elements.

There are two situations, however, when nodal assembly can be advantageous. The first can occur with an extremely large problem on a conventional computer and the second with a problem of only moderate size on a minicomputer. In both of these cases, there is a limitation on computer storage but not on additional arithmetic computation. The choice of an *indirect* (iterative) solution algorithm with its minimal storage is therefore indicated (Chapter 10). With the simple iterative schemes, only several rows of the system matrix are needed in core at any one time and nodal assembly is compatible with this requirement. Up to the present, iterative methods have been seldom incorporated in the larger finite element program packages [2–4], but the projected continuing decrease in the price/performance ratio of minicomputers is likely to soon bring these machines into standard design office use, with a consequent demand for large-scale programs suited to their characteristics. The development of new finite element formulations, such as the absolute–relative displacement method [5], that appear suited to iterative solution may well accelerate the utilization of minicomputers.

With assembly by elements, the element **k** matrices are obtained for all elements in turn, and inserted into the system **K** matrix. The element parameters need to be computed only once for each element. When local coordinates are used, the necessary transformations can be handled expeditiously on an element–matrix basis as shown earlier. For these and other reasons, element assembly has been widely adopted and most large-scale program packages use this approach.

The computer programs listed in the previous chapters used, for the solution of the system equation, a library subroutine that required that the system **K** matrix be held in full storage mode with its obvious disadvantage. This drawback can be overcome in a number of ways, one of the simplest of which, called tridiagonalization, is presented next. Other, more efficient approaches, such as the frontal solution method (see Section 10.2.4) and simultaneous assembly and reduction (see Section 3.3.8), are described in detail in the literature [2–4].

6.4 TRIDIAGONALIZATION

When the system $\mathbf{K}$ matrix is too large to be in full storage mode or in symmetric band storage mode, partitioning can be used to divide the system matrix equation into a sequence of submatrix equations, each of which is of manageable size. This approach may be used to tridiagonalize the system matrix, allowing a solution to be obtained from submatrix equations. Although there are other procedures that can be adopted, tridiagonalization can still be a useful approach, which offers a significantly reduced core requirement at the price of some additional housekeeping.

Consider the partitioned region shown in Fig. 6.2a and the corresponding partitioned system matrix in Fig. 6.2b. Using this partitioning, the system matrix equation can be written as

$$\begin{bmatrix} \mathbf{K}_1 & \mathbf{C}_1 \\ \mathbf{C}_2 & \mathbf{K}_2 \end{bmatrix} \begin{bmatrix} \boldsymbol{\phi}_1 \\ \boldsymbol{\phi}_2 \end{bmatrix} = \begin{bmatrix} \mathbf{R}_1 \\ \mathbf{R}_2 \end{bmatrix}, \tag{6.12}$$

where $\mathbf{K}_1$ and $\mathbf{K}_2$ are the partition $\mathbf{K}$ matrices, $\boldsymbol{\phi}_1$ and $\boldsymbol{\phi}_2$ are the partition nodal vectors, and $\mathbf{R}_1$ and $\mathbf{R}_2$ are the right-hand side matrices, all respectively of partitions I and II. The matrix $\mathbf{C}_1$ is the influence of partition II on partition I and $\mathbf{C}_2$ is the influence of partition I on partition II. Due to the inherent symmetry of the system $\mathbf{K}$ matrix, $\mathbf{C}_1$ and $\mathbf{C}_2$ must contain respectively equal elements in symmetric locations. Hence Eq. (6.12) can be written as

$$\begin{bmatrix} \mathbf{K}_1 & \mathbf{C}_1 \\ \mathbf{C}_1^{\mathsf{T}} & \mathbf{K}_2 \end{bmatrix} \begin{bmatrix} \boldsymbol{\phi}_1 \\ \boldsymbol{\phi}_2 \end{bmatrix} = \begin{bmatrix} \mathbf{R}_1 \\ \mathbf{R}_2 \end{bmatrix}. \tag{6.13}$$

From Eq. (6.13) the two submatrix equations are

$$\mathbf{K}_1 \boldsymbol{\phi}_1 + \mathbf{C}_1 \boldsymbol{\phi}_2 = \mathbf{R}_1, \tag{6.14}$$

$$\mathbf{C}_1^{\mathsf{T}} \boldsymbol{\phi}_1 + \mathbf{K}_2 \boldsymbol{\phi}_2 = \mathbf{R}_2. \tag{6.15}$$

Inverting the matrix $\mathbf{K}_1$, denoting this inverse by $\mathbf{K}_1^{-1}$, and premultiplying Eq. (6.14) by $\mathbf{K}_1^{-1}$ result in

$$\boldsymbol{\phi}_1 = \mathbf{K}_1^{-1} \mathbf{R}_1 - \mathbf{K}_1^{-1} \mathbf{C}_1 \boldsymbol{\phi}_2. \tag{6.16}$$

Substituting Eq. (6.16) into Eq. (6.15) gives after some rearrangement

$$\boldsymbol{\phi}_2 = (\bar{\mathbf{K}}_2)^{-1} \bar{\mathbf{R}}_2, \tag{6.17}$$

where

$$\bar{\mathbf{K}}_2 = \mathbf{K}_2 - \mathbf{C}_1^{\mathsf{T}} \mathbf{K}_1^{-1} \mathbf{C}_1, \qquad \bar{\mathbf{R}}_2 = \mathbf{R}_2 - \mathbf{C}_1^{\mathsf{T}} \mathbf{K}_1^{-1} \mathbf{R}_1. \tag{6.18}$$

Since the submatrices in Eqs. (6.18) are explicitly known, Eq. (6.17) can be solved for $\boldsymbol{\phi}_2$. Substitution of this solution for $\boldsymbol{\phi}_2$ into Eq. (6.16) allows the solution for $\boldsymbol{\phi}_1$ to be obtained The whole $\mathbf{K}$ matrix need not be stored but only the submatrices $\mathbf{K}_1, \mathbf{K}_2, \mathbf{C}_1$ need to be kept in core. The matrices $\mathbf{K}_1$ and $\mathbf{K}_2$ are of different size, with the matrix $\mathbf{C}_1$ not being square in consequence, which is a common occurrence in practice. In this particular example, because of the poor numbering sequence of Fig. 6.2a and the use of only two partitions, there is no advantage compared with band symmetric storage of $\mathbf{K}$.

Consider then the numbering sequence and the partitioning of Fig. 6.3a. The corresponding banded matrix is shown in its partitioned form in Fig. 6.3b. Assuming that the system $\mathbf{K}$ matrix is symmetric, the system matrix equation can be written in its submatrix form as

$$\begin{bmatrix} \mathbf{K}_1 & \mathbf{C}_1 & & \\ \mathbf{C}_1{}^{\mathrm{T}} & \mathbf{K}_2 & \mathbf{C}_2 & \\ & \mathbf{C}_2{}^{\mathrm{T}} & \mathbf{K}_3 & \mathbf{C}_3 \\ & & \mathbf{C}_3{}^{\mathrm{T}} & \mathbf{K}_4 \end{bmatrix} \begin{bmatrix} \boldsymbol{\phi}_1 \\ \boldsymbol{\phi}_2 \\ \boldsymbol{\phi}_3 \\ \boldsymbol{\phi}_4 \end{bmatrix} = \begin{bmatrix} \mathbf{R}_1 \\ \mathbf{R}_2 \\ \mathbf{R}_3 \\ \mathbf{R}_4 \end{bmatrix}, \tag{6.19}$$

where $\mathbf{K}_i$, $\boldsymbol{\phi}_i$, and $\mathbf{R}_i$ are, respectively, the partition $\mathbf{K}$ matrix, the partition nodal vector, and the right-hand side matrix of partition i (where $i = 1, 2, 3, 4$). $\mathbf{C}_i$ is the influence of partition $i + 1$ on partition i, where $i = 1, 2, 3$. It is seen that the coefficient matrix in Eq. (6.19) is a tridiagonal banded matrix. Partitioning the region into any number of partitions (equal to or greater than three) will always result in a tridiagonal form for the system matrix equation.

The solution procedure for Eq. (6.19) follows in a similar fashion to the previous case, using as the generalized equivalents to Eqs. (6.18),

$$\bar{\mathbf{K}}_{i+1} = \mathbf{K}_{i+1} - \mathbf{C}_i{}^{\mathrm{T}}(\bar{\mathbf{K}}_i)^{-1}\mathbf{C}_i, \qquad i = 1, 2, \ldots, N - 1, \tag{6.20a}$$
$$\bar{\mathbf{R}}_{i+1} = \mathbf{R}_{i+1} - \mathbf{C}_i{}^{\mathrm{T}}(\bar{\mathbf{K}}_i)^{-1}\mathbf{R}_i, \qquad i = 1, 2, \ldots, N - 1, \tag{6.20b}$$

where N is the number of partitions. For $i = 0$, the following apply

$$\bar{\mathbf{K}}_1 = \mathbf{K}_1, \qquad \bar{\mathbf{R}}_1 = \mathbf{R}_1, \tag{6.20c,d}$$

The first submatrix equation of Eq. (6.19) is

$$\mathbf{K}_1\boldsymbol{\phi}_1 + \mathbf{C}_1\boldsymbol{\phi}_2 = \mathbf{R}_1, \tag{6.21}$$

which, using Eqs. (6.20) and the inverse of $\bar{\mathbf{K}}_1$, denoted $(\bar{\mathbf{K}}_1)^{-1}$, becomes

$$\boldsymbol{\phi}_1 = (\bar{\mathbf{K}}_1)^{-1}\bar{\mathbf{R}}_1 - (\bar{\mathbf{K}}_1)^{-1}\mathbf{C}_1\boldsymbol{\phi}_2. \tag{6.22}$$

The second submatrix equation of Eq. (6.19) is

$$\mathbf{C}_1{}^{\mathrm{T}}\boldsymbol{\phi}_1 + \mathbf{K}_2\boldsymbol{\phi}_2 + \mathbf{C}_2\boldsymbol{\phi}_3 = \mathbf{R}_2, \tag{6.23}$$

which, after substitution for ϕ_1 from Eq. (6.22), can be rearranged as

$$\phi_2 = (\bar{\mathbf{K}}_2)^{-1}\bar{\mathbf{R}}_2 - (\bar{\mathbf{K}}_2)^{-1}\mathbf{C}_2\phi_3, \tag{6.24}$$

where use is made of Eqs. (6.20).

From Eqs. (6.22) and (6.24), the general form of the solution for ϕ_i will have become evident, and it can easily be shown that

$$\phi_i = \bar{\mathbf{K}}_i^{-1}\bar{\mathbf{R}}_i - \bar{\mathbf{K}}_i^{-1}\mathbf{C}_i\phi_{i+1}, \qquad i = 1, 2, \ldots, N - 1. \tag{6.25}$$

From the final submatrix equation there is obtained

$$\mathbf{C}_3^{\mathrm{T}}\phi_3 + \mathbf{K}_4\phi_4 = \mathbf{R}_4, \tag{6.26}$$

which, using Eqs. (6.20) and (6.25), can be rearranged as

$$\phi_4 = (\bar{\mathbf{K}}_4)^{-1}\bar{\mathbf{R}}_4. \tag{6.27}$$

This result can be generalized to give

$$\phi_N = (\bar{\mathbf{K}}^{-1})_N\bar{\mathbf{R}}_N, \tag{6.28}$$

which, together with Eqs. (6.25), yields a set of equations for all the ϕ_i, $i = 1, 2, \ldots, N$. Equation (6.27) can be solved for ϕ_4 since the submatrices are explicitly available. Then successive back substitution, using Eqs. (6.25) with $i = 3, 2$, and 1, yields the solutions for ϕ_3, ϕ_2, and ϕ_1.

Illustrative Example 6.1   Consider the heat transfer problem of Fig. 2.1, using the finite element mesh of Fig. 2.4. Let the region be partitioned as shown in Fig. 6.4. The system matrix equation for this problem has been derived

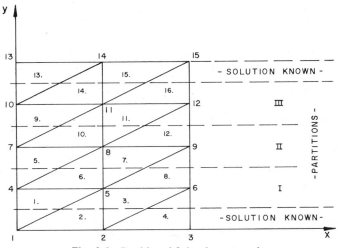

Fig. 6.4   Partitioned finite element mesh.

previously and is given in Eq. (2.36). As indicated in Fig. 6.4, the solution is known at the nodes 1, 2, 3 and 13, 14, 15. Consequently, the corresponding nodal equations can be eliminated according to the procedure outlined in Section 3.3.4. It can be shown that the system $\mathbf{K}$ matrix then reduces to

$$
\begin{bmatrix}
10 & -2 & 0 & -4 & 0 & 0 & & & \\
-2 & 20 & -2 & 0 & -8 & 0 & & & \\
0 & -2 & 10 & 0 & 0 & -4 & & & \\
-4 & 0 & 0 & 10 & -2 & 0 & -4 & 0 & 0 \\
0 & -8 & 0 & -2 & 20 & -2 & 0 & -8 & 0 \\
0 & 0 & -4 & 0 & -2 & 10 & 0 & 0 & -4 \\
& & & -4 & 0 & 0 & 10 & -2 & 0 \\
& & & 0 & -8 & 0 & -2 & 20 & -2 \\
& & & 0 & 0 & -4 & 0 & -2 & 10
\end{bmatrix}
\begin{bmatrix}
\bar{T}_4 \\ \bar{T}_5 \\ \bar{T}_6 \\ \bar{T}_7 \\ \bar{T}_8 \\ \bar{T}_9 \\ \bar{T}_{10} \\ \bar{T}_{11} \\ \bar{T}_{12}
\end{bmatrix}
=
\begin{bmatrix}
200 \\ 400 \\ 200 \\ 0 \\ 0 \\ 0 \\ 400 \\ 800 \\ 400
\end{bmatrix}, \quad (6.29)
$$

where the partitioning lines of Fig. 6.4 have been inserted and where it is noted that the resulting system $\mathbf{K}$ matrix is now symmetric. Equation (6.29) can also be written as

$$
\begin{bmatrix}
\mathbf{K}_1 & \mathbf{C}_1 & \\
\mathbf{C}_1^{\mathrm{T}} & \mathbf{K}_2 & \mathbf{C}_2 \\
& \mathbf{C}_2^{\mathrm{T}} & \mathbf{K}_3
\end{bmatrix}
\begin{bmatrix}
\mathbf{T}_1 \\ \mathbf{T}_2 \\ \mathbf{T}_3
\end{bmatrix}
=
\begin{bmatrix}
\mathbf{R}_1 \\ \mathbf{R}_2 \\ \mathbf{R}_3
\end{bmatrix}, \quad (6.30)
$$

where the various submatrices are defined through Eq. (6.29).

Using Eqs. (6.20), $\bar{\mathbf{K}}_i$ and $\bar{\mathbf{R}}_i$ for $i = 1, 2,$ and 3 are obtained as

$$
\bar{\mathbf{K}}_1 = \begin{bmatrix} 10 & -2 & 0 \\ -2 & 20 & -2 \\ 0 & -2 & 10 \end{bmatrix}, \qquad \bar{\mathbf{R}}_1 = \begin{bmatrix} 200 \\ 400 \\ 200 \end{bmatrix}, \quad (6.31)
$$

$$
\bar{\mathbf{K}}_2 = \begin{bmatrix} 8.36667 & -2.33333 & -0.03333 \\ -2.33333 & 16.66667 & -2.33333 \\ -0.03333 & -2.33333 & 8.36667 \end{bmatrix}, \qquad \bar{\mathbf{R}}_2 = \begin{bmatrix} 100 \\ 200 \\ 100 \end{bmatrix}, \quad (6.32)
$$

$$
\bar{\mathbf{K}}_3 = \begin{bmatrix} 8.0060 & -2.5833 & -0.0893 \\ -2.5833 & 15.8333 & -2.5833 \\ -0.0893 & -2.5833 & 8.0060 \end{bmatrix}, \qquad \bar{\mathbf{R}}_3 = \begin{bmatrix} 466.6667 \\ 933.3333 \\ 466.6667 \end{bmatrix}, \quad (6.33)
$$

The inverses of $\bar{\mathbf{K}}_1$, $\bar{\mathbf{K}}_2$, and $\bar{\mathbf{K}}_3$ can then be computed as

$$
\bar{\mathbf{K}}_1^{-1} = \frac{1}{1920} \begin{bmatrix} 196 & 20 & 4 \\ 20 & 100 & 20 \\ 4 & 20 & 196 \end{bmatrix}, \quad (6.34)
$$

$$\bar{\mathbf{K}}_2^{-1} = \frac{1}{1075.2000} \begin{bmatrix} 134 & 19.6 & 6 \\ 19.6 & 70 & 19.6 \\ 6 & 19.6 & 134 \end{bmatrix}, \tag{6.35}$$

$$\bar{\mathbf{K}}_3^{-1} = \frac{1}{906.6788} \begin{bmatrix} 120.0880 & 20.9126 & 8.0874 \\ 20.9126 & 64.0881 & 20.9126 \\ 8.0874 & 20.9126 & 120.0880 \end{bmatrix}. \tag{6.36}$$

The solution for $\mathbf{T}_3$ is now obtained from $\mathbf{T}_3 = (\bar{\mathbf{K}}_3)^{-1}\mathbf{R}_3$ as

$$\mathbf{T}_3 = \frac{1}{906.6788} \begin{bmatrix} 120.0880 & 20.9126 & 8.0874 \\ 20.9126 & 64.0881 & 20.9126 \\ 8.0874 & 20.9126 & 120.0880 \end{bmatrix} \begin{bmatrix} 466.6667 \\ 933.3333 \\ 466.6667 \end{bmatrix} = \begin{bmatrix} 87.50 \\ 87.50 \\ 87.50 \end{bmatrix}. \tag{6.37}$$

The solution for $\mathbf{T}_2$ then follows from Eq. (6.25), with $i = 2$, as

$$\mathbf{T}_2 = \frac{1}{1075.20} \begin{bmatrix} 134 & 19.6 & 6 \\ 19.6 & 70 & 19.6 \\ 6 & 19.6 & 134 \end{bmatrix} \left\{ \begin{bmatrix} 100 \\ 200 \\ 100 \end{bmatrix} - \begin{bmatrix} -4 & 0 & 0 \\ 0 & -8 & 0 \\ 0 & 0 & -4 \end{bmatrix} \begin{bmatrix} 87.50 \\ 87.50 \\ 87.50 \end{bmatrix} \right\}$$

$$= \begin{bmatrix} 75.00 \\ 75.00 \\ 75.00 \end{bmatrix}. \tag{6.38}$$

Finally, $\mathbf{T}_1$ is computed calculated from Eq. (6.25), with $i = 1$, as

$$\mathbf{T}_1 = \frac{1}{1920} \begin{bmatrix} 196 & 20 & 4 \\ 20 & 100 & 20 \\ 4 & 20 & 196 \end{bmatrix} \left\{ \begin{bmatrix} 200 \\ 400 \\ 200 \end{bmatrix} - \begin{bmatrix} -4 & 0 & 0 \\ 0 & -8 & 0 \\ 0 & 0 & -4 \end{bmatrix} \begin{bmatrix} 75.00 \\ 75.00 \\ 75.00 \end{bmatrix} \right\} = \begin{bmatrix} 62.50 \\ 62.50 \\ 62.50 \end{bmatrix}. \tag{6.39}$$

From Eqs. (6.37)–(6.39), the solution can therefore be tabulated as

$$\begin{array}{lll} \bar{T}_1 = 50, & \bar{T}_2 = 50, & \bar{T}_3 = 50, \\ \bar{T}_4 = 62.5, & \bar{T}_5 = 62.5, & \bar{T}_6 = 62.5, \\ \bar{T}_7 = 75, & \bar{T}_8 = 75, & \bar{T}_9 = 75, \\ \bar{T}_{10} = 87.5, & \bar{T}_{11} = 87.5, & \bar{T}_{12} = 87.5, \\ \bar{T}_{13} = 100, & \bar{T}_{14} = 100, & \bar{T}_{15} = 100, \end{array} \tag{6.40}$$

which is identical to that given previously in Eq. (2.37).

Exercise 6.1   Prepare a flow chart and develop a computer program based on the tridiagonalization procedure developed in Section 6.4.

REFERENCES

1. J. A. George, On the density of finite element matrices, *Internat. J. Numer. Methods Engrg.* **5**, 297–300 (1972).
2. K-J. Bathe and E. L. Wilson. "Numerical Methods in Finite Element Analysis." Prentice-Hall, Englewood Cliffs, New Jersey, 1976.
3. S. J. Fenves, N. Perrone, A. R. Robinson, and W. C. Schnobrich (eds.), "Numerical and Computer Methods in Structural Mechanics." Academic Press, New York, 1973.
4. D. H. Norrie and G. de Vries, "Finite Element Bibliography." Plenum, New York, 1976.
5. E. L. Wilson, Special numerical and computer techniques for the analysis of finite element systems, *Proc. U.S.–Germany Symp. Formulations and Computational Algorithms in Finite Element Analysis, 9–13 August 1976, MIT, Cambridge, Massachusetts.*

7

THE VARIATIONAL CALCULUS AND ITS APPLICATION

In this chapter, some of the basic concepts of the classical calculus of variations and their applications to field problems are briefly reviewed. Further details on the variational calculus may be found in standard works [1-3]. Applications of the variational calculus to physical phenomena have been extensive and some of the more important variational principles are presented at the end of this chapter.

7.1 MAXIMA AND MINIMA OF FUNCTIONS

7.1.1 Functions of One Independent Variable

A fundamental theorem of calculus states that any function $y = f(x)$ that is continuous[†] in a region $a \leq x \leq b$ attains a maximum and a minimum value in that region. Moreover, if $f(x)$ attains its minimum (or maximum value) at $x = x_0$, where $a < x_0 < b$, then x_0 can be found as the

---

[†] It is assumed throughout this chapter that the necessary conditions of existence and continuity of the function and its derivatives are satisfied.

solution $x = x_0$ to the equation

$$\frac{df(x)}{dx} = 0. \tag{7.1}$$

provided the first derivative of $f(x)$ exists at x_0.

The statement leading to Eq. (7.1) can be expressed in the following alternative form. If the continuous function $y = f(x)$, defined in the region $a \leq x \leq b$, attains its maximum or minimum value at the point $x = x_0$, then the first derivative of $f(x)$ with respect to x must vanish at that point, and consequently the first variation of f, denoted by δf, and sometimes referred to as the differential df, must vanish there for arbitrary changes δx in x, that is,

$$\delta f = \left(\frac{df}{dx}\right)\delta x = 0. \tag{7.2}$$

Further differentiation of the function $y = f(x)$ determines whether $f(x)$ has a relative minimum, maximum, or minimax condition at the point $x = x_0$, according to the criteria:

$$\text{minimum} \quad \text{if} \quad \frac{d^2f}{dx^2} > 0 \quad \text{at} \quad x = x_0, \tag{7.3a}$$

$$\text{maximum} \quad \text{if} \quad \frac{d^2f}{dx^2} < 0 \quad \text{at} \quad x = x_0, \tag{7.3b}$$

$$\text{minimax} \quad \text{if} \quad \frac{d^2f}{dx^2} = 0 \quad \text{at} \quad x = x_0. \tag{7.3c}$$

Equation (7.1) [or the alternative form, Eq. (7.2)] is a necessary condition for $f(x)$ to be a minimum at the point x_0, although Eqs. (7.3) show that this is not a sufficient condition. Equation (7.2) is, however, a necessary and sufficient condition for $f(x)$ to be stationary at the point $x = x_0$. A function $f(x)$ is said to be stationary at $x = x_0$ if it attains a minimum, maximum, or a minimax condition at that point.

7.1.2   Functions with Two Independent Variables

The necessary conditions for a stationary value of the continuous function

$$z = f(x, y), \tag{7.4}$$

will now be considered. Since z depends on both x and y, it is now necessary to differentiate with respect to both variables and to solve the resulting equations simultaneously. Thus, if z is stationary at the point (x_0, y_0), then x_0 and y_0 are the solutions to the system of equations

$$\partial f/\partial x = 0, \tag{7.5a}$$

$$\partial f/\partial y = 0. \tag{7.5b}$$

Alternatively, if $z = f(x, y)$ is stationary at the point $(x, y) = (x_0, y_0)$, the variation of this function must vanish for arbitrary variations δx and δy; that is,

$$\delta f = \left(\frac{\partial f}{\partial x}\right)\delta x + \left(\frac{\partial f}{\partial y}\right)\delta y = 0, \tag{7.6}$$

at the point $(x, y) = (x_0, y_0)$. Equations (7.5a) and (7.5b) follow from Eq. (7.6) as a consequence of the arbitrariness of δx and δy.

7.1.3 Functions with n Independent Variables

The procedure of the previous section may be generalized to functions with n independent variables. Consider a function $f(x_1, x_2, \ldots, x_n)$ that is continuous in a closed domain. It is desired to find the point $(x_1, x_2, \ldots, x_n) = (x_1{}^0, x_2{}^0, \ldots, x_n{}^0) = X^0$ at which the function $f(x_i)$ is stationary. Extending Eq. (7.6) shows that the necessary and sufficient condition for $f(x_i)$ to be stationary at X^0 is that the first variation of $f(x_i)$ must vanish at X^0, for arbitrary variations δx_i, $i = 1, 2, \ldots, n$; that is,

$$\delta f = \left(\frac{\partial f}{\partial x_1}\right)\delta x_1 + \left(\frac{\partial f}{\partial x_2}\right)\delta x_2 + \cdots + \left(\frac{\partial f}{\partial x_i}\right)\delta x_i + \cdots + \left(\frac{\partial f}{\partial x_n}\right)\delta x_n = 0. \tag{7.7}$$

Equation (7.7) can be written more simply as

$$\delta f = \left\{\sum_{j=1}^{n} \frac{\partial f(x_i)}{\partial x_j}\bigg|_{X^0} \delta x_j\right\} = \sum_{j=1}^{n} \frac{\partial f(X^0)}{\partial x_j}\delta x_j = 0, \tag{7.8}$$

where

$$\frac{\partial f(X^0)}{\partial x_j} = \frac{\partial f(x_i)}{\partial x_j}\bigg|_{X^0}, \qquad i = 1, 2, \ldots, n, \tag{7.9}$$

is the partial derivative of $f(x_i)$ with respect to x_j, evaluated at the point X^0.

Since Eq. (7.7) is true for arbitrary variations $\delta x_j, j = 1, 2, \ldots, n$, all but one of these may be chosen as zero. Let the independent variable for which there is a nonzero change be x_k, that is,

$$\delta x_j = 0, \qquad j = 1, 2, \ldots, n, \qquad j \neq k. \tag{7.10}$$

Substituting Eq. (7.10) into Eq. (7.7) results in

$$\delta f = \frac{\partial f(X^0)}{\partial x_k}\delta x_k = 0, \tag{7.11}$$

for arbitrary δx_k. Consequently, it follows that

$$\frac{\partial f(X^0)}{\partial x_k} = 0. \tag{7.12}$$

Since the independent variable x_k was chosen arbitrarily, the result obtained in Eq. (7.12) must be true for all x_k. The necessary and sufficient conditions for $f(x_i)$ to be stationary at the point X^0 reduce, therefore, to the system of simultaneous equations

$$\frac{\partial f(X^0)}{\partial x_i} = 0, \qquad i = 1, 2, \ldots, n. \tag{7.13}$$

7.2 LAGRANGE MULTIPLIERS

The previous sections dealt with necessary and sufficient conditions for a function of independent variables to be stationary. In many applications, however, the variables are not all independent. Consider, for example, necessary and sufficient conditions for the continuous function $f(x_i)$, $i = 1, 2, \ldots, n$, defined in a given region, to be stationary, if the x_i are related through the *m equations of constraint*

$$g_k(x_i) = 0, \qquad k = 1, 2, \ldots, m, \qquad m < n. \tag{7.14}$$

One possible method of solution is as follows. The m equations of constraint can, at least in principle, be solved for m of the variables in terms of the remaining $n - m$ variables. The relations so obtained can be substituted into the function $f(x_i)$, $i = 1, 2, \ldots, n$, to yield a new function $F(x_{m+1}, x_{m+2}, \ldots, x_n)$. In this function, all the variables $x_{m+1}, x_{m+2}, \ldots, x_n$ are independent since the original dependency has been taken care of. The criteria developed in Section 7.1.3 may thus be applied directly to $F(x_{m+1}, x_{m+2}, \ldots, x_n)$ to yield the set of $n - m$ simultaneous equations

$$\frac{\partial F(x_{m+1}, x_{m+2}, \ldots, x_n)}{\partial x_j} = 0, \qquad j = m + 1, m + 2, \ldots, n. \tag{7.15}$$

The set of $n - m$ equations represented by Eq. (7.15), together with the m equations in Eq. (7.14) comprise a system of n equations from which the required n variables $x_1^0, x_2^0, \ldots, x_n^0$ may be obtained.

Although the above procedure will yield the correct solution, it is somewhat cumbersome in its application, especially if functions of three or more variables are considered. A more elegant approach is the *method of Lagrange multipliers*, which is outlined in the following.

It has already been established that a necessary and sufficient condition for $f(x_i)$ to be stationary at the point X^0 is that the first variation of $f(x_i)$ at that point must vanish for arbitrary variations δx_j, that is,

$$\delta f(X^0) = \sum_{j=1}^{n} \frac{\partial f(X^0)}{\partial x_j} \delta x_j = 0. \tag{7.16}$$

In the present case, however, not all the x_i are independent, and hence it cannot be concluded that all the partial derivatives $\partial f(X^0)/\partial x_j$ vanish simultaneously.

The equations of constraint given by Eqs. (7.14) yield

$$\delta g_k(X^0) = \sum_{j=1}^{n} \frac{\partial g_k(X^0)}{\partial x_j} \delta x_j = 0, \qquad k = 1, 2, \ldots, m, \quad m < n. \tag{7.17}$$

Into these relations, m new variables, the Lagrange multipliers $\lambda_1, \lambda_2, \ldots, \lambda_m$, are introduced as follows. Each of the equations in (7.17) is multiplied by its appropriate λ_k and the results added to Eq. (7.16), to give

$$\delta f(X^0) + \sum_{k=1}^{m} \lambda_k \delta g_k(X^0) = \sum_{j=1}^{n} \left[\frac{\partial f(X^0)}{\partial x_j} + \sum_{k=1}^{m} \lambda_k \frac{\partial g_k(X^0)}{\partial x_j} \right] \delta x_j = 0. \tag{7.18}$$

Since the λ_k are arbitrary, they may be chosen such that the first m of the expressions in the square brackets in Eq. (7.18) vanish. The remaining $n - m$ variables x_j in Eq. (7.18) are independent, hence their variations δx_j are arbitrary and the remaining $n - m$ expressions in the square brackets in Eq. (7.18) must also vanish. Consequently, these two conditions yield the set of n simultaneous equations

$$\frac{\partial f(X^0)}{\partial x_j} + \sum_{k=1}^{m} \lambda_k \frac{\partial g_k(X^0)}{\partial x_j} = 0, \qquad j = 1, 2, \ldots, n, \tag{7.19}$$

which, together with the m equations of constraint [Eqs. (7.14)] comprise a set of $m + n$ equations from which the $m + n$ unknowns $\lambda_1, \lambda_2, \ldots, \lambda_m$, $x_1{}^0, x_2{}^0, \ldots, x_n{}^0$, can be derived.

The procedure outlined above can also be developed from another point of view. Let a new function $F(x_1, x_2, \ldots, x_n; \lambda_1, \lambda_2, \ldots, \lambda_m)$ be defined as

$$F(x_i; \lambda_j) = f(x_i) + \sum_{k=1}^{m} \lambda_k g_k(x_i), \qquad i = 1, 2, \ldots, n,$$

$$j = 1, 2, \ldots, m, \quad m < n. \tag{7.20}$$

Since the m equations of constraint [Eqs. (7.14)] are included in the expression for $F(x_i; \lambda_j)$, [Eq. (7.20)], it is evident that $F(x_i; \lambda_j)$ is a function of $n + m$ independent variables. From Section 7.1.3, necessary and sufficient conditions for $F(x_i; \lambda_j)$ to be stationary are therefore

$$\frac{\partial F(x_i; \lambda_j)}{\partial x_j} = 0, \qquad j = 1, 2, \ldots, n, \tag{7.21a}$$

$$\frac{\partial F(x_i; \lambda_j)}{\partial \lambda_k} = 0, \qquad k = 1, 2, \ldots, m. \tag{7.21b}$$

Substituting $F(x_i; \lambda_j)$ from Eq. (7.20) into Eqs. (7.21) then results in the set of $m + n$ equations

$$\frac{\partial f(x_i)}{\partial x_j} + \sum_{k=1}^{m} \lambda_k \frac{\partial g_k(x_i)}{\partial x_j} = 0, \qquad j = 1, 2, \ldots, n, \tag{7.22}$$

$$g_k(x_i) = 0, \qquad k = 1, 2, \ldots, m, \tag{7.23}$$

which are seen to be the same as Eqs. (7.19) and (7.14), respectively.

It is seen from the above that the use of Lagrange multipliers is a powerful tool when dealing with equations of constraint. As a particular case, the boundary conditions of a field problem could be considered as equations of constraint and the method of Lagrange multipliers applied.

7.3 MAXIMA AND MINIMA OF FUNCTIONALS

In previous sections, the problem of finding the stationary values of an explicit function $f(x_i)$ was considered. In many applications, however, the stationary value of an integral is required, instead of the stationary value of a simple function. Such an integral is known as a functional, and the necessary conditions for this to be stationary are now considered.

7.3.1 One Independent Variable and Several Dependent Variables—Euler's Equations

In this section, functionals with one independent variable and several dependent variables will be examined. Consider the functional

$$\chi = \int_a^b F\left[y_i(x), \frac{dy_i(x)}{dx} \right] dx, \qquad i = 1, 2, \ldots, m, \tag{7.24}$$

for which the conditions

$$y_i(a) = a_i, \qquad y_i(b) = b_i, \tag{7.25}$$

apply, and where the function F depends not only on x, but also on the m functions $y_i(x)$ and their first derivatives $dy_i(x)/dx$. The trial functions $y_i(x)$ in Eq. (7.24) must be *admissible*, that is, they should not violate any of the requirements of the variational process. For this problem, the class of admissible functions includes those functions that are continuous and have piecewise continuous first derivatives in $a \le x \le b$.

Consider now the variation of χ, which from Eq. (7.24) becomes[†] for a stationary value of χ

$$\delta\chi = \int_a^b \left[\frac{\partial F}{\partial y_i} \delta y_i + \frac{\partial F}{\partial(dy_i/dx)} \delta\left(\frac{dy_i}{dx}\right) \right] dx = 0. \tag{7.26}$$

Writing

$$y_i' = \frac{dy_i}{dx} \tag{7.27}$$

and noting that

$$\delta\left(\frac{dy_i}{dx}\right) = \frac{d}{dx}(\delta y_i) = (\delta y_i)' \tag{7.28}$$

allows Eq. (7.26) to be written as

$$\delta\chi = \int_a^b \left[\frac{\partial F}{\partial y_i} \delta y_i + \frac{\partial F}{\partial y_i'} (\delta y_i)' \right] dx = 0. \tag{7.29}$$

Integrating the second right-hand side term by parts gives

$$\delta\chi = \int_a^b \left[\frac{\partial F}{\partial y_i} - \frac{d}{dx}\left(\frac{\partial F}{\partial y_i'}\right) \right] \delta y_i \, dx + \frac{\partial F}{\partial y_i'} \delta y_i \Big|_a^b = 0. \tag{7.30}$$

If the trial functions $y_i(x)$ satisfy the principal boundary conditions [Eqs. (7.25)], then Eq. (7.30) reduces to

$$\delta\chi = \int_a^b \left[\frac{\partial F}{\partial y_i} - \frac{d}{dx}\left(\frac{\partial F}{\partial y_i'}\right) \right] \delta y_i \, dx = 0. \tag{7.31}$$

Since the expression in square brackets within the integrand of Eq. (7.31) is continuous and the variations δy_i, $i = 1, 2, \ldots, m$, are arbitrary,

$$\frac{\partial F}{\partial y_i} - \frac{d}{dx}\left(\frac{\partial F}{\partial y_i'}\right) = 0, \qquad i = 1, 2, \ldots, m. \tag{7.32}$$

Equations (7.32) are ordinary differential equations of second order and are called the *Euler equations* for this problem. As the above has shown, the particular trial functions $y_i(x)$ for which χ has a stationary value also satisfy the Euler equations.

[†] Using the summation convention (p. 33).

7.3.2   One Dependent Variable and Several Independent Variables

In many problems, the functional involves only one dependent variable but several independent variables. To illustrate, necessary and sufficient conditions for the functional

$$\chi = \frac{1}{2}\int_D\left[\left(\frac{\partial T}{\partial x}\right)^2 + \left(\frac{\partial T}{\partial y}\right)^2\right]dD + \int_S pT\,dS, \qquad (7.33)$$

to be stationary will be investigated, where p is a function of position along the boundary S enclosing the domain D, and where, to be admissible, the trial functions $T(x, y)$ are required to be continuous with piecewise continuous first derivatives in $D + S$.

As before, the necessary condition for χ to be stationary is that the variation of χ, $\delta\chi$, vanishes; that is,

$$\delta\chi = \int_D\left[\frac{\partial T}{\partial x}\delta\left(\frac{\partial T}{\partial x}\right) + \frac{\partial T}{\partial y}\delta\left(\frac{\partial T}{\partial y}\right)\right]dD + \int_S p\delta T\,dS = 0. \qquad (7.34)$$

Noting that

$$\delta\left(\frac{\partial T}{\partial x}\right) = \frac{\partial}{\partial x}(\delta T), \qquad (7.35a)$$

$$\delta\left(\frac{\partial T}{\partial y}\right) = \frac{\partial}{\partial y}(\delta T), \qquad (7.35b)$$

allows Eq. (7.34) to be written as

$$\delta\chi = \int_D\left[\frac{\partial T}{\partial x}\frac{\partial}{\partial x}(\delta T) + \frac{\partial T}{\partial y}\frac{\partial}{\partial y}(\delta T)\right]dD + \int_S p\delta T\,dS = 0. \qquad (7.36)$$

Green's identity can be written in the form

$$\int_D\left(\frac{\partial u}{\partial x}\frac{\partial v}{\partial x} + \frac{\partial u}{\partial y}\frac{\partial v}{\partial y}\right)dD = -\int_D v\left(\frac{\partial^2 u}{\partial x^2} + \frac{\partial^2 u}{\partial y^2}\right)dD$$

$$+ \int_S v\left(\frac{\partial u}{\partial x}n_x + \frac{\partial u}{\partial y}n_y\right)dS \qquad (7.37)$$

where n_x, n_y are the x, y components of the unit outward normal to S, denoted by $\mathbf{n}$. The functions $u = u(x, y)$ and $v = v(x, y)$ are required to be continuous in $D + S$ but can possess piecewise continuous second derivatives in $D + S$. The first derivatives of u are required to be continuous but those of v can be piecewise continuous [4].

The case where T is continuous and has continuous first derivatives in $D + S$ is first considered. Since the continuity of δT is the same as that of T,

Green's identity can be used to rewrite the domain integral in the form

$$\int_D \left[\frac{\partial T}{\partial x} \frac{\partial}{\partial x}(\delta T) + \frac{\partial T}{\partial y} \frac{\partial}{\partial y}(\delta T) \right] dD = -\int_D \delta T \nabla^2 T \, dD$$

$$+ \int_S \left(\frac{\partial T}{\partial x} n_x + \frac{\partial T}{\partial y} n_y \right) \delta T \, dS$$

$$= -\int_D \delta T \nabla^2 T \, dD + \int_S \frac{\partial T}{\partial n} \delta T \, dS, \quad (7.38)$$

where

$$\nabla^2 T = \frac{\partial^2 T}{\partial x^2} + \frac{\partial^2 T}{\partial y^2}. \quad (7.39)$$

Substituting Eq. (7.38) into Eq. (7.36) finally results in

$$\delta \chi = -\int_D \nabla^2 T \delta T \, dD + \int_S \left(p + \frac{\partial T}{\partial n} \right) \delta T \, dS = 0. \quad (7.40)$$

Since the domain and surface integral in Eq. (7.40) are independent, it follows that both

$$\int_D \nabla^2 T \delta T \, dD = 0 \quad (7.41a)$$

and

$$\int_S \left(p + \frac{\partial T}{\partial n} \right) \delta T \, dS = 0. \quad (7.41b)$$

From the arbitrariness of δT in D, it follows from Eq. (7.41a) that

$$\nabla^2 T = 0 \quad \text{in } D, \quad (7.42)$$

and from Eq. (7.41b) that

$$\frac{\partial T}{\partial n} + p = 0 \quad \text{on } S. \quad (7.43)$$

Alternatively, if $T(x, y)$ is prescribed as $T = g$ on that part of the boundary denoted S_1, so that

$$T = g \quad \text{on } S_1, \quad (7.44)$$

and not on the remaining part, denoted $S_2^{\dagger}$, where

$$S = S_1 + S_2, \quad (7.45)$$

and, furthermore, if the trial functions $T(x, y)$ are chosen to satisfy the

$\dagger$ The surface integral in Eq. (7.53) and subsequent equations is correspondingly restricted to S_2.

boundary conditions in Eq. (7.44), then the variations of T must vanish on S_1; that is,

$$\delta T = 0 \qquad \text{on } S_1. \tag{7.46}$$

Eq. (7.41b) now becomes

$$\int_{S_2} \left(p + \frac{\partial T}{\partial n} \right) \delta T \, dS = 0, \tag{7.47}$$

which on account of the arbitrariness of δT on S_2 gives

$$\frac{\partial T}{\partial n} + p = 0 \qquad \text{on } S_2. \tag{7.48}$$

Consider now the case where T is continuous with piecewise continuous first derivatives. The surfaces of discontinuity of the first derivatives divide the domain D into subdomains. Green's identity can be applied to each subdomain and hence Eq. (7.38) is valid for a subdomain. Adding these subdomain equations yields a domain equation identical to Eq. (7.38) except for an additional term, an interface integral, appearing on the right-hand side. This interface integral carries down into Eq. (7.40), being added to the middle expression of the equation. However, by virtue of δT being arbitrary over $D + S$, the domain integral, the boundary surface integral and the interface integral in Eq. (7.40) must be independent and separately equal to zero. The subsequent equations, (7.41)–(7.48), therefore, remain valid when T is continuous with piecewise continuous first derivatives.

In summary, the above shows that that function[†] $T(x, y)$ making the functional of Eq. (7.33) stationary is also the solution to the field equation

$$\nabla^2 T = 0 \qquad \text{in } D, \tag{7.42}$$

subject to the Dirichlet boundary condition

$$T = g \qquad \text{on } S_1, \tag{7.44}$$

and the Neumann boundary condition

$$\frac{\partial T}{\partial n} + p = 0 \qquad \text{on } S_2. \tag{7.48}$$

In this example, the Dirichlet boundary condition given in Eq. (7.44) is the prescribed or principal boundary condition, whereas the Neumann given in Eq. (7.48) is the natural boundary condition. The latter name follows from the condition being a natural consequence of the variational procedure.

[†] Within the class of admissible functions, namely, those functions which are continuous and have piecewise continuous first derivatives.

In passing, it should be pointed out that in certain cases a natural boundary condition may be changed by modifying the functional to suit the need of the problem under investigation.

7.4   ADMISSIBILITY AND FINITE ELEMENTS

It is important that the role played by the admissibility conditions imposed on trial functions in a variational problem be clearly understood. The functional for such a problem can, in general, be written in the form

$$\chi = \int_D F(y_1, y_2, \ldots, y_m) \, dD \tag{7.49}$$

where the y_i, $i = 1, 2, \ldots, m$, are functions of the independent variables. For a variational procedure such as outlined in Section 7.3.1 or 7.3.2 to be valid, it will be found necessary to place restrictions on the continuity of the trial functions $y_1, y_2, \ldots, y_m$ to be used in Eq. (7.49). These continuity restrictions are the *admissibility conditions* for the problem. For simplicity, the case of a single dependent variable, that is, $m = 1$ in Eq. (7.49), will be considered, but the discussion can easily be extended to two or more dependent variables.

It should be noted that the continuity conditions imposed on a trial function by the governing *differential equation*, the *functional*, and the *variational development*[†] are not, in general, identical. Consider, for example, the problem of Section 7.3.2 for which the relevant differential equation, (7.42), is of second order. In a physical problem governed by this equation, the *physical solution* is, in general, continuous with continuous first and second derivatives. The *functional* given in Eq. (7.33) involves only first derivatives and can be evaluated if the trial function is continuous with piecewise continuous first derivatives. If the function were itself piecewise continuous, the first derivatives would be indefinite[‡] at the points of discontinuity and the integral would correspondingly have an indefinite value. The *variational development* between Eqs. (7.35) and (7.38) requires the trial function to be continuous with continuous first derivatives although, as noted, the formulation can be extended to allow the trial function to be continuous with piecewise continuous first derivatives. The latter are the least restrictive continuity conditions allowed by the variational procedure, and are thus adopted as the admissibility conditions for the problem.

[†] The admissibility conditions specify the continuity permitted by a variational procedure. For convenience, in this section, the continuity allowed by the functional and by the (remainder of the) variational development are considered separately. Since the latter may be more restrictive than the former, it is the latter that determines the admissibility conditions.

[‡] In the literature, such indefinite derivatives are sometimes described as *infinite derivatives*. Within the framework of the standard calculus, this is incorrect. By defining the discontinuity in terms of the δ-function, however, the description can be accommodated [5, 6].

It will be noted that the continuity conditions allowed by the variational development are the same as those derived from the functional. In fact, the continuity conditions associated with a variational development cannot be more general than those derived by considering the functional alone. In the next chapter, a quite large class of common engineering problem is defined through Eq. (8.3). For this class, the most general continuity conditions allowed by a variational procedure are the same as those that derive from the functional. This is not necessarily true for other problems as Section 8.8 shows.

For the problem of Section 7.3.2, the continuity conditions to be considered are the following:

Origin	*Required continuity for trial function*
(a) Differential equation	Continuous with first and second derivatives continuous
(b) Functional	Continuous with first derivatives piecewise continuous
(c) Variational development	Continuous with first and second derivatives piecewise continuous

The class of trial functions specified by (a) above are included within the classes defined by (b) and (c) since a continuous function can be regarded as a special case of a piecewise continuous function. Consider now the situation when the solution to the differential equation is to be found by searching among the allowable trial functions for that particular function that makes the functional stationary. If no additional constraint is placed on the continuity of the solution, the same solution will be found irrespective of whether the search is made among the restricted set of trial functions specified by (a) or among the broader field allowed by (c). This solution will be continuous with continuous first and second derivatives. As indicated by Courant and John [7], a variational procedure will always yield a solution that is continuous with continuous derivatives, if this continuity is included within the allowable continuity of the trial functions.

When a trial function is specified in terms of finite elements, however, the type of element chosen may preclude interelement continuity of one or more derivatives. To fix ideas, consider the use of the linear three-node triangular element in the problem of Section 7.3.2. The trial function for the temperature T will in consequence be continuous with first derivatives that are strictly[†] piecewise continuous. Since under (b) and (c) the variational

---

[†] That is, the first derivatives are, in general, prevented from being continuous.

procedure is valid for such continuity, that particular trial function from among the class of admissible functions can be found, which makes the functional stationary. For this particular problem, it can be shown that the stationary value obtained by the variational process is, in fact, a minimum. With the chosen finite element, this minimum value will not, however, be as small as the minimum[†] that would be obtained if the trial function were not only continuous but allowed to possess continuous derivatives. The finite element solution obtained is thus an approximation to the "exact" solution[‡] to the differential equation.

As the foregoing indicates, the finite element trial function that is selected for a problem should not violate the variational procedure; that is, its continuity should satisfy the admissibility conditions.[§]

7.5 VARIATIONAL PRINCIPLES IN PHYSICAL PHENOMENA

For many physical problems, true variational principles have been developed giving rise to a functional whose stationary conditions yield the solution to the problem. Some of the more commonly used functionals in engineering have been conveniently referenced by Desai and Abel [8]. As noted by Finlayson and Scriven [9], true functionals are generally available only where the governing equation is linear and self-adjoint. For other problems, various so-called variational principles have been proposed, one of the earliest being Onsager's [10] variational principle, followed by those due to Rosen [11–13], Chambers [14], Herivel [15], Hays [16], Glandsdorff and Prigogine (who introduced the local potential) [17, 18], and Biot [19–22]. These cited variational principles have been reviewed critically by Finlayson and Scriven [9] who demonstrate that these are not true variational principles but are equivalent to a Galerkin or similar residual method which can in fact be applied more expeditiously. Other variational principles have also been proposed by Lieber *et al.* [23, 24], Visser [25], and Gurtin [26]. The last mentioned reference has been used for finite element analyses of propagation problems [27–29] but has been shown [30], in one case at least, to be equivalent to the Galerkin method.

From the above, it is difficult to avoid the conclusion that, except for linear self-adjoint problems, there is no particular value in attempting a

---

[†] This would be the smallest value of the functional that is possible within the continuity constraints of a variational procedure for this problem.

[‡] That is, the solution that is continuous with continuous first and second derivatives.

[§] It is sometimes possible to violate these conditions and still obtain a solution that is not only a satisfactory approximation to the exact solution but also converges to the exact solution as element size tends to zero. See the paragraph on nonconforming elements in Section 8.4.

variational finite element solution as the much simpler residual method will suffice. As has been said by Finlayson and Scriven [9]:

these approximation schemes are far more readily set up as the straight-forward Galerkin method or another closely related version of the method of weighted residuals. That these direct approximation procedures avoid completely the effort and embellishment of a variational formulation has not been emphasized adequately in the literature . . . . Apart from self-adjoint, linear systems which are comparatively rare, there is no practical need for variational formalism. When approximate solutions are in order the applied scientist and engineer are better advised to turn immediately to direct approximation methods for their problems, rather than search for or try to understand quasi-variational formulations and restricted variational principles.

REFERENCES

1. R. S. Schecter, "The Variational Method in Engineering." McGraw-Hill, New York, 1967.
2. M. J. Forray, "Variational Calculus in Science and Engineering." McGraw-Hill, New York, 1968.
3. S. G. Miklin, "Variational Methods in Mathematical Physics." Pergamon, Oxford, 1964.
4. W. J. Sternberg and T. L. Smith, "The Theory of Potential and Spherical Harmonics." Univ. of Toronto Press, Toronto, 1944; rev. ed. 1946.
5. Lynn, P. P., Arya, S. K., Finite elements formulated by the weighted least squares criterion, *Int. J. Num. Meth. Engrg.* **8**, 71–90 (1974).
6. Strang, G., Variational crimes in the finite element method, *in* "Mathematical Foundations of the Finite Element Method with Applications to Partial Differential Equations" (A. K. Aziz, ed.), pp. 689–710. Academic Press, New York, 1972
7. Courant, R., John, F., "Calculus and Analysis," Vol. 2. Wiley (Interscience), New York, 1974.
8. C. S. Desai and J. F. Abel, "Introduction to the Finite Element Method." Van Nostrand-Reinhold, Princeton, New Jersey, 1972.
9. B. A. Finlayson and L. E. Scriven, On the search for variational principles, *Internat J. Heat Mass Transfer* **10**, 799–821 (1967).
10. L. Onsager, Reciprocal relations in irreversible processes I, *Phys. Rev.* **37**, 405–426 (1931).
11. P. Rosen, On variational principles for irreversible processes, *J. Chem. Phys.* **21**, No. 7, 1220–1221 (1953).
12. P. Rosen, Use of restricted variational principles for the formation of differential equations, *J. Appl. Phys.* **25**, 336–338 (1954).
13. P. Rosen, Variational approach to magneto-hydrodynamics, *Phys. Fluids* **1**, 251 (1958).
14. L. C. Chambers, A variational principle for the conduction of heat, *Quart. J. Mech. Appl. Math.* **9**, 234–235 (1956).
15. J. W. Herivel, A general virational principle for dissipative systems I & II, *Proc. Roy. Irish Acad.* **56**, Sect. A, 37–44, 67–75 (1954).
16. D. F. Hays, Variational formulation of the heat equation: Temperature-dependent thermal conductivity, *in* "Non-Equilibrium Thermodynamics, Variational Techniques and Stability" (R. J. Donnelly, R. Herrman, and I. Prigogine, eds.), pp. 17–43. Univ. of Chicago Press, Chicago, Illinois, 1966.

17. P. Glansdorff and I. Prigogine, On a general evolution criterion in mascroscopic physics, *Physica* **30**, 351–374 (1964).
18. P. Glansdorff and I. Prigogine, Sur les propriétés différentielles de la production d'entropie, *Physica Grav.* **20**, 773–780 (1954).
19. M. A. Biot, Variational principles in irreversible thermodynamics with application to viscoelasticity, *Phys. Rev.* **97**, 1463–1469 (1955).
20. M. A. Biot, New Methods in heat flow analysis with application to flight structures, *J. Aeronaut. Sci.* **24**, 857–873 (1957).
21. M. A. Biot, Further developments of new methods in heat-flow analysis, *J. Aerospace Sci.* **26**, 367–381 (1959).
22. M. A. Biot, "Variational Principles in Heat Transfer." Oxford Univ. Press, London and New York, 1970.
23. P. Lieber and W. Koon-Sang, A principle of minimum dissipation for real fluids, *Proc. Internat. Congr. Appl. Mech., 9th, Brussels* pp. 114–126 (1957).
24. P. Lieber, O. Anderson, and W. Koon-Sang, A principle of virtual displacements for real fluids, *Proc. Internal. Congr. Appl. Mech., 9th Brussels* pp. 106–113 (1957).
25. W. Visser, A finite element method for the determination of nonstationary temperature distribution and temperature distortion, *Proc, Conf. Matrix Methods Struct. Mech., 1st, Wright-Patterson AFB, Ohio, 26–28 October 1965* (AFFDL-TR-66-80), pp. 925–943 (November 1966).
26. M. Gurtin, Variational principles for linear initial value problems, *Quart. Appl. Math.* **22**, No. 3, 252–256 (1964).
27. E. L. Wilson and R. E. Nickell, Application of the finite element method to heat conduction analysis, *Nucl. Engrg. Design* **4**, 276–286 (1966).
28. R. A. Brocci, Analysis of Axi-symmetric Linear Heat Conduction Problems by Finite Element method. Paper 69-WA/HT-37, ASME Winter Annual Meeting, Los Angeles, California (November 16–20, 1969).
29. I. Javandel and P. A. Witherspoon, Application of the finite element method to transient flow in porous media, *Soc. Pet. Engrg. J.* 241–252 (September 1968) [also in the *Transactions* **343** (1968)].
30. E. C. Lemmon and H. S. Heaton, Accuracy, Stability and Oscillation Characteristics of Finite Element Method for Solving Heat Conduction Equation. Paper 69-WA/HT-35, ASME Winter Annual Meeting, Los Angeles, California (November 16–20, 1969).

8

CONVERGENCE, COMPLETENESS, AND CONFORMITY

In general, a finite element solution will be an approximation to the true or exact solution. How close this computed solution is to the exact solution and whether or not it converges to the exact solution are both important questions. This chapter briefly considers the accuracy and convergence of the finite element method, with the aid of heuristic arguments.

8.1 ACCURACY, STABILITY, AND CONVERGENCE IN NUMERICAL COMPUTATION

When a numerical procedure is being selected, the accuracy, stability, and convergence of the computing method should be among the characteristics considered. Accuracy is a measure of the closeness of the numerical solution to the exact or true solution. Stability[†] refers to the growth of error as a particular computation proceeds, an unstable computation being

---

[†] The *stability* analysis of the finite element method is beyond the scope of this textbook and will not be considered here. It is assumed throughout the discussion however that the solutions are stable.

one in which truncation, round-off, or other errors accumulate in an un-
bounded fashion so that the true solution is soon swamped by the error.
Convergence is the ever-closer approach of successive computed solutions
to the limiting solution as some computational parameter, such as element
size or number of terms in the trial solution, is refined. The term convergence
is also applied in the same sense to an iterative procedure in which some
or all of the results from one computation become the input for the next
(repeated) calculation, with a convergent procedure being one in which the
difference between successive results continues to decrease, tending to zero
in the limit. These three terms are illustrated in Fig. 8.1. More precise definitions can be found in textbooks on numerical analysis and computation
[1–3]. It should be noted that the desirable situation is for each computation
to be stable, with successive calculations converging rapidly to the exact
solution.

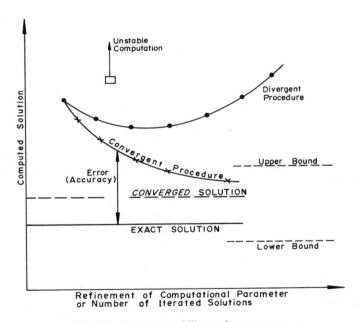

Fig. 8.1 Accuracy, stability, and convergence.

It will be noted from Fig. 8.1 that as the computational procedure is
refined, the accuracy increases if the process is convergent and decreases
if not convergent. Thus, an examination of the errors in the finite element
method and their causes leads naturally into a consideration of convergence.

8.2 ERRORS IN THE FINITE ELEMENT METHOD

In addition to the usual rounding and truncation errors associated with any computational brocedure, there are those errors that relate to the finite element technique itself [4–6]. These can be subdivided into:

(a) *discretization errors*, resulting from geometric differences between the boundaries of the region and its finite element approximation; and

(b) *trial function* or *shape function errors*, due to the difference between the true solution and its trial function representation.

Discretization errors can be reduced by using smaller elements or by placing curved elements along the boundaries and, in any event, tend to zero as the element size tends to zero. Trial function errors do not necessarily decrease as the element size reduces and may thus prevent convergence to the exact solution or even cause divergence.

In the following sections, trial function errors are considered in relation to the behavior of the trial function across the element, discontinuities at the interelement boundaries, and admissibility of the trial function.

8.3 TRIAL FUNCTION ERROR AND COMPLETENESS

Consider a set of linearly independent functions, denoted by ϕ_r. Such a set is said to be *complete* if the linear combination $\sum_{r=1}^{M} \alpha_r \phi_r$ converges in some specified sense to an arbitrary function f as M tends to infinity, where the α_r are suitably chosen constants. Although various definitions of completeness are possible [7–9], depending on how convergence is specified, only completeness in the sense of convergence to the mean need be considered here. For this situation, a set of functions ϕ_r is complete if any function f can be approximated to any desired degree of accuracy by taking a sufficient number of terms in the linear combination $\sum_{r=1}^{M} \alpha_r \phi_r$ [9], with the limiting case being

$$\lim_{M \to \infty} \sum_{r=1}^{M} \alpha_r \phi_r \to f. \tag{8.1}$$

A polynomial series, in one or more variables, with all the terms in the sequence present, can be shown to be complete in this sense.

From the preceding, it is evident that a trial function that is a polynomial series can match the true solution exactly only over an element of finite size, in general, if the polynomial is complete and of infinite degree. Since in the practical case, a finite number of terms must be used, a polynomial trial function representation cannot be other than an approximation to the true solution. Alternatively expressed, the trial function error will, in general, be nonzero for a polynomial trial representation of finite degree over an element of practical size. The conditions under which this trial function error will tend to zero as the element size tends to zero will now be investigated.

To illustrate, a one-dimensional problem is considered, but the conclusions can be extended to two- and three-dimensional cases. Let it be assumed that the functional contains $\hat{u}$ and its derivatives up to and including order p. The polynomial series for $\hat{u}$ must be of at least pth degree if the pth derivative is to have nonzero representation. Choosing the polynomial to be of pth degree gives the following representations over an element

$$\hat{u} = \alpha_0 + \alpha_1 x + \alpha_2 x^2 + \alpha_3 x^3 + \cdots \qquad\qquad + \alpha_p x^p,$$

$$\frac{d\hat{u}}{dx} = \alpha_1 + 2\alpha_2 x + 3\alpha_3 x^2 + \cdots \qquad\qquad + p\alpha_p x^{p-1},$$

$$\frac{d^2\hat{u}}{dx^2} = 2\alpha_2 + 6\alpha_3 x \cdots \qquad\qquad + \quad p(p-1)\alpha_p x^{p-2},$$

$$\vdots \qquad\qquad\qquad\qquad\qquad\qquad\qquad\qquad (8.2)$$

$$\frac{d^{p-1}\hat{u}}{dx^{p-1}} = (p-1)!\alpha_{p-1} + p!\alpha_p x,$$

$$\frac{d^p\hat{u}}{dx^p} = p!\alpha_p.$$

It is noted from Eq. (8.2) that since the polynomial for $\hat{u}$ is complete, each of the derivatives has a representation that includes a constant term. As the element size tends to zero, each derivative representation will tend to its exact value and consequently the element contribution χ^e will also tend to its exact value. The functional χ will similarly tend to its exact value and the finite element solution will therefore tend to the exact solution. The procedure is thus convergent.

The foregoing allows the following *restricted* convergence criterion to be deduced:

Criterion I(R) A condition for convergence is that a complete polynomial, of order at least p, be used for the representation of the variable within an element, where p is the order of the highest derivative of the variable appearing in the functional.

Where a complete polynomial of degree greater than the minimum is used for the variable representation, the approximation can be expected to be more accurate and the trial function error of smaller magnitude, with more rapid convergence as a consequence. Numerical experiments have confirmed these predictions [10], at least for conforming elements.

So far, no consideration has been given to physical parameters such as conductivity, modulus of elasticity, and the like, which may be present in the functional. If such a parameter is not a constant across the region, it can be approximated across each element using a polynomial series, with at least the first (constant) term of the series being present to ensure that the representation tends to exactness as the element size tends to zero.

It is worth noting that the completeness criterion enunciated above is a particular case of the more general requirement:

Criterion I A necessary condition for convergence is that the representation of a variable and any of its derivatives appearing in the functional must tend to their exact values, over each element, as the element size tends to zero.

8.4 TRIAL FUNCTION ERROR AND CONFORMITY

In solid mechanics, with the governing equations formulated in terms of strains and the solution being sought in terms of displacements, it became customary to describe the displacement field as *compatible* if the displacements were continuous across the region, in which case the strains would be piecewise continuous. This definition has been carried over into the finite element area to describe a trial function representation that is similarly continuous across the region. The more general term *conforming* was used, apparently for the first time in 1966 [11], by Bazely, Cheung, Irons, and Zienkiewics [12]. The trial function is described as conforming if the variable and its derivatives up to order $p - 1$ are continuous across interelement boundaries, where p is the order of the highest derivative of the variable appearing in the functional.

For the class of problem

$$\mathscr{L}u = f, \tag{8.3}$$

where $\mathscr{L}$ is a linear self-adjoint positive-bounded-below[†] differential operator and u and f are functions of the independent variables, a variational formulation is possible [9, 11] with the functional being of quadratic-linear form. This class, which includes many engineering problems, can be extended by considering u to be a vector, in which case $\mathscr{L}$ becomes a matrix and f a vector. To be admissible, the trial function must, in general, be continuous

---

[†] For the definition of *positive-bounded-below* and *self-adjoint*, see [7, 9, 11]. The term *symmetric* is sometimes used in place of *self-adjoint*.

and have continuous derivatives up to and including order $p - 1$, where p is the order of the highest derivative appearing in the functional. For the Ritz finite element formulation of such problems, the condition of admissibility therefore requires that conformable elements be used.

For problems in linear elasticity (which are a subclass of that above, for which $\mathcal{L}$ is required to be positive definite), convergence of the Ritz finite element formulation based on the principle of minimum potential energy can be investigated for *conforming* elements (i.e., admissible trial functions) by using a Taylor series expansion for solution u within each element. Such an approach has been used by McLay [13, 14], Cowper [10, 15], and others, and the results can be summarized as follows. If the strain energy contains derivatives of u whose highest order is p, convergence is assured if in each element a complete polynomial of order at least p is used for the trial function $\hat{u}$. More rapid convergence can be achieved by selecting higher order polynomials. If such higher order polynomials are complete only up to and including order p, the error will be greater and the rate of convergence less than if the polynomial were fully complete. These results are consistent with the discussion of Section 8.3 and with criterion I(R).

A more general approach to convergence acknowledges that in practice the functional is written as a sum of element contributions. As a consequence, the derivatives in the functional are evaluated not by differentiating across the domain but by separate differentiation within each element so that the problem of nonexistent derivatives at interelement discontinuities is bypassed. The convergence of even nonconforming elements can thus be investigated by considering whether the functional (calculated in the manner described) tends to its true value as the element size tends to zero. The convergence analysis is most conveniently formulated using Hilbert spaces and energy norms. With this latter approach, Oliviera [16] demonstrated that for the class of problem considered earlier, convergence of the Ritz finite element method is assured if the element approximation for **u** is a polynomial complete at least to order p (where p is the order of the highest derivative in the functional), provided that the conformity condition is satisfied. That completeness[†] and conformity are sufficient conditions for convergence was subsequently reaffirmed by Oden [11] in a more general analysis of the same class of problem.

The foregoing allows the following restricted convergence criterion to be stated for the class of problem considered.

Criterion II(R)   A condition for convergence is that the elements be conforming, that is, the representations of the variable and its derivatives

---

† This term is used here and subsequently to mean that the element approximation is a polynomial complete at least up to and including order p.

up to and including order $p - 1$ must be continuous across interelement boundaries, where p is the order of the highest derivative appearing in the functional.

Oliviera's conclusion is thus that criterion I(R) and criterion II(R) are sufficient conditions for convergence of the Ritz variational finite element method, for this same class of problem.

8.5 TRIAL FUNCTION ERROR AND NONCONFORMITY

Using the Hilbert space approach, Patterson [17] has recently shown that sufficient conditions for convergence of the variational finite element method are the completeness criterion and a weak "conformity" criterion, which is usually satisfied in practical applications. This "conformity" or interelement criterion requires that the difference or discontinuity in $\hat{u}$ across each interelement boundary shrink to zero faster than the diameter of the largest subdomain, for increasingly finer subdivisions. In two-dimensional problems, this criterion is satisfied if the difference in $\hat{u}$ vanishes at least twice on each subdomain interface. The class of problems for which these results were obtained is the same as that considered earlier, except that the more general relationship $\mathscr{L}\mathbf{u} = \mathbf{f}$ was used, with $\mathscr{L}$ being constrained to be positive definite. Homogeneous boundary conditions were considered in [17], but it was also stated that the same results had been obtained for a general class of nonhomogeneous boundary conditions.

Patterson's interelement criterion can be generalized by realizing that a condition for convergence to the exact solution must be that the trial function $\hat{u}$ tend to the exact solution as the element mesh is increasingly refined. This leads to the following more general convergence criterion:

Criterion II A condition for convergence is that the representation of the variable and its derivatives in the functional must tend to the same continuity[†] as the exact solution, as the element size tends to zero.

8.6 NONCONFORMITY, NONCOMPLETENESS, AND ACCURACY

For the class of problem considered, it has been demonstrated that criterion I(R) (completeness) and criterion II(R) (conformity) are sufficient conditions for convergence. More generally, either of criteria I/I(R) together

---

[†] In this section, the exact solution is assumed to be continuous, but there are physical problems where the solution will be discontinuous, for example, in regions where shock waves or cracks are present.

with either of criteria II/II(R) are sufficient conditions for convergence of the variational finite element method. The following are also worth noting:

(1) If criterion I is satisfied, criterion II will be satisfied in consequence, hence criterion I is a necessary and sufficient condition for convergence.

(2) Completeness is a stronger requirement than conformity, as Patterson's work has shown, and in many practical cases, completeness is a sufficient condition for convergence.

It is also evident that criterion I and criterion II are sufficient conditions for convergence for other finite element methods such as the weighted residual and least squares procedures, provided the word *functional* in these criteria is replaced by *defining* or *key integral*. In all finite element methods, as has been emphasized recently by Zienkiewicz [18], there is such a defining integral derived from the governing equations for the problem, by a procedure characteristic of the particular method.

It might be thought from the preceding that all element representations for which convergence was assured would be equally valuable, but nothing can be further from the truth. It cannot be emphasized too strongly that what is of concern in practice is *accuracy*. To have an element representation that yields a result convergent to the exact solution in the limit as the element size tends to zero is little consolation if the result is grossly inaccurate at the finite size of element dictated by computational economics. How can the accuracy of a computed solution be determined in the practical case? The answer is that, in general, it cannot. A sufficient indication of the accuracy can, however, often be obtained in one of two ways. The first is to solve a problem of the same type for which the exact solution is known analytically, using the same element and as similar a configuration as practical. The error obtained can be used to estimate the order of the error for the practical problem. The second method requires that the type of convergence be first determined for the particular finite element formulation and problem. If the convergence is known to be monotonic[†] as the finite element mesh is refined, the problem can be solved several times with successively smaller elements and the results extrapolated to give some estimate of the converged solution.

As has been demonstrated by Oden [11], a Ritz finite element formulation is monotonically convergent to the exact solution if

(1) the element representation satisfies the completeness and conformity conditions of criteria I(R) and II(R);

[†] Note that monotonic convergence describes only the manner in which the sequence converges. It does not necessarily mean that the converged result is the true solution.

(2)   the mesh refinement is such that the element at each level of refinement represent a subdivision of the elements at the previous level;

(3)   the subfields at each level of refinement are contained in the subfields of the previous level.

Consider, for example, a linear representation of a variable over a triangular element and suppose the completeness and conformity conditions of (1) are satisfied. Subdivision of each triangle by joining the midpoints of each side would satisfy condition (2) and since each successive subfield would be linear, condition (3) would also be satisfied. It may not be easy, however, to satisfy the requirements for monotonic convergence for other than the simplest formulations. In particular, refining the mesh by the required procedure may yield more elements after one or two subdivisions than can be accommodated by the available computer facilities.

When neither completeness nor conformity are satisfied but the more general criteria I and II are, so that convergence is assured in the limit, it is often assumed that at any size of finite element, a unique solution will be obtained. At large element sizes, however, the computation may be unstable or multiple solutions may exist; but, as the element size is decreased, a critical point will be reached below which a stable converging solution is obtained. The only difficulty is that this critical size may be uneconomically small. Even where a unique solution is obtained for each size of finite element, the errors over the practical range of size may be large and, if convergence is not monotonic, may also vary in an unexpected manner.

The preceding difficulties do not mean that nonconforming and/or noncomplete elements should not be used. Valuable elements exist that are nonconforming and in some cases also noncomplete, but which yield high accuracy and rapid convergence. Indeed, the performance of such elements can be superior to conforming complete elements of the same polynomial order. As Zienkiewicz [18] has recently pointed out, nonconforming or incompatible elements are the "best elements for practical use" in certain areas of application. Nonconforming elements should certainly not be overlooked but are not recommended for the inexperienced practitioner.

8.7   THE PATCH TEST

A practically oriented approach to convergence can be found in the *patch test* of Irons [19, 20], which is here outlined for a solid mechanics problem. In the simplest form of the test, a patch of elements (with at least one noninternal node completely surrounded by elements) is loaded at the boundary by forces corresponding to a constant strain throughout the patch. If the formulation is to be convergent, the patch test requires that the values of displacement, strain, and stress computed in the finite element solution be consistent with the imposed state of constant strain. The test can alternatively be performed using applied displacements corresponding to a state of constant strain across the patch. A higher order patch test is also available, which requires that the solution throughout the patch be consistent with a more complex loading prescribed at the boundary.

The patch test is not restricted to conforming or complete elements but can also be used to determine whether elements that do not satisfy these criteria will give a converging solution. Although developed through engineering intuition, the test has been confirmed mathematically by Strang [21] as being a necessary and sufficient test for convergence, for the following cases: (a) when nonconforming elements are used, (b) when the formulation incorporates numerical integration. The test can be extended to problems other than those of solid mechanics, as indicated recently by Oliviera [22].

8.8   ADMISSIBILITY

For the class of problem specified by Eq. (8.3), the operator $\mathcal{L}$ is of order $2p$, that is, the highest derivative in the operator is of order $2p$. The quadratic-linear functional for the problem involves derivatives whose highest order is p. The *principal* boundary conditions involve derivatives up to and including order $p - 1$, whereas the *natural* boundary conditions involve derivatives of order $p, p + 1, \ldots, 2p - 1$ [11]. For a trial function $\hat{u}$ to be *admissible*, in general, it must be continuous and possess continuous derivatives up to order $p - 1$ over the region. As shown in Section 7.4 the variational procedure is valid only if the trial functions belong to the class of admissible functions.

It will be noted that the condition of admissibility requires that the element be conforming, *for the class of problem whose governing equation is* Eq. (8.3). It is sometimes thought that admissibility and conformity are equivalent for other problems as well, but this is not true in general, as the following example indicates. Consider the governing field equations for two-dimensional inviscid incompressible fluid flow, in terms of the x and y components of velocity, u and v respectively,

$$\frac{\partial u}{\partial x} + \frac{\partial v}{\partial y} = 0 \qquad \text{in } D, \tag{8.4a}$$

$$\frac{\partial v}{\partial x} - \frac{\partial u}{\partial y} = 0 \qquad \text{in } D. \tag{8.4b}$$

Using a least-squares formulation, the functional is derived as

$$\chi = \int\int_D \left[\left(\frac{\partial u}{\partial x} + \frac{\partial v}{\partial y} \right)^2 + \left(\frac{\partial v}{\partial x} - \frac{\partial u}{\partial y} \right)^2 \right] dD. \tag{8.5}$$

Conformity would require continuity only in the trial functions $\hat{u}$ and $\hat{v}$ and not in their derivatives. The condition of admissibility for this problem has been shown [23] to be that $\hat{u}$ and $\hat{v}$ be continuous with continuous first derivatives and piecewise continuous second derivatives.

8.9 PHYSICAL EQUIVALENTS TO COMPLETENESS AND CONFORMITY

The mathematical requirements of completeness and conformity are sometimes equivalent to meaningful physical conditions. For example, consider the displacement functions u and v in the x and y directions, respectively, in a two-dimensional plane strain problem. Let the region of interest be subdivided into a finite number of three-node triangular elements and let the approximations for both u and v vary linearly within each elements.

The functional for plane strain (see Chapter 11) involves the functions u and v and their first partial derivatives, so that $p = 1$ for this problem. The conformity condition is thus that $\hat{u}$ and $\hat{v}$ be continuous, and this is satisfied by the element chosen. The completeness condition is also satisfied for both u and v since their trial functions are complete first-order polynomials. Rigid body displacements, that is, uniform displacements of the element, are possible due to the constant terms in the u and v representations. Strains, being first derivatives of displacements, are approximated by constants across each element and hence piecewise constant strains across the region are representable. When higher order complete polynomials are used, even in the more general three-dimensional elasticity problem, the presence of the constant and linear terms in the trial functions still ensures that rigid body displacements and piecewise constant strains are representable.

In the early finite element literature, the conditions that the displacement function be capable of representing rigid body motion and uniform strain over an element were proposed as necessary for convergence. As Section 8.3 and the preceding discussion indicate, these conditions are encompassed within the completeness criterion. It is important to note, however, that it has been assumed in the above that rectilinear coordinates are being used. If the polynomial representations are expressed in terms of curvilinear coordinates, as for example would be natural for curved plates and shells, then the constant and linear terms no longer correspond to rigid body motion and uniform strain [10, 11].

8.10 COMPLETENESS AND GEOMETRIC ISOTROPY

The representation of a variable across an element should be independent of the coordinate system used, or more particularly, the representation should be *geometrically invariant* under *orthogonal* transformations of the coordinate system. Alternative descriptions are *spatial isotropy* and *geometric isotropy*, with the latter becoming more common. Apart from invariance, geometric isotropy also ensures that the representation along any boundary or edge of an element is a complete polynomial of the same order as across the subdomain of the element [24].

As shown in previous chapters, it is often convenient to derive the element matrix equations with reference to a local coordinate system and then to transform these to the global system. In such cases, it is essential that the elements possess geometric isotropy, otherwise the transformation could destroy convergence requirements previously satisfied.

When a complete polynomial is chosen as the trial function, it can be shown that the corresponding element possesses geometric isotropy. If some of the terms are to be deleted from the series, this should be done in such a way that the element corresponding to this incomplete polynomial still retains geometric isotropy. In deciding which terms to drop, it is clear that a symmetric pair (such as x^3, y^3 or x^5y^2, x^2y^5) does not introduce any preferential bias to either coordinate. Indeed, it can be shown that a polynomial series, complete except for one (or more) symmetric pair(s), gives a geometrically invariant representation, and hence possesses geometric isotropy, provided that the order of the original complete polynomial has not been reduced.

To illustrate the deletion of symmetric pair(s) from a complete polynomial, consider the ten-term complete cubic polynomial representation for ϕ in two dimensions,

$$\hat{\phi} = \alpha_1 + \alpha_2 x + \alpha_3 y + \alpha_4 x^2 + \alpha_5 xy + \alpha_6 y^2 + \alpha_7 x^3 + \alpha_8 x^2 y + \alpha_9 xy^2 + \alpha_{10} y^3.$$
$$(8.6)$$

An element based on this complete polynomial possesses geometric isotropy, but so also do those using the following incomplete cubic series,

$$\hat{\phi} = \alpha_1 + \alpha_2 x + \alpha_3 y + \alpha_5 xy + \alpha_7 x^3 + \alpha_8 x^2 y + \alpha_9 xy^2 + \alpha_{10} y^3, \qquad (8.7)$$

$$\hat{\phi} = \alpha_1 + \alpha_2 x + \alpha_3 y + \alpha_5 xy + \alpha_7 x^3 + \alpha_{10} y^3. \qquad (8.8)$$

For an element based on any one of the above series, substitution of the boundary equation $y = ax + b$ into the appropriate equation generates a complete cubic polynomial in terms of x. A complete cubic polynomial is thus also obtained in terms of s, where s is measured along any boundary of the element, providing verification of the statement in the first paragraph of this section.

REFERENCES

1. S. H. Crandell, "Engineering Analysis." McGraw-Hill, New York, 1956.
2. W. F. Ames, "Numerical Methods of Partial Differential Equations," 2nd ed. Academic Press, New York, 1977.
3. G. E. Forsythe and W. R. Wasow, "Finite Difference Methods for Partial Differential Equations." Wiley, New York, 1960.
4. E. Schrem, Computer implementation of the finite element method, in "Numerical Computer Methods in Structural Mechanics" (S. J. Fenves et al., eds.), pp. 79–122. Academic Press, New York, 1973.
5. G. von Fuchs and J. R. Roy, Solution of the Stiffness Matrix Equations in ASKA, Rep. No. 50, Inst. für Statik und Dynamik Univ. Stuttgart, 1968.
6. J. R. Roy, Numerical error in structural solutions, J. Struct. Div., ASCE 97, 1039–1054 (1971).
7. K. Rektorys, "A Survey of Applicable Mathematics." Iliffe Books, London, 1969.
8. S. G. Miklin, "Variational Methods of Mathematical Physics." Pergamon Press, Oxford, 1964.
9. D. H. Norrie and G. de Vries, "The Finite Element Method." Academic Press, New York, 1973.
10. G. R. Cowper, Variational procedures and convergence, in "Numerical and Computer Methods in Structural Mechanics" (S. J. Fenves, et al., eds.), pp. 1–12, Academic Press, New York, 1973.
11. J. T. Oden, "Finite Elements of Non-Linear Continua." McGraw-Hill, New York, 1972.
12. G. P. Bazely, Y. K. Cheung, B. M. Irons, and O. C. Zienkiewicz, Triangular elements in plate bending—conforming and non-conforming solutions, Proc. Conf. Matrix Methods Struct. Mech., Wright-Patterson AFB, Ohio, October 26–28, 1965 (AFFDL-TR-66-80), pp. 547–576 (November, 1965).
13. M. W. Johnson, Jr., and R. W. McLay, Convergence of the finite element method in the theory of elasticity, J. Appl. Mech. 35, 274–278 (1968).
14. R. W. McLay, Completeness and convergence properties of finite element displacement functions, A.I.A.A. 5th Aerospace Sci. Meeting, New York, Paper No. 67–143 (January 1967).
15. G. R. Cowper, E. Kosko, G. M. Lindberg, and M. D. Olson, A high precision triangular plate bending element, National Research Council of Canada, Aero. Rep. LR-514 (December 1969).
16. E. R. de Arantes e Oliviera, Theoretical foundations of the finite element method, Internat. J. Solids and Structures 4, 929–952 (1968).
17. C. Patterson, Sufficient conditions for convergence in the finite element method for any solution of finite energy, in "The Mathematics of Finite Elements and Applications," (J. R. Whiteman, ed.), pp. 213–224. Academic Press, 1973.
18. O. C. Zienkiewicz, Recent developments, trends, and applications of finite element methods, Proc. Internat. Conf. Finite Element Methods in Engrg., Dept. of Civil Engrg., Univ. of Adelaide, December 6–8, 1976, pp. 1.1–1.38.
19. B. M. Irons, The patch test for engineers, Proc. Finite Element Symp. Atlas Computer Lab., Chilton, Didcot, England, 26–28 March, 1974, pp. 171–192.
20. B. M. Irons, The superpatch theorem and other propositions relating to the patch test, Proc. Canad. Congress Appl. Mech., 5th, University of New Brunswick, 26–30 May, 1975, pp. 651–652.
21. G. Strang, Variational crimes in the finite element method, in "The Mathematical Foundations of the Finite Element Method" (A. K. Aziz, ed.), pp. 689–710. Academic Press, New York, 1972.

22. E. R. de Arantes e Oliviera, Convergence and accuracy in the finite element method, *Proc. World Congr. Finite Element Methods Struct. Mech., Bournemouth, England, 12–17 October, 1975*, pp. 0.1–0.24. Robinson and Associates, Verwood, Dorset, England.

23. G. de Vries, T. Labrujere, and D. H. Norrie, A least-square finite element solution for potential flow, Dept. of Mechanical Engineering, University of Calgary, Rep. 86 (December, 1976).

24. K. H. Huebner, "The Finite Element Method for Engineers," Wiley, New York, 1975.

9

ELEMENTS AND THEIR PROPERTIES

In previous chapters several different kinds of element were used in the formulation and application of the finite element method. Each element trial function represented the unknown function as a linear combination of its nodal values. These latter consisted of the values of the function at the nodes together with, in the case of derivative elements, the values of their derivatives. In principle, an approximation involving a nonlinear dependency on these nodal values could be used, but any improvement in accuracy would be unlikely to outweigh the additional complexity of such an approach.

It was noted earlier that the trial function must be compatible with the corresponding element, to enable the coefficients α_i in the trial function to be determined *uniquely*. More generally, it can be shown that the trial function, the location and number of the element nodes, as well as the number of unknowns per node[†] cannot all be specified independently. Moreover, the type and order of the governing equation and the convergence requirements of the variational procedure must also be taken into account when selecting elements and their trial functions.

[†] The degrees of freedom per node.

Within the constraints mentioned above, a variety of acceptable elements have been developed and some of the more important of these are considered in this chapter. Further details may be found in various surveys on elements [1–4], as well as in standard texts on the finite element method [5–17]. A recent finite element bibliography [18] also references many element types.

9.1 ELEMENT CLASSIFICATION

The most obvious grouping of elements is into one-, two-, and three-dimensional categories and the subsequent treatment in this chapter will be on this basis. These categories can be subdivided further, according to whether the nodal values involve only the function (Lagrangian elements) or also include derivatives (Hermitian elements).

9.2 ELEMENT SHAPE FUNCTIONS

The general trial function representation over any element e is the linear form

$$\hat{u}^e = \mathbf{N}^e \mathbf{u}^e, \tag{9.1}$$

where $\mathbf{N}$ is the *shape function matrix*, and $\mathbf{u}$ is the *element nodal vector*.[†]

For a Lagrangian element, there is only one degree of freedom per node, namely the value of the function, and hence Eq. (9.1) can be written as

$$\hat{u} = \begin{bmatrix} N_1 & N_2 & \cdots & N_k & \cdots & N_s \end{bmatrix} \begin{bmatrix} \bar{u}_1 \\ \bar{u}_2 \\ \vdots \\ \bar{u}_k \\ \vdots \\ \bar{u}_s \end{bmatrix}, \tag{9.2}$$

where $1, 2, \ldots, k, \ldots, s$ are the node identifiers, and s is the total number of nodes of element e.

For a Hermitian element, each of the $\bar{u}_k$ must be regarded as being a column matrix itself since the derivative(s) of the function now appear as variables at the nodes also. If each of the s nodes of element e has q degrees of freedom, then the entries $\bar{u}_k$ in Eq. (9.2) are to be considered as column

---

† In the remainder of this chapter, the superscript e will be dropped, for convenience.

matrices

$$\bar{\mathbf{u}}_k = \begin{bmatrix} \bar{u}_{k1} \\ \bar{u}_{k2} \\ \vdots \\ \bar{u}_{kq} \end{bmatrix}, \qquad k = 1, 2, \ldots, s, \tag{9.3}$$

and similarly, the shape functions N_k are to be considered as row matrices

$$\mathbf{N}_k = \begin{bmatrix} N_{k1} & N_{k2} & \cdots & N_{kq} \end{bmatrix}, \qquad k = 1, 2, \ldots, s. \tag{9.4}$$

Equation (9.2) is therefore the general expression for both Lagragian and Hermitian representations, provided that the proper shape functions and nodal values are adopted.

In previous chapters the shape functions obtained were found to be functions of the independent global (or local) coordinates x, y (or ξ, η) and the coordinates of the nodes. It is, in fact, always true that the shape functions are functions of the independent variables and the nodal coordinates. To derive the shape functions for any selected element, two methods are available, the first of which uses generalized coordinates and the second interpolation functions. These procedures are briefly considered in the next two sections.

9.2.1 Deriving Shape Functions from Generalized Coordinates

This approach is particularly suitable for those simple elements based on low-order polynomials which are complete.[†] The manipulations become tedious for the more complex elements and the interpolation method is then usually resorted to.

The procedure is illustrated using the rectangular element e with sides parallel to the global system Oxy in Fig. 9.1. The simplest trial function for the element contains only four unknown constants α_i corresponding to the element's four nodes. This trial function $\hat{u}$ can be obtained by deleting two terms from the complete second-order polynomial which contains six constants. Deleting the symmetric pair x^2, y^2, in order to retain geometric isotropy, yields the trial function

$$\hat{u} = \alpha_1 + \alpha_2 x + \alpha_3 y + \alpha_4 xy, \tag{9.5}$$

which is seen to vary linearly along the element boundaries. The coefficients

---

[†] Complete polynomial elements will be subsequently referred to as *linear, quadratic,* etc., thus identifying the order of the polynomial.

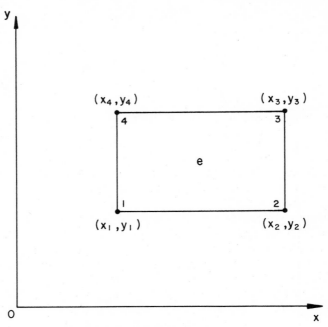

Fig. 9.1   Rectangular four-node element e.

α_i in Eq. (9.5) are the generalized coordinates of the element. Applying Eq. (9.5) to the four nodes, in turn, yields the equations

$$
\begin{aligned}
\bar{u}_1 &= \alpha_1 + \alpha_2 x_1 + \alpha_3 y_1 + \alpha_4 x_1 y_1, \\
\bar{u}_2 &= \alpha_1 + \alpha_2 x_2 + \alpha_3 y_2 + \alpha_4 x_2 y_2, \\
\bar{u}_3 &= \alpha_1 + \alpha_2 x_3 + \alpha_3 y_3 + \alpha_4 x_3 y_3, \\
\bar{u}_4 &= \alpha_1 + \alpha_2 x_4 + \alpha_3 y_4 + \alpha_4 x_4 y_4.
\end{aligned}
\tag{9.6}
$$

In matrix notation Eqs. (9.6) can be written as

$$
\mathbf{u} = \mathbf{A}\boldsymbol{\alpha},
\tag{9.7}
$$

where $\mathbf{u}$ is the element nodal vector

$$
\mathbf{u} = [u_i] = \begin{bmatrix} \bar{u}_1 \\ \bar{u}_2 \\ \bar{u}_3 \\ \bar{u}_4 \end{bmatrix},
\tag{9.8}
$$

and

$$
\mathbf{A} = [a_{ij}] = \begin{bmatrix} 1 & x_1 & y_1 & x_1y_1 \\ 1 & x_2 & y_2 & x_2y_2 \\ 1 & x_3 & y_3 & x_3y_3 \\ 1 & x_4 & y_4 & x_4y_4 \end{bmatrix}, \quad \boldsymbol{\alpha} = [\alpha_i] = \begin{bmatrix} \alpha_1 \\ \alpha_2 \\ \alpha_3 \\ \alpha_4 \end{bmatrix}. \tag{9.9}
$$

Equations (9.9) allows Eq. (9.5) to be written as

$$
\hat{u} = \mathbf{X}\boldsymbol{\alpha}, \tag{9.10}
$$

where

$$
\mathbf{X} = \begin{bmatrix} 1 & x & y & xy \end{bmatrix}. \tag{9.11}
$$

The matrix $\mathbf{A}$ in Eqs. (9.9) can be inverted provided that its determinant does not vanish. It can be shown that

$$
\det \mathbf{A} = -\Delta^2, \tag{9.12}
$$

where Δ is the area of the rectangle, which never vanishes. Consequently, $\det \mathbf{A}$ never equals zero, implying that the matrix $\mathbf{A}$ is nonsingular. Inverting $\mathbf{A}$, denoting this inverse by $\mathbf{A}^{-1}$, and premultiplying Eq. (9.7) by $\mathbf{A}^{-1}$ give

$$
\boldsymbol{\alpha} = \mathbf{A}^{-1}\mathbf{u}. \tag{9.13}
$$

Substituting Eq. (9.13) into Eq. (9.10) then results in

$$
\hat{u} = \mathbf{X}\mathbf{A}^{-1}\mathbf{u}, \tag{9.14}
$$

which can also be written in the form

$$
\hat{u} = \mathbf{N}\mathbf{u}, \tag{9.15}
$$

where the shape function matrix $\mathbf{N}$ is given by

$$
\mathbf{N} = \mathbf{X}\mathbf{A}^{-1}. \tag{9.16}
$$

The above procedure can also be formulated in terms of a local coordinate system $\bar{O}\xi\eta$, where, for example, $\bar{O}$ is chosen at node 1 and the ξ and η axes are chosen parallel to the x and y axes, respectively. The subsequent manipulations are simplified using the notation

$$
a = x_2 - x_1 = x_3 - x_4, \qquad b = y_4 - y_1 = y_3 - y_2. \tag{9.17}
$$

For Lagrangian elements, when the form of the polynomial approximation is known explicitly, the shape functions can, in principle, always be calculated using the above procedure. Where the algebra becomes extensive, numerical evaluation of the shape functions is resorted to. For Hermitian elements, the formulation requires appropriate modification.

9.2.2 Deriving Shape Functions from Interpolation Functions

Consider again the rectangular element e shown in Fig. 9.1, but suppose now that the trial function representation of Eq. (9.5) is not known. The shape function relation for $\hat{u}$ is, however, always known in the general form of Eq. (9.2), that is,

$$\hat{u} = \mathbf{Nu} = \begin{bmatrix} N_1 & N_2 & N_3 & N_4 \end{bmatrix} \begin{bmatrix} \bar{u}_1 \\ \bar{u}_2 \\ \bar{u}_3 \\ \bar{u}_4 \end{bmatrix}. \tag{9.18}$$

Each shape function N_k must have the value unity at node k and zero at every other node, so that $\hat{u}$ will reduce to $\bar{u}_k$ when Eq. (9.18) is applied to node k. This property allows interpolating formulas to be used to derive the shape functions, as the following sections show.

9.2.2.1 Lagrangian Elements

Consider the approximation of the function $u(x)$ by a p'th order polynomial where the values of $u(x)$ are given as $u_1, u_2, \ldots, u_p$ at the $p + 1$ points $x_1, x_2, \ldots, x_p$. From numerical analysis, it is known that the function $u(x)$ can be written as the pth order polynomial

$$u(x) = \sum_{i=1}^{p+1} L_i(x)u_i, \tag{9.19}$$

where the $L_i(x)$ are the Lagrange polynomials defined by

$$L_i(x) = \prod_{j=1, j \neq i}^{p+1} \frac{x - x_j}{x_i - x_j}. \tag{9.20}$$

It is worth noting that the so-called *base points* $x_1, x_2, \ldots, x_p$ need not be spaced equally, although this is often a convenience.

Applying Eqs. (9.19) and (9.20) to side 1-2 of the rectangle e in Fig. 9.1 allows $\hat{u}$ along this side to be written as

$$\hat{u}\big|_{1\text{-}2} = L_1(x)\bar{u}_1 + L_2(x)\bar{u}_2, \tag{9.21}$$

where

$$L_1(x) = \frac{x - x_2}{x_1 - x_2}, \quad L_2(x) = \frac{x - x_1}{x_2 - x_1}. \tag{9.22}$$

Similarly along side 4-3, there is obtained

$$\hat{u}\big|_{4\text{-}3} = L_1(x)\bar{u}_3 + L_2(x)\bar{u}_4, \tag{9.23}$$

where $L_1(x)$ and $L_2(x)$ are defined in Eqs. (9.22).

The representations in Eqs. (9.21) and (9.23) apply for the constant values of y, $y = y_1$ and $y = y_4$, respectively. Lagrange's interpolating formula can be applied again, this time in the y direction to give the interpolation over the element as

$$\hat{u} = L_1(y)\hat{u}|_{1\text{-}2} + L_2(y)\hat{u}|_{4\text{-}3}, \tag{9.24}$$

where

$$L_1(y) = \frac{y - y_4}{y_1 - y_4}, \qquad L_2(y) = \frac{y - y_1}{y_4 - y_1}. \tag{9.25}$$

Substituting Eqs. (9.21) and (9.23) into Eq. (9.24) finally allows the trial function $\hat{u}$ within element e to be written as

$$\hat{u} = L_1(x)L_1(y)\bar{u}_1 + L_2(x)L_1(y)\bar{u}_2 + L_1(x)L_2(y)\bar{u}_3 + L_2(x)L_2(y)\bar{u}_4, \tag{9.26}$$

where the various Lagrange polynomials are given by Eqs. (9.22) and (9.25). Comparison of Eq. (9.26) with (9.18) yields the shape functions as

$$\begin{aligned} N_1 &= L_1(x)L_1(y), & N_2 &= L_2(x)L_1(y), \\ N_3 &= L_1(x)L_2(y), & N_4 &= L_2(x)L_2(y). \end{aligned} \tag{9.27}$$

It can also be shown, which is left as an exercise, that the shape functions obtained by substituting Eqs. (9.22) and (9.25) into Eq. (9.27) are identical to those which may be obtained from Eq. (9.15), showing that the two methods are equivalent.

9.2.2.2   Hermitian Elements

The shape functions for Hermitian elements can be derived in a manner similar to that above, but using *Hermite polynomials* in place of Lagrange polynomials. The nodal vector will now, in addition to nodal values of the function, also include nodal values of the derivatives.

To illustrate, consider an s-node one-dimensional element e, in which the nodes are not necessarily equally spaced. Let each node have two degrees of freedom, consisting of the function u and its first derivative $\partial u / \partial x$. Consequently, the trial function within element e can be written as

$$\hat{u} = \sum_{i=1}^{s} \left[N_{0i}(x)\bar{u}_i + N_{1i}(x)\frac{\partial \bar{u}_i}{\partial x} \right]. \tag{9.28}$$

In a shape function N_{ij} of Eq. (9.28), the first subscript denotes the order of differentiation of the accompanying nodal variable, and the second identifies the node.

For the representation in Eq. (9.28) to yield $\bar{u}_k$ and $\partial \bar{u}_k / \partial x$ at node k, the functions $N_{0i}(x)$ and $N_{1i}(x)$ must (for $i \neq j$) satisfy the relations

$$
\begin{aligned}
N_{0i}(x_i) &= 1, & N_{1i}(x_i) &= 0, \\
N'_{0i}(x_i) &= 0, & N'_{1i}(x_i) &= 1, \\
N_{0i}(x_j) &= 0, & N_{1i}(x_j) &= 0, \\
N'_{0i}(x_j) &= 0, & N'_{1i}(x_j) &= 0.
\end{aligned}
\tag{9.29}
$$

Equations (9.29) can be satisfied by the Hermitian polynomials [19]

$$
N_{0i}(x) = \prod_{j=1, j \neq i}^{s} \frac{(x - x_j)^2}{(x_i - x_j)^2} \left[1 + 2 \sum_{j=1, j \neq i}^{s} \frac{x_i - x}{x_i - x_j} \right].
\tag{9.30a}
$$

and

$$
N_{1i}(x) = \prod_{j=1, j \neq i}^{s} \frac{(x - x_j)^2}{(x_i - x_j)^2} (x - x_i),
\tag{9.30b}
$$

As a particular example, consider the case when $s = 2$. Equation (9.28) then becomes

$$
\hat{u} = N_{01}(x)\bar{u}_1 + N_{11}(x)\frac{\partial \bar{u}_1}{\partial x} + N_{02}(x)\bar{u}_2 + N_{12}(x)\frac{\partial \bar{u}_2}{\partial x},
\tag{9.31}
$$

where the shape functions are obtained from Eqs. (9.30) and (9.31) as

$$
\begin{aligned}
N_{01}(x) &= \frac{(x_2 - x)^2}{(x_2 - x_1)^2} \left[1 + 2\left(\frac{x - x_1}{x_2 - x_1}\right) \right], \\[2mm]
N_{02}(x) &= \frac{(x - x_1)^2}{(x_2 - x_1)^2} \left[1 + 2\left(\frac{x_2 - x}{x_2 - x_1}\right) \right], \\[2mm]
N_{11}(x) &= \frac{(x_2 - x)^2}{(x_2 - x_1)^2} (x - x_1), \\[2mm]
N_{12}(x) &= \frac{(x - x_1)^2}{(x_2 - x_1)^2} (x - x_2).
\end{aligned}
\tag{9.32}
$$

If the local coordinate $\xi = (x - x_1)/L$ is used, where $L = x_2 - x_1$, Eqs. (9.31) and (9.32) become

$$
\hat{u} = N_{01}(\xi)\bar{u}_1 + N_{11}(\xi)\frac{\partial \bar{u}_1}{\partial x} + N_{02}(\xi)\bar{u}_2 + N_{12}(\xi)\frac{\partial \bar{u}_2}{\partial x}.
\tag{9.33}
$$

$$
\begin{aligned}
N_{01}(\xi) &= 1 - 3\xi^2 + 2\xi^3, & N_{02}(\xi) &= 3\xi^2 - 2\xi^3, \\
N_{11}(\xi) &= \xi L (1 - \xi)^2, & N_{12}(\xi) &= L(\xi^3 - \xi^2).
\end{aligned}
\tag{9.34}
$$

The nodal derivatives $\partial \bar{u}_1/\partial x$, $\partial \bar{u}_2/\partial x$ in Eq. (9.33) may be replaced by $\partial \bar{u}_1/\partial \xi$, $\partial \bar{u}_2/\partial \xi$, respectively, through the relation $\partial/\partial x = L^{-1}\,\partial/\partial \xi$, in which case L drops out of the second and fourth component equations of (9.34).

The procedure can be [10, 11, 17] extended to include, in addition to the function u and its first derivatives, higher order derivatives of u as well. For two-dimensional elements, the interpolation is applied twice, first in the x direction and secondly in the y direction as in Section 9.2.2.1, yielding shape functions that are products of the one-dimensional shape functions. The simplest such rectangular element will have four degrees of freedom per node, namely u, $\partial u/\partial x$, $\partial u/\partial y$, and $\partial^2 u/\partial x\,\partial y$, and is described in Section 9.5.2.3.

9.3 NATURAL COORDINATES

When an arbitrary global coordinate system is used, the nodal coordinates are limited in value only by the boundaries of the region. It would be a useful simplification if the extremum values of these coordinates were restricted to $-1, 0$, or 1. This can be accomplished by choosing a local coordinate system peculiar to the element, such that the coordinates vary linearly between the normalized nodal coordinates. A frame of reference of this kind is known as a natural coordinate system.

One advantage of natural coordinates is that an integration over the element, in a finite element formulation, can often be expressed in a standard analytical form.

9.3.1 Natural Coordinates in One Dimension

Consider the one-dimensional element e, with node identifiers 1 and 2, shown in Fig. 9.2. The coordinates in the global system Ox are x_1 and x_2 for nodes 1 and 2, respectively. Introducing the local system $\bar{O}\xi_1$ with origin at x_1 and the ξ_1 axis along the x axis, allows an x coordinate within the element to be written as

$$\xi_1 = x - x_1, \tag{9.35}$$

or, dividing by the length $(x_2 - x_1)$ of the element, as

$$\bar{\xi}_1 = \frac{x - x_1}{x_2 - x_1}. \tag{9.36a}$$

In Eq. (9.36a) and the remainder of this section, a bar is used to indicate that the coordinate is normalized.

If another local system $\bar{\bar{O}}\xi_2$ is chosen so that its origin $\bar{\bar{O}}$ coincides with x_2, with the ξ_2 axis along the x axis (but in the opposite sense), then for any

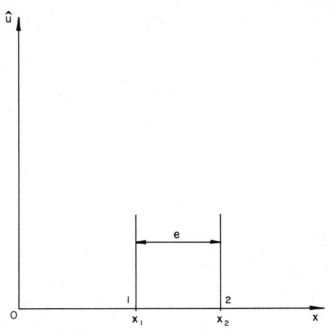

Fig. 9.2 One-dimensional element e.

x within element e,

$$\overline{\xi}_2 = \frac{x_2 - x}{x_2 - x_1}.$$ (9.36b)

From Eq. (9.36a) it is noted that $\overline{\xi}_1 = 0$ at $x = x_1$ and $\overline{\xi}_1 = 1$ at $x = x_2$. Similarly, from Eq. (9.36b), $\overline{\xi}_2 = 1$ at $x = x_1$ and $\overline{\xi}_2 = 0$ at $x = x_2$. It is easily verified that $\overline{\xi}_1$ and $\overline{\xi}_2$ are identical to $L_2(x)$ and $L_1(x)$ of Eqs. (9.22), respectively. Both of the coordinates $\overline{\xi}_1$ and $\overline{\xi}_2$ vary linearly with x as may be seen from Eqs. (9.36a) and (9.36b). That only one of the coordinates $\overline{\xi}_1$ and $\overline{\xi}_2$ is independent is shown by the relationship

$$\overline{\xi}_1 + \overline{\xi}_2 = L_2(x) + L_1(x) = 1,$$ (9.37)

which is easily proven.

The natural coordinates $\overline{\xi}_1$ and $\overline{\xi}_2$ (or $L_2(x)$ and $L_1(x)$) are functions of the independent variable x and the nodal coordinates x_1 and x_2, and have the value 1 at one node and 0 at the other node. The approximation $\hat{u}$ across the element e can thus be interpolated as

$$\hat{u} = \overline{\xi}_2 \bar{u}_1 + \overline{\xi}_1 \bar{u}_2$$ (9.38)

or

$$\hat{u} = L_1(x)\bar{u}_1 + L_2(x)\bar{u}_2. \tag{9.39}$$

Comparison of Eqs. (9.39) with Eqs. (9.2) shows that the shape functions N_1 and N_2 are given by

$$N_1 = L_1(x), \qquad N_2 = L_2(x). \tag{9.40}$$

It may also be shown, using the generalized coordinate method of the previous section, that the trial function

$$\hat{u} = \alpha_1 + \alpha_2 x \tag{9.41}$$

yields the same shape functions as given in Eqs. (9.40). For this element, the shape functions therefore are determinable either indirectly through Eq. (9.15) using the generalized coordinate method, or directly using the interpolation method. Although both routes are quite straightforward in this particular case, the interpolation method is generally advantageous for elements which are more complex.

When evaluating an element contribution, derivatives such as $\partial u/\partial x$, $\partial u/\partial y$, and product terms such as $x\,\partial u/\partial x$ will commonly be encountered. With natural coordinates, element contributions generally can be expressed as a product of nodal values and integrals of the form $\int_e L_1{}^a(x)L_2{}^b(x)\,dx$, where a and b are integer powers. The integration can be carried out analytically using the relation

$$\int_e L_1{}^a(x)L_2{}^b(x)\,dx = \frac{a!\,b!}{(a+n+1)!}\,h^e, \tag{9.42}$$

where h^e is the length of element e.

9.3.2   Natural Coordinates in Two Dimensions

In the two-dimensional case, the area coordinate is analogous to the length coordinate in the one-dimensional case. For the point P in a three-node triangular element e, such a coordinate is defined by taking the area subtended by P and its appropriate base or datum, and dividing this by the area of the whole triangle. Thus, as shown in Fig. 9.3, the area coordinate L_1 is given by

$$L_1 = \frac{A_1}{\Delta}, \tag{9.43}$$

where A_1 is the area of the triangle formed by the point P and the L_1 datum or base, and Δ is the area of the triangular element. The L_2 and L_3 coordinates are defined similarly. The triangle bases correspond to $L_1 = 0$, $L_2 = 0$, and

$L_3 = 0$, with the opposite vertex nodes being equal to $L_1 = 1$, $L_2 = 1$, and $L_3 = 1$, respectively.

From simple geometry, it can be shown that the relationship between the Cartesian and area coordinates is given by

$$\begin{bmatrix} x \\ y \\ 1 \end{bmatrix} = \begin{bmatrix} x_1 & x_2 & x_3 \\ y_1 & y_2 & y_3 \\ 1 & 1 & 1 \end{bmatrix} \begin{bmatrix} L_1 \\ L_2 \\ L_3 \end{bmatrix}, \qquad (9.44)$$

where the third component equation is easily obtained from Fig. 9.3 and the definitions of L_1, L_2, and L_3.

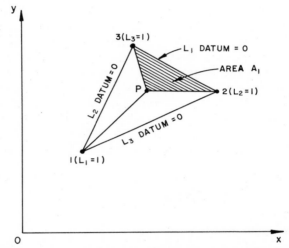

Fig. 9.3 Area coordinates for a typical three-node triangular element e.

The area coordinates L_1, L_2, and L_3 are similar to shape functions in that they have the values of 0 and 1 at the nodal points. The approximation for $\hat{u}$ across the (linear) element e can therefore be written in the form

$$\hat{u} = L_1 \bar{u}_1 + L_2 \bar{u}_2 + L_3 \bar{u}_3. \qquad (9.45)$$

Comparison of Eq. (9.45) with Eq. (9.2) shows that the shape functions N_1, N_2, and N_3 for this element are given by

$$N_1 = L_1, \qquad N_2 = L_2, \qquad N_3 = L_3. \qquad (9.46)$$

An element contribution originally derived with reference to the global system can be transformed to the natural system of area coordinates through Eq. (9.44). In general, the element contributions will then involve integrals of the form $\int_e L_1{}^a L_2{}^b L_3{}^c \, dD_e$, which can be evaluated analytically through the relation

$$\int_e L_1{}^a L_2{}^b L_3{}^c \, dD_e = \frac{a!b!c!}{(a+b+c+2)!} 2\Delta. \tag{9.47}$$

It may be noted that natural coordinates can also be defined for quadrilateral elements [6].

9.3.3 Natural Coordinates in Three Dimensions

For the three-dimensional case, the natural coordinates are volume ratios or volume coordinates. For the point P in the four-node tetrahedral element e shown in Fig. 9.4, the volume coordinate L_1 is defined as

$$L_1 = \frac{V_1}{V}, \tag{9.48}$$

where V_1 is the volume subtended by the point P and the face opposite node 1, with V being the volume of the whole tetrahedron. Volume coordinates L_2, L_3, and L_4 are similarly defined. The relationship between the Cartesian coordinates x, y, z and the volume coordinates L_1, L_2, L_3, L_4 will be found to be

$$\begin{bmatrix} x \\ y \\ z \\ 1 \end{bmatrix} = \begin{bmatrix} x_1 & x_2 & x_3 & x_4 \\ y_1 & y_2 & y_3 & y_4 \\ z_1 & z_2 & z_3 & z_4 \\ 1 & 1 & 1 & 1 \end{bmatrix} \begin{bmatrix} L_1 \\ L_2 \\ L_3 \\ L_4 \end{bmatrix}. \tag{9.49}$$

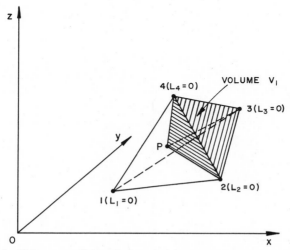

Fig. 9.4 Volume coordinates for a typical four-node tetrahedral element e.

For the four-node tetrahedron, it can be verified that the shape functions N_i derived from the linear interpolation

$$\hat{u}^e = \alpha_1 + \alpha_2 x + \alpha_3 y + \alpha_4 z \tag{9.50}$$

are identical to the respective volume coordinates; hence for this element

$$N_i = L_i, \qquad i = 1, 2, 3, 4. \tag{9.51}$$

By inverting Eq. (9.49) the volume coordinates can be obtained in terms of global coordinates as

$$L_i = \frac{1}{6V}(a_i + b_i x + c_i y + d_i z), \qquad i = 1, 2, 3, 4, \tag{9.52}$$

where

$$6V = \begin{bmatrix} 1 & x_1 & y_1 & z_1 \\ 1 & x_2 & y_2 & z_2 \\ 1 & x_3 & y_3 & z_3 \\ 1 & x_4 & y_4 & z_4 \end{bmatrix}, \tag{9.53}$$

and the a_i, b_i, c_i are obtained by cyclic permutation, as necessary, of the subscripts 1, 2, 3, 4 in the following:

$$
a_1 = \begin{bmatrix} x_2 & y_2 & z_2 \\ x_3 & y_3 & z_3 \\ x_4 & y_4 & z_4 \end{bmatrix}, \qquad b_1 = -\begin{bmatrix} 1 & y_2 & z_2 \\ 1 & y_3 & z_3 \\ 1 & y_4 & z_4 \end{bmatrix},
$$
$$
c_1 = -\begin{bmatrix} x_2 & 1 & z_2 \\ x_3 & 1 & z_3 \\ x_4 & 1 & z_4 \end{bmatrix}, \qquad d_1 = -\begin{bmatrix} x_2 & y_2 & 1 \\ x_3 & y_3 & 1 \\ x_4 & y_4 & 1 \end{bmatrix}. \tag{9.54}
$$

Note that the signs in these equations depend on the sense in which the nodes are identified. The above relations apply for a right-handed Cartesian coordinate system when the nodes 1, 2, 3 are ordered counterclockwise when viewed from node 4.

With volume coordinates integrals occurring in the element contributions can be evaluated using the relation

$$\iiint_e L_1^a L_2^b L_3^c L_4^d \, dV_e = \frac{a! \, b! \, c! \, d!}{(a + b + c + d + 3)!} \, 6V, \tag{9.55}$$

where V is the volume of the element e.

It is worth noting that natural coordinates can also be defined for general hexahedral elements [6].

9.4 ONE-DIMENSIONAL ELEMENTS

9.4.1 Shape Functions

Although other one-dimensional elements exist, the simple ones described in Section 9.3, together with the presentation in Chapter 1, will suffice for this book.

9.5 TWO-DIMENSIONAL ELEMENTS

Triangles are the simplest polygonal[†] figures into which any two-dimensional region can be subdivided, and this accounts in part for the popularity of the triangular finite element. The next possible choice is that of rectangular, or more generally, quadrilateral elements, which is also quite widely used. Higher order polygons are not normally used.

9.5.1 Triangular Elements

Of the various triangular elements available, the simple three-node Lagrangian element corresponding to a linear trial function has been used the most extensively. Elements based on higher order polynomial trial functions are now, however, coming into wider use.

9.5.1.1 The Lagrangian Triangular Family

Triangular elements of this family can be simply formed by selecting a sufficient number of nodes to allow a unique solution for the coefficients in the chosen polynomial trial function. A complete polynomial of order n contains $\frac{1}{2}(n+1)(n+2)$ coefficients and an s-node Lagrangian triangular element, based on this polynomial, must contain the same number of nodes, hence

$$s = \tfrac{1}{2}(n+1)(n+2). \tag{9.56}$$

Although other possibilities exist for the location of the s nodes, the arrangements shown in Table 9.1 lead to relatively simple shape functions.

The shape functions for these complete polynomial elements can be derived by the method of generalized coordinates presented earlier, although the algebra becomes more difficult as the order increases. Alternatively, the method of interpolation can be used, choosing each shape function to be the product of three Lagrange interpolating formulas [20, 6, 11].

[†] Only regular (straight-sided) polygons are considered in this section. Curved-sided elements are described in Section 9.7.

TABLE 9.1

The Lagrangian Triangular Family

Element	Type	Order of polynomial used as trial functions	Number of terms in trial function
	Linear	1	3
	Quadratic	2	6
	Cubic	3	10
	Quartic	4	15
	Quintic	5	21

The Lagrangian elements, shown in Table 9.1, all possess continuity of the function across interelement boundaries and hence throughout the region of the problem. This can be verified by noting that the complete polynomials cause the corresponding elements to possess geometric isotropy (see Section 8.11). It also follows from the fact that the number of nodes on any side of an element is the same as the number of coefficients in the polynomial along that side, thus enabling these coefficients to be determined uniquely. Since the polynomial on the side of the adjoining element is uniquely pegged on the same nodal values, there must be continuity of the trial function across the interelement boundary.

9.5.1.2 Four-Node Cubic Triangular Element

In Chapter 5 the four-node triangular Hermitian element with a complete cubic trial function was introduced. The geometrics of this four-node triangular element are the same as for the three-node element except for the additional fourth node chosen at the centroid. It will be recalled that in addition to specifying the function and its first derivatives (with respect to x and y) as nodal parameters at each of the three vertex nodes, the function is also specified at the centroidal node. These ten nodal values can be seen to be sufficient to determine the complete cubic trial function uniquely.

A complete listing of the shape functions for this element has been given by Fellipa and Clough [21]. As shown in Chapter 5, however, the explicit use of the shape functions can be bypassed with a resulting simplification in the formulation. The same chapter also showed how the element matrix equations could be reduced in size, by deleting the centroidal nodal value from the formulation, using condensation. Alternatively, this nodal parameter can be eliminated as described by Tocher, Holand, and others [22–24].

The interelement compatibility of this element is of interest. Along any side the trial function $\hat{u}$ is represented by a cubic polynomial in s where s is measured along the side. At each vertex node u, $\partial u/\partial x$, and $\partial u/\partial y$ are specified as nodal values. The derivative $\partial u/\partial s$ in the s direction can be obtained as a linear combination of $\partial u/\partial x$ and $\partial u/\partial y$, and hence is known at each vertex node. Along any side, therefore, the values of u and $\partial u/\partial s$ at its end points are known, giving sufficient conditions for solving for the four coefficients of the cubic polynomial. Since these cubic representations along the sides of adjoining elements are uniquely pegged on the same nodal values, interelement continuity of the trial function exists. In similar fashion, it can be shown that the first derivatives of the cubic trial function are not continuous across interelement boundaries. If such continuity is required, a higher order interpolation is needed such as is described in the next section.

9.5.1.3 Six-Node Quintic Triangular Element

This element, shown in Fig. 9.5, was first proposed by Fraeijs de Veubeke [25] and is based on the complete fifth-order polynomial representation. Its nodal parameters are the values of u, $\partial u/\partial x$, $\partial u/\partial y$, $\partial^2 u/\partial x^2$, $\partial^2 u/\partial x\,\partial y$, $\partial^2 u/\partial y^2$ at each vertex node, together with the nodal value of $\partial u/\partial n$ at each midside node. There are thus 21 conditions available to determine the 21 coefficients of the complete fifth-order polynomial uniquely. It is left as an exercise to show, in a similar manner to that in the previous section, that interelement continuity exists both for the trial function and its derivative normal to the element boundary. It follows that the trial function and its first derivatives with respect to x and y are continuous throughout the region.

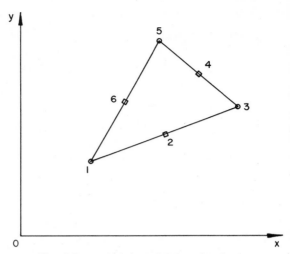

Fig. 9.5   Typical six-node triangular element.

The midside nodes can be eliminated to yield an 18 degree-of-freedom element by constraining $\partial u/\partial n$ to vary cubically along the sides [26–30]. Interelement continuity in the trial function and its first derivatives is retained in this process.

The quintic elements not only give a continuous solution surface with continuous first derivatives but also allow the second derivatives to be obtained nodalwise as part of the solution. Boundary conditions specified in terms of second-order derivatives can therefore be prescribed as equivalent Dirichlet conditions (Section 5.4), if so desired.

For higher order trial functions, such as the quintic, the use of a local coordinate system $\bar{O}\xi\eta$, as shown in Chapter 5, is advantageous.

9.5.1.4   Other Triangular Elements

Some of the other commonly used triangular elements are given in Table 9.2. Further details may be found in the references cited.

The term *constrained* in the table needs explanation. A constrained polynomial, in the present sense, is a complete polynomial for which one or more equations of constraint exist, relating the coefficients. The constrained cubic in Table 9.2 has a complete cubic polynomial (10-term) representation, with one equation of constraint. Only nine nodal parameters are required for the element, these being sufficient to allow determination of the ten polynomial coefficients.

TABLE 9.2

Some Triangular Elements

Element	Polynomial	Degrees of freedom	Nodal Parameters at ○	Nodal Parameters at □	Interelement continuity	Reference
	Complete quadratic	6	u	$\dfrac{\partial u}{\partial n}$	u	31
	Complete cubic	10	$u, \dfrac{\partial u}{\partial x}, \dfrac{\partial u}{\partial y}$	$\dfrac{\partial u}{\partial n}$	$u, \dfrac{\partial u}{\partial n}$	32
	Constrained cubic	9	$u, \dfrac{\partial u}{\partial x}, \dfrac{\partial u}{\partial y}$		u in the non-conforming version, $u, \partial u/\partial n$ in the conforming version	33, 34
	Complete quartic	15	$u, \dfrac{\partial u}{\partial x}, \dfrac{\partial u}{\partial y}$	$u, \dfrac{\partial u}{\partial n}$	u	35

9.5.2 Rectangular and Quadrilateral Elements

Rectangular elements on their own are not well suited to irregular two-dimensional regions, but are often used in combination with the more extensively adopted triangular elements. Quadrilateral elements are better adapted for irregular regions but have not received as wide application as triangular elements. Only a brief description of rectangular and quadrilateral elements will therefore be given, but it should be remembered that for certain applications such elements may be used to advantage.

9.5.2.1 Lagrangian Rectangular Elements

Although there are other rectangular elements which are Lagrangian in the sense that the nodal parameters comprise only values of the variables, this particular family of elements is often identified as *the* Lagrangian rectangular family because of its direct derivation from the Lagrange polynomials.

Consider the two-dimensional element e shown in Fig. 9.6, in which there are m equally spaced nodes in each row and n equally spaced nodes in each column. For a node ij, the shape function is defined by the product of two Lagrange polynomials as

$$N_{ij} = L_j^m(x)L_i^n(y), \tag{9.57}$$

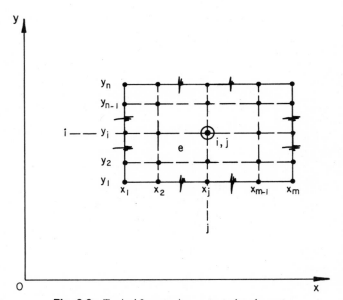

Fig. 9.6 Typical Lagrangian rectangular element e.

where $L_j^m(x)$, $L_i^n(y)$ are defined by Eq. (9.20) and the superscripts m and n are used to indicate the order of the polynomial.

From the properties of Lagrange polynomials $N_{ij}(x, y)$ has the value of unity at x_i, y_j and zero at every other nodal position, as required by a shape function. The Lagrange's interpolation relation $\hat{u}$ for this mn Lagrangian element can thus be written as

$$
\begin{aligned}
\hat{u} = \;& N_{11}\bar{u}_{11} + N_{12}\bar{u}_{12} + \cdots + N_{1m}\bar{u}_{1m} \\
& + N_{21}\bar{u}_{21} + N_{22}\bar{u}_{22} + \cdots + N_{2m}\bar{u}_{1m} + \cdots \\
& + N_{n1}\bar{u}_{n1} + N_{n2}\bar{u}_{n2} + \cdots + N_{nm}\bar{u}_{nm}
\end{aligned}
\tag{9.58}
$$

or

$$
\hat{u} = \sum_{i=1}^{n} \sum_{j=1}^{m} N_{ij}\bar{u}_{ij}.
\tag{9.59}
$$

Lagrange elements of this family have interelement continuity in $\hat{u}$ only and have an incomplete interpolation polynomial. It can be shown that geometric isotropy is present only if equal numbers of nodes are used in the x and y direction. Except for the first member of this family, the *bilinear* element, the Lagrangian elements suffer from the disadvantages of having internal nodes and giving poor curve fitting, especially for higher order polynomials. Consequently, other than the bilinear element, they are rarely used.

Illustrative Example 9.1 The shape functions for the *bilinear* element with node identifiers 1, 2, 3, and 4, shown in Fig. 9.7, are derived in the following

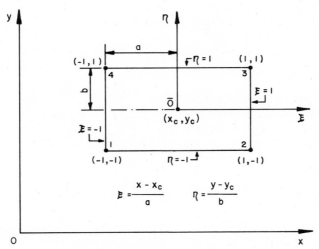

Fig. 9.7 Typical bilinear element e.

with reference to the local coordinate system $\bar{O}\xi\eta$. For node 1, the shape function N_1 is given by Eq. (9.57), which after dropping the superscripts m and n for convenience, becomes

$$N_1 = N_{11} = L_1(\xi)L_1(\eta). \tag{9.60}$$

Substitution into Eq. (9.60) from Eq. (9.20) for $m, n = 2$ then gives

$$N_1 = \left(\frac{\xi - \xi_2}{\xi_1 - \xi_2}\right)\left(\frac{\eta - \eta_4}{\eta_1 - \eta_4}\right) = \left(\frac{\xi - 1}{-1 - 1}\right)\left(\frac{\eta - 1}{-1 - 1}\right)$$

$$= \tfrac{1}{4}(\xi - 1)(\eta - 1). \tag{9.61a}$$

Similarly, it can be shown that

$$N_2 = -\tfrac{1}{4}(\xi + 1)(\eta - 1), \tag{9.61b}$$
$$N_3 = \tfrac{1}{4}(\xi + 1)(\eta + 1), \tag{9.61c}$$
$$N_4 = -\tfrac{1}{4}(\xi - 1)(\eta + 1). \tag{9.61d}$$

The trial function $\hat{u}$ can then be written in the general form

$$\hat{u} = N_1\bar{u}_1 + N_2\bar{u}_2 + N_3\bar{u}_3 + N_4\bar{u}_4, \tag{9.62}$$

where the shape functions are evaluated through Eqs. (9.61).

9.5.2.2   Serendipity Elements

The first three members of this family of elements are shown in Table 9.3. It should be noted that the description of these elements as linear, quadratic, cubic refers to the variation of the trial function in the ξ direction at constant η, or in the η direction at constant ξ. The trial functions for these elements are incomplete quadratic, cubic, and quartic polynomials, respectively, in ξ and η.

The shape functions for the first three serendipity elements were originally found by inspection and are given in Table 9.4 in terms of the local coordinates ξ and η. These shape functions can also be obtained by using the following incomplete polynomial trial functions.

Linear: $\hat{u} = \alpha_1 + \alpha_2\xi + \alpha_3\eta + \alpha_4\xi\eta$ $\qquad\qquad$ (9.63a)

Quadratic: $\hat{u} = \alpha_1 + \alpha_2\xi + \alpha_3\eta + \alpha_4\xi^2 + \alpha_5\xi\eta + \alpha_6\eta^2$
$\qquad\qquad + \alpha_7\xi^2\eta + \alpha_8\xi\eta^2$ $\qquad\qquad\qquad\qquad\qquad$ (9.63b)

Cubic: $\hat{u} = \alpha_1 + \alpha_2\xi + \alpha_3\eta + \alpha_4\xi^2 + \alpha_5\xi\eta + \alpha_6\eta^2 + \alpha_7\xi^3$
$\qquad\qquad + \alpha_8\xi^2\eta + \alpha_9\xi\eta^2 + \alpha_{10}\eta^3 + \alpha_{11}\xi^3\eta + \alpha_{12}\xi\eta^3$ $\quad$ (9.63c)

TABLE 9.3

The Serendipity Family of Elements

Element	Type	Number of nodes
	Linear	4
	Quadratic	8
	Cubic	12

It will be noted that in Eqs. (9.63) symmetric pairs of terms have been omitted from the complete polynomials to retain geometric isotropy (see Section 8.11).

As originally developed the serendipity elements possessed an equal number of nodes in the x and y directions. More recently, an algorithm has been developed [36] for constructing serendipity elements with *either* equal *or* unequal numbers of nodes in the two (or three) directions. The same reference also shows how to produce *modified serendipity* elements that have complete polynomial trial functions, without additional nodes in two dimensions, but with a midface node in three dimensions. The modified elements are "more efficient computationally" than either Lagrangian rectangular elements or complete triangular elements. Moreover, use of elements with a different number of nodes along each side allows *low-order elements* in regions where the solution is not expected to change drastically, to be matched with *higher order elements* in other regions.

Along element boundaries, the trial function of a serendipity element is a complete polynomial, and hence there exists interelement continuity of the trial function.

Serendipity elements form a useful class of rectangular elements, which, if combined with triangular elements, can be used quite effectively in regions with curved boundaries.

TABLE 9.4

Shape Functions for Serendipity Elements

Element type	Corner nodes — Coordinates of points ●	Corner nodes — Shape function	Side nodes — Coordinates of points □	Side nodes — Shape function
	ξ_i, η_i	$N_i = \frac{1}{4}(1 + \xi\xi_i)(1 + \eta\eta_i)$		
	ξ_i, η_i	$N_i = \frac{1}{4}(1 + \xi\xi_i)(1 + \eta\eta_i)(\xi\xi_i + \eta\eta_i - 1)$	$\xi_i = 0$ $\eta_i = 0$	$N_i = \frac{1}{2}(1 - \xi^2)(1 + \eta\eta_i)$ $N_i = \frac{1}{2}(1 + \xi\xi_i)(1 - \eta^2)$
	ξ_i, η_i	$N_i = \frac{1}{32}(1 + \xi\xi_i)(1 + \eta\eta_i)[9\xi^2 + \eta^2 - 10]$	$\xi_i = \pm 1$ $\eta_i = \pm\frac{1}{3}$ $\xi_i = \pm\frac{1}{3}$ $\eta_i = \pm 1$	$N_i = \frac{9}{32}(1 + \xi\xi_i)(1 - \eta^2)(1 + 9\eta\eta_i)$ $N_i = \frac{9}{32}(1 + \eta\eta_i)(1 - \xi^2)(1 + 9\xi\xi_i)$

9.5.2.3 Hermitian Elements

Shape functions for rectangular Hermitian elements can be obtained by forming products of the Hermitian functions in each coordinate direction, in a fashion generally similar to that for the Lagrangian rectangular elements of Section 9.5.2.1. Consider, for example, the first-order Hermitian rectangle shown in Fig. 9.8, where the local coordinates $\xi = (x - x_1)/a$ and $\xi = (y - y_1)/b$ are used and where the parameters at each node are the nodal values of u, $\partial u/\partial x$, $\partial u/\partial y$, and $\partial^2 u/\partial x \partial y$.

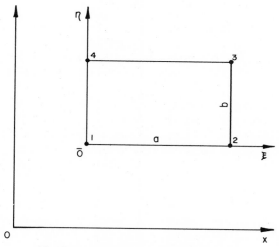

Fig. 9.8 Hermitian rectangle of first order.

It has been shown in Section 9.2.2.2 that for the one-dimensional two-node Hermitian element involving first derivatives as nodal variables, the element approximation $\hat{u}$ could be written as

$$\hat{u} = \sum_{i=1}^{2} \left(N_{0i}(\xi)\bar{u}_i + N_{1i}(\xi)\frac{\partial \bar{u}_i}{\partial x} \right). \tag{9.64}$$

This interpolation can be extended into the two directions x and y in a manner similar to that used for Lagrange's interpolation. Using the coordinate system $\bar{O}\xi\eta$, it can then be shown [37] for the rectangular element in Fig. 9.8 that

$$\hat{u} = \sum_{i=1}^{4} \bar{N}_{1i}\bar{u}_i + \bar{N}_{2i}\frac{\partial \bar{u}_i}{\partial x} + \bar{N}_{3i}\frac{\partial \bar{u}_i}{\partial y} + \bar{N}_{4i}\frac{\partial^2 \bar{u}_i}{\partial x \partial y}, \tag{9.65}$$

where

$$\bar{N}_{1i} = N_{0i}(\xi)N_{0i}(\eta), \qquad \bar{N}_{2i} = N_{0i}(\xi)N_{1i}(\eta),$$
$$\bar{N}_{3i} = N_{1i}(\xi)N_{0i}(\eta), \qquad \bar{N}_{4i} = N_{1i}(\xi)N_{1i}(\eta). \tag{9.66}$$

The functions N_{ij} in Eqs. (9.66)[†] are given by

$$N_{01}(\xi) = N_{04}(\xi) = 1 - 3\xi^2 + 2\xi^3, \qquad N_{01}(\eta) = N_{02}(\eta) = 1 - 3\eta^2 + 2\eta^3,$$
$$N_{11}(\xi) = N_{14}(\xi) = a(\xi - 2\xi^2 + \xi^3), \qquad N_{11}(\eta) = N_{12}(\eta) = b(\eta - 2\eta^2 + \eta^3),$$
$$N_{02}(\xi) = N_{03}(\xi) = 3\xi^2 - 2\xi^3, \qquad N_{04}(\eta) = N_{03}(\eta) = 3\eta^2 - 2\eta^3, \tag{9.67}$$
$$N_{12}(\xi) = N_{13}(\xi) = -a(\xi^2 - \xi^3), \qquad N_{14}(\eta) = N_{13}(\eta) - b(\eta^2 - \eta^3).$$

The trial function is an incomplete sixth-order polynomial in terms of ξ and η. It can also be shown that the trial function [Eq. (9.65)] possesses continuity in both the function and its first derivatives. The element stiffness matrix for this first-order element in plate bending is available in the literature [37].

Higher order trial functions, based on Hermitian elements with more than two nodes per side, and using derivatives of order greater than two, have been developed [38, 39], but become rather complex and are seldom used. A comparison of results from several higher order Hermitian elements, when applied to plate bending, has been given by Smith [40].

9.5.2.4 Twelve-Term Rectangle

For an element similar to that shown in Fig. 9.7, but having u, $\partial u/\partial x$, and $\partial u/\partial y$ as nodal parameters, an incomplete quartic polynomial with 12 terms can be used to determine the shape functions as shown in Section 9.2.1.2. A complete polynomial of fourth order contains 15 terms, so three of these must be deleted. The usual choice is $\xi^2\eta^2$ plus the symmetric pair ξ^4, η^4, giving for the interpolation

$$\hat{u} = \alpha_1 + \alpha_2\xi + \xi_3\eta + \alpha_4\xi^2 + \alpha_5\xi\eta + \alpha_6\eta^2 + \alpha_7\xi^3 + \alpha_8\xi^2\eta$$
$$+ \alpha_9\xi\eta^2 + \alpha_{10}\eta^3 + \alpha_{11}\xi^3\eta + \alpha_{12}\xi\eta^3. \tag{9.68}$$

The trial function, in shape function form, can be written as

$$\hat{u} = \sum_{i=1}^{4} N_{1i}\bar{u}_i + N_{2i}\left(\frac{\partial\bar{u}}{\partial x}\right)_i + N_{3i}\left(\frac{\partial\bar{u}}{\partial y}\right)_i, \tag{9.69}$$

where it can be shown that the shape functions are as tabulated in Table 9.5, when the local coordinate axes ξ, η of Fig. 9.7 are used.

[†] The shape functions N_{ij} of Eqs. (9.66) and (9.67) can be identified with the one-dimensional shape functions N_{ij} of Eqs. (9.33) and (9.34).

TABLE 9.5

Shape Functions for the Twelve-Term Rectangle

i	Shape functions N_{1i}	N_{2i}	N_{3i}	Variables e, f, g, h
1	$\frac{1}{8}eg(\frac{1}{2}eg - \frac{1}{2}fh - ef - gh)$	$\frac{a}{8}eg^2h$	$-\frac{b}{8}e^2gh$	$e = \eta - 1$
2	$\frac{1}{8}eh(\frac{1}{2}eh - \frac{1}{2}fg + ef + gh)$	$\frac{a}{8}eh^2g$	$\frac{b}{8}e^2fh$	$f = \eta + 1$
3	$\frac{1}{8}fh(\frac{1}{2}fh - \frac{1}{2}eg - ef - gh)$	$\frac{a}{8}fh^2g$	$\frac{b}{8}ef^2h$	$g = \xi - 1$
4	$\frac{1}{8}fg(\frac{1}{2}fg - \frac{1}{2}eh + ef + gh)$	$\frac{a}{8}fg^2h$	$-\frac{b}{8}ef^2g$	$h = \xi + 1$

9.5.3 Quadrilateral Elements

One approach to quadrilateral elements is to form such elements from triangular elements. Figure 9.9 illustrates how averaging can be used with simple linear triangles to obtain a quadrilateral element. First, the quadrilateral is divided by one diagonal, and secondly, by the other diagonal. For each divided quadrilateral, the element k matrix is found from the linear representations within the component triangular elements. The two element k matrices are then averaged to obtain the final quadrilateral element k matrix. Higher order triangles can be used to form higher order quadrilaterals.

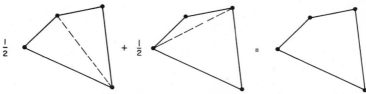

Fig. 9.9 Quadrilateral element—averaging.

Figure 9.10a shows the quadrilateral element developed by de Veubeke [41] from four complete cubic polynomial triangles. There are 16 degrees of freedom for the element, three at each vertex node, namely u, $\partial u/\partial x$, and $\partial u/\partial y$, and one, $\partial u/\partial n$, at each midside node.

The Clough and Felippa quadrilateral element [42] illustrated in Fig. 9.10b is also developed from four triangles, each of which is constructed

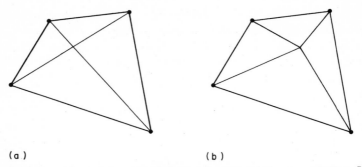

(a)   (b)

Fig. 9.10   Quadrilateral elements: (a) de Veubeke [41]; (b) Clough and Felippa [42].

from three triangular subregions. There are 12 degrees of freedom for the element, three at each vertex node.

An alternative approach to the quadrilateral element, shown in Fig. 9.11, uses a transformation from natural coordinates to global coordinates, to generate a quadrilateral from a square. The relationship between the natural coordinates (ξ, η) and the global coordinates (x, y) in this case is given by

$$x = \tfrac{1}{4}[(1 - \xi)(1 - \eta)x_1 + (1 + \xi)(1 - \eta)x_2 + (1 + \xi)(1 + \eta)x_3 \\ + (1 - \xi)(1 + \eta)x_4], \tag{9.70a}$$

$$y = \tfrac{1}{4}[(1 - \xi)(1 - \eta)y_1 + (1 + \xi)(1 - \eta)y_2 + (1 + \xi)(1 + \eta)y_3 \\ + (1 - \xi)(1 + \eta)y_4]. \tag{9.70b}$$

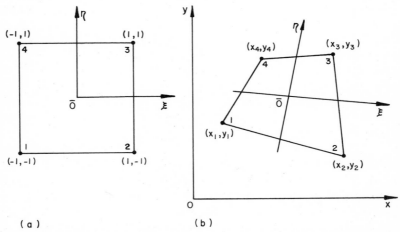

(a)   (b)

Fig. 9.11   Typical quadrilateral element: (a) natural coordinates; (b) global and local coordinates.

The trial function in the $\bar{O}\xi\eta$ system can be transferred to the quadrilateral on a point-by-point basis since the trial function has the same value at corresponding ξ, η and x, y points. For example, the trial function $\hat{u}$ for the particular four-node rectangle obtained in Section 9.5.2.1. [Eq. (9.62)] can be transferred to the general quadrilateral element using the transformation of Eqs. (9.70). The above procedure illustrates the mapping of a simple parent element into an element of more general shape. This approach is considered further in Section 9.7.

9.6 THREE-DIMENSIONAL ELEMENTS

Three-dimensional problems inherently involve a large number of degrees of freedom. For example, in solid mechanics, the three displacements u, v, and w, and their first derivatives in the x, y, and z directions, respectively, cause the simple tetrahedral element to have 48 degrees of freedom. Even with a reasonable number of elements, the resulting system matrix equation can easily have several thousand unknowns. It is not surprising, therefore, that three-dimensional structures, even of the *shell type* such as used in aircraft, automobiles, and ships, can involve tens of thousands of unknowns. Three-dimensional finite element formulations in other areas, for example, pollutant distributions in tidal estuaries, also involve a large number of unknowns.

In general, there is an advantage in concentrating the nodal parameters at the vertex nodes since the vertex nodes are common to a larger number of elements than the edge or face nodes. For a fixed number of nodal parameters per element, this decreases the number of nodal parameters in the system, thus decreasing the size of the system matrix. Since a face-node is common to only two elements, the use of such nodes is to be avoided. The penalty with using edge-nodes is considerably less, particularly if a frontal solution method is adopted. In the foregoing, internal nodes have not been taken into account since they can be simply eliminated by condensation.

In view of the above, elements with vertex nodes will be emphasized in the following sections. All the elements which are considered have inter-element continuity in the variable. Other elements and their relative performances are described in the literature [18, 43–46].

9.6.1 Tetrahedral Elements

The most popular of the three-dimensional elements are the tetrahedral, although occasionally it is difficult to subdivide a region into only this type of element.[†] For this reason tetrahedral elements are often mixed with hexadral (brick-shaped) elements.

[†] Accuracy is likely to suffer if the elements become long and thin.

9.6.1.1 Lagrangian Tetrahedra

The first three elements of this family, namely the four-node, 10-node, and 20-node elements, are shown in Fig. 9.12 and correspond respectively to complete linear, quadratic, and cubic polynomial trial functions. Each node has only one degree of freedom, namely the value of the function $\hat{u}$ at that node. The shape functions can be obtained by the generalized coordinate method, although for other than the linear element are more easily obtained as products of interpolation functions [47].

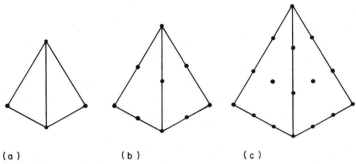

(a)   (b)   (c)

Fig. 9.12   Lagrangian tetrahedra: (a) four node; (b) ten node; (c) 20 node.

For the four-node element, the trial function is linear, that is,

$$\hat{u} = \alpha_1 + \alpha_2 x + \alpha_3 y + \alpha_4 z. \tag{9.71}$$

The shape functions for this element can be quite simply shown to be

$$N_1 = L_1, \qquad N_2 = L_2, \qquad N_3 = L_3, \qquad N_4 = L_4, \tag{9.72}$$

where L_1, L_2, L_3, and L_4 are the volume coordinates, defined previously in Section 9.3.3.

9.6.1.2 Other Tetrahedra

As indicated earlier, there is an advantage with higher order elements in concentrating the nodal parameters at the vertex nodes. A valuable tetrahedral element with only vertex parameters is illustrated in Fig. 9.13. The trial function within this element is an incomplete cubic polynomial in x, y, z [46]. The nodal parameters are the function u and its first derivatives in the x, y, and z directions, giving a total of 16 degrees of freedom. Since a complete cubic polynomial in three dimensions has 20 terms, four have to be deleted to allow a unique determination of the shape functions. The evaluation of the shape functions for this element and the derivation of the stiffness matrix can be found in the literature [43, 48]. In elasticity problems,

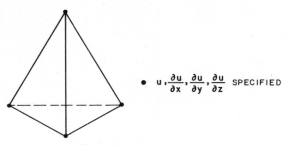

$$\bullet \quad u, \frac{\partial u}{\partial x}, \frac{\partial u}{\partial y}, \frac{\partial u}{\partial z} \quad \text{SPECIFIED}$$

Fig. 9.13 T48 tetrahedron.

where there are three displacements u, v, and w, in the x, y, and z directions, respectively, to be considered at any point, the resulting 48 nodal parameters give the element its common name of T48.

For other tetrahedral elements appropriate references [6, 43–46] can be consulted.

9.6.2 Hexahedral Rectangular Elements

The hexahedral family of elements are sometimes known as brick elements, for obvious reasons. The most commonly used hexahedral elements belong to the Lagrangian and serendipity families, and these are considered in the following.

9.6.2.1 Lagrangian Elements

As in the two-dimensional case, three-dimensional Lagrangian elements have shape functions that are products of the Lagrange interpolation functions. The first element of this family has eight nodes as shown in Fig. 9.14.

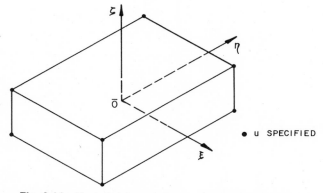

Fig. 9.14 Typical eight-node Lagrangian hexahedral element.

At each node the value of the function is specified, giving a total of eight nodal parameters. Higher order elements, in addition to the corner nodes, may have edge, face, and interior nodes, but these elements are not widely used. For the first Lagrangian element, using normalized orthogonal coordinates ξ, η, ζ and choosing the origin $\bar{O}$ at the centroid of the element [see Fig. 9.14], it can be shown that the shape functions are

$$N_i = \tfrac{1}{8}(1 + \xi\xi_i)(1 + \eta\eta_i)(1 + \zeta\zeta_i), \qquad i = 1, 2, \ldots, 8. \tag{9.73}$$

The trial function $\hat{u}$ within element e can be written in terms of these N_i, in the usual way, as

$$\hat{u} = \sum_{i=1}^{8} N_i \bar{u}_i. \tag{9.74}$$

9.6.2.2 Serendipity Elements

As was the case for serendipity elements in two dimensions, the three-dimensional serendipity elements do not contain internal nodes. The first three elements of the family are shown in Fig. 9.15, where it is seen that they possess 8, 20, and 32 nodes, respectively, at each of which the nodal value of the function is specified. For the same local coordinate system $\bar{O}\xi\eta\zeta$ as used in Fig. 9.14, the shape functions for the serendipity elements have been shown [49] to be the following:

"Linear" Element

$$N_i = \tfrac{1}{8}(1 + \xi\xi_i)(1 + \eta\eta_i)(1 + \zeta\zeta_i). \tag{9.75}$$

"Quadratic" Element
  Vertex Node

$$N_i = \tfrac{1}{8}(1 + \xi\xi_i)(1 + \eta\eta_i)(1 + \zeta\zeta_i)(\xi\xi_i + \eta\eta_i + \zeta\zeta_i - 2). \tag{9.76}$$

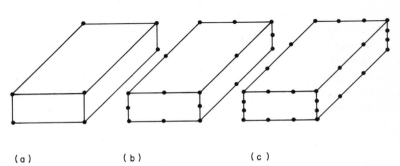

(a)   (b)   (c)

Fig. 9.15 Serendipity hexahedral elements: (a) linear; (b) quadratic; (c) cubic.

Typical Midside Node

$$\xi_i = 0, \quad \eta_i = \pm 1, \quad \zeta_i = \pm 1,$$
$$N_i = \tfrac{1}{4}(1 - \xi^2)(1 + \eta\eta_i)(1 + \zeta\zeta_i). \tag{9.77}$$

"Cubic" Element
  Vertex Node

$$N_i = \tfrac{1}{64}(1 + \xi\xi_i)(1 + \eta\eta_i)(1 + \zeta\zeta_i)[9(\xi^2 + \eta^2 + \zeta^2) - 19]. \tag{9.78}$$

Typical Midside Node

$$\xi_i = \pm\tfrac{1}{3}, \quad \eta_i = \pm 1, \quad \zeta_i = \pm 1,$$
$$N_i = \tfrac{9}{64}(1 - \xi^2)(1 + 9\xi\xi_i)(1 + \eta\eta_i)(1 + \zeta\zeta_i). \tag{9.79}$$

The shape functions for the linear element of this family are seen to be identical to those of the linear Lagrangian element. All the above elements have considerable practical use.

9.6.3 Pentahedral Elements

These elements, in the shape of triangular prisms, are used quite often in conjunction with hexahedral elements. The shape functions are generated by forming products of triangle interpolation functions with Lagrange or serendipity functions in the remaining dimension [49].

9.7 ISOPARAMETRIC ELEMENTS

In problems that have curved boundaries it is necessary to use many straight-sided (faced) elements along the boundaries in order to achieve a reasonable geometric representation of these boundaries. The number of elements needed can be reduced appreciably if curved elements are used, with a consequent reduction in the total number of variables in the system. For three-dimensional problems, where the total number of variables is inherently large, such a reduction can be very important.

While other methods of creating curved elements exist, the only method used extensively in practice involves mapping from regular (straight-edged or -sided) elements. Since the shape functions of the regular parent element are known with respect to a local coordinate system, those of the generated curvilinear element can also be determined. As developed by Irons and Zienkiewicz *et al.* [49–51], the mapping from local coordinates ξ, η, ζ to Cartesian coordinates x, y, z is through the shape function relationships:

$$x = \mathbf{N}_m\mathbf{x}, \tag{9.80a}$$
$$y = \mathbf{N}_m\mathbf{y}, \tag{9.80b}$$
$$z = \mathbf{N}_m\mathbf{z}. \tag{9.80c}$$

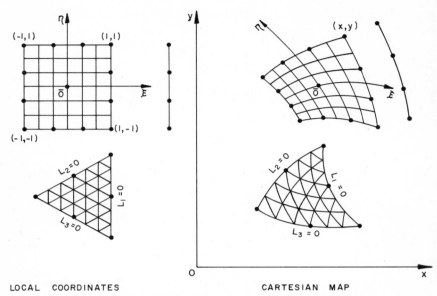

LOCAL   COORDINATES   CARTESIAN   MAP

Fig. 9.16a   Two-dimensional isoparametric element mapping.

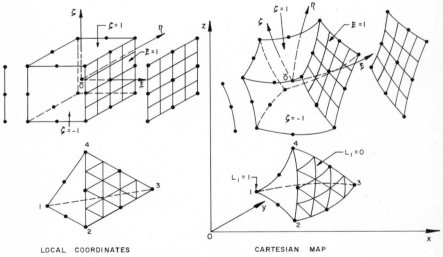

LOCAL   COORDINATES   CARTESIAN   MAP

Fig. 9.16b   Three-dimensional isoparametric element mapping.

In Eqs. (9.80), the entries in the shape function matrix $\mathbf{N}_m$ are functions of ξ, η, ζ, and the column matrices $\mathbf{x}$, $\mathbf{y}$, and $\mathbf{z}$ list the nodal coordinates with respect to the global system. The trial function $\hat{u}$ can be written with respect to the local coordinate system as

$$\hat{u} = \mathbf{N}\mathbf{u}, \qquad (9.81)$$

where the elements of the shape function matrix $\mathbf{N}$ are functions of ξ, η, and ζ.

Referring to Fig. 9.16, it is seen that for any point in the parent element with local coordinates ξ, η, ζ, a corresponding point with global coordinates x, y, z can be generated in the derived element, using Eqs. (9.80). The value of the trial function at the point x, y, z is the same as its value at the corresponding ξ, η, ζ and this latter can be evaluated through Eq. (9.81).

It is convenient to choose the shape function matrices $\mathbf{N}_m$ and $\mathbf{N}$ identical in form, in which case the generated element is termed isoparametric. If the shape function matrix $\mathbf{N}_m$ is a lower order representation than the matrix $\mathbf{N}$, the generated curvilinear element is subparametric, and if $\mathbf{N}_m$ is a higher order representation, the element is superparametric.

If linear shape functions are used in Eqs. (9.80) and (9.81), a two-dimensional rectangle will be mapped into an arbitrary quadrilateral and a three-dimensional brick will become a hexahedron with plane, but not parallel, faces. Higher order mapping functions, such as quadratic and cubic, can be used to produce curvilinear elements.

When developing element matrix equations, using isoparametric elements, it is necessary to calculate derivatives of shape functions with respect to the $\bar{O}\xi\eta\zeta$ system from the corresponding derivatives in the $Oxyz$ system. These sets of derivatives can conveniently be related through the Jacobian matrix of transformation.

For two-dimensional regions, several curved triangular elements have been devised. A cubic curved triangle, originally developed for potential flow problems, for example, is described by Isaacs [52]. Further details of isoparametric elements and their application can be found in the references already cited.

9.8 TRANSFORMATION FROM LOCAL TO GLOBAL COORDINATES

It has become evident that many shape functions can most simply be expressed with reference to a particular local coordinate system. In such cases, the resulting element matrix equations will contain the unknown variables relative to the local system. These element matrix equations must be transformed to their corresponding equations with respect to the global system before assembly into the system matrix equation.

For problems involving nonisotropic media, an element matrix equation has its simplest form if derived with respect to those local axes that are respectively parallel to the principal axes of the medium at that point. Again, transformation from the local to the global system is necessary before assembly.

The transformation is most easily effected by using the appropriate matrix of transformation as is shown in the following. Consider the element contribution, derived with respect to the local system, given by

$$\chi^e = (\mathbf{u}_L)^\mathrm{T} \mathbf{k}_L \mathbf{u}_L, \tag{9.82}$$

where $\mathbf{k}_L$ is the element $\mathbf{k}$ matrix and $\mathbf{u}_L$ the element nodal vector, both with reference to the local system. Let the transformation from the local system to the global system be given by[†]

$$\mathbf{u}_L = \mathbf{Tu}, \tag{9.83}$$

where $\mathbf{u}$ is the element nodal vector with respect to the global system and $\mathbf{T}$ is the transformation matrix between the two systems. Substituting Eq. (9.83) into Eq. (9.82) gives

$$\chi^e = (\mathbf{u})^\mathrm{T} \mathbf{T}^\mathrm{T} \mathbf{k}_L \mathbf{Tu}. \tag{9.84}$$

Consequently, the element $\mathbf{k}$ matrix with respect to the global system, denoted $\mathbf{k}$, becomes

$$\mathbf{k} = \mathbf{T}^\mathrm{T} \mathbf{k}_L \mathbf{T}. \tag{9.85}$$

Since the transformation from the local to the global system has been illustrated in Chapter 5, no further examples will be presented in this section.

9.9 SELECTION OF AN ELEMENT

In the preceding sections, a selection of the more common elements has been described. For a particular problem, the question of which element to choose is not an insignificant one. The complexity of programming, the total computation effort and cost, and the accuracy of the solution are strongly affected by the element selected. Unfortunately there are no iron-clad rules for choosing the best element since the selection depends on the type of problem, the geometry of the boundaries, the boundary conditions, the accuracy required, the size of the available computer, the maximum allowable computing cost, as well as other factors.

To aid in the selection of an element, there are however certain guidelines available. To begin, the trial function must be able to represent all the

---

[†] Where the transformation is a simple rotation, the transformation matrix has been previously denoted by $\boldsymbol{R}$.

derivatives that occur in the functional. The simplest way to satisfy the convergence requirements is to adopt an element based on a complete polynomial and, moreover, to use only trial functions that are admissible, which for a large class of problems corresponds to using conforming elements. Elements that do not satisfy both the completeness and conformity conditions should be used only after a careful examination of their performance. Since such elements may be very efficient, they should however not be excluded from consideration when an element is being selected.

If the boundaries of the problem are regular, elements of simple geometry are usually selected, whereas for curved boundaries, the decision is more difficult since both regular and curved elements have been used successfully for such cases. To match irregular boundaries, the choice is between many regular elements or few, more complex, isoparametric elements.

There is considerable advantage, as previously indicated, in choosing elements that have their nodal parameters concentrated at the vertices. Derivative elements can be valuable where the solution involves derivatives since the latter do not then need to be estimated by subsequent interpolation.

REFERENCES

1. J. H. Argyris, Continua and discontinua, *Proc. Conf. Matrix Methods Struct. Mech., 1st, Wright-Patterson AFB, Ohio October 26–28, 1965* (AFFDL-TR-66-80) (November 1966).
2. R. W. Clough, Comparison in three-dimensional finite elements, *in* "Finite Element Methods in Stress Analyses" (I. Holand and K. Bell, eds.). Tapir Press, Trondheim, Norway, 1969.
3. O. C. Zienkiewicz, Isoparametric and allied numerically integrated elements—a review, *Proc. Symp. Num. Comput. Methods Struct. Mech.* Univ. of Illinois, Urbana, Illinois (September 1971).
4. A. R. Mitchell, Element types and base functions, *in* "Numerical Solution of Partial Differential Equations" (J. G. Gram, ed.), pp. 107–150. Reidel, Dordrecth, 1973.
5. D. H. Norrie and G. de Vries, "The Finite Element Method." Academic Press, New York, 1973.
6. K. H. Huebner, "The Finite Element Method for Engineers." Wiley, New York, 1975.
7. R. T. Fenner, "Finite Element Methods for Engineers." MacMillan, New York, 1975.
8. O. C. Zienkiewicz, "The Finite Element Method in Engineering Science." McGraw-Hill, New York, 1971.
9. O. Ural, "The Finite Element Method." Intext Educational Publ., New York, 1973.
10. H. C. Martin and G. F. Carey, "Introduction to Finite Element Analysis." McGraw-Hill, New York, 1973.
11. R. H. Gallagher, "Finite Element Analysis." Prentice-Hall, Englewood Cliffs, New Jersey, 1975.
12. J. H. Robinson, "Integrated Theory of Finite Element Methods." Wiley (Interscience), New York, 1973.
13. K. C. Rockey, H. R. Evans, D. W. Griffiths, and D. A. Nethercot, "The Finite Element Method." Crosby Lockwood Staples, London, 1975.

14.   C. A. Brebbia and J. J. Connor, "Fundamentals of Finite Element Techniques," Halsted Press, 1974.
15.   C. S. Desia and J. F. Abel, "Introduction to the Finite Element Method." Van Nostrand-Reinhold, Princeton, New Jersey, 1972.
16.   R. D. Cook, "Concepts and Applications of Finite Element Analysis." Wiley, New York, 1974.
17.   Pin Tong and J. N. Rossettos, "The Finite Element Method." MIT Press, Cambridge, Massachusetts, 1977.
18.   D. H. Norrie and G. de Vries, "A Finite Element Bibliography." Plenum Press, New York, 1976.
19.   R. W. Hamming, "Numerical Methods for Scientists and Engineers." McGraw-Hill, New York, 1962.
20.   P. Silvester, Higher-order polynomial triangular elements for potential problems, *Internat. J. Engrg. Sci.* 7(8), 849–861 (1969).
21.   C. A. Felippa and R. W. Clough, The finite element method in solid mechanics, *in* "Numerical Solution of Field Problems in Continuum Physics" (*SIAM–AMS Proc.*), Vol. 2, 210–252. Amer. Math. Soc., Providence, Rhode Island 1970.
22.   J. L. Tocher and B. J. Hartz, Higher-order finite element for plane stress, *Proc. ASCE, J. Engrg. Mech. Div.* 93, No. EM4, 149–174 (1967).
23.   I. Holand and P. G. Bergan, Discussion of higher-order finite element for plane stress, *Proc. ASCE, J. Engrg. Mech. Div.* 94, No. EM2, 698–702, (April 1968).
24.   I. Holand, The finite element method in plane stress analysis, *in* "The Finite Element Method in Stress Analysis" (I. Holand and K. Bell, eds.), Chapter 2. Tapir Press, Trondheim, Norway, 1969.
25.   B. Fraeijs de Veubeke, Displacement and equilibrium models in the finite element method, *Symp. Numer. Methods in Elasticity, University College of Swansea, January 1964, in* "Stress Analysis" (O. C. Zienkiewicz and G. Holister, eds.), Chapter 9, pp. 145–147. Wiley, New York, 1965.
26.   C. Brebbia and J. Connor, Plate bending, *in* "Finite Element Techniques in Structural Mechanics" (H. Tottenham and C. Brebbia, eds.), Chapter 4, 112–114. Stress Analysis Publ., Southampton, 1971.
27.   K. Bell, A refined triangular plate bending finite element, *Int. J. Numer. Methods Engrg.* 1, No. 1, 101–122 (1969).
28.   G. R. Cowper, E. Kosko, G. Lindberg, and M. Olson, Static and dynamic applications of a high precision triangular plate bending element, *AIAA J.* 7, No. 10, 1957–1965 (1969).
29.   G. Butlin and R. Ford, A compatible triangular plate bending finite element, *Internat. J. Numer. Methods Engrg.* 6, 323–332 (1970).
30.   K. Bell, Triangular plate bending elements, *in* "Finite Element Methods in Stress Analysis" (I. Holand and K. Bell, eds.), Chapter 7. Tapir Press, Trondheim, Norway, 1969.
31.   L. S. D. Morley, The constant-moment plate bending element, *J. Strain Anal.* 6, No. 1, 2–24 (1971).
32.   R. W. Clough and J. L. Tocher, Finite element stiffness matrices for analysis of plate bending, *Proc. Conf. Matrix Methods. Struct. Mech., 1st, Wright-Patterson AFB, Ohio, 26–28 October, 1965* (AFFDL-TR-66-80), pp. 515–546 (November 1966).
33.   G. Bazeley, Y. K. Cheung, B. Irons, and O. Zienkiewicz, Triangular elements in plate bending—conforming and non-forming solutions, *Proc. Conf. Matrix Methods. Struct. Mech., 1st, Wright-Patterson AFB, Ohio, 26–28 October 1965* (AFFDL-TR-66-80). pp. 547–576 (November 1966).
34.   D. G. Harrison and Y. K. Cheung, A higher-order triangular finite element for the solution of field problems in orthotropic media, *Internat. J. Numer. Methods. Engrg.,* 7, 287–295

(1973). (*Note* that there is an error on p. 294. The first element of the S_{y_1} submatrix should read $10c_1{}^2 - 4c_{I+1}e_{I-1}$ instead of the printed version, which omits the exponent 2 in the first term.)

35. K. Bell, Analysis of Thin Plates in Bending Using Triangular Finite Elements, Div. of Struct. Mech., Technical Univ. of Norway, Trondheim, Norway (February 1968).

36. R. L. Taylor, On completeness of shape functions for finite element analysis, *Internat. J. Numer. Methods. Engrg.* **4**, No. 1, 17–22 (1972).

37. F. K. Bogner, R. L. Fox, and L. A. Schmidt, The generation of interelement-compatible stiffness and mass matrices by the use of interpolation formulas, *Proc. Conf. Matrix Methods. Struct. Mech., 1st, Wright-Patterson AFB, Ohio, 26–28 October, 1965* (AFFDL-TR-66-80), (November 1966).

38. G. Birkhoff, M. H. Schultz, and R. S. Varga, Piece-wise hermitian interpolation in one and two variables with application to differential equations, *Numer. Math.* **2**, 232–256 (1968).

39. I. M. Smith and W. Duncan, The effectiveness of nodal continuities in finite element analysis of thin rectangular and skew plates in bending, *Internat. J. Numer, Methods. Engrg.* **2**, 253–258 (1970).

40. I. M. Smith, A finite element analysis for moderately thick rectangular plates in bending, *Internat. J. Mech. Sci.* **10**, 563–570 (1968).

41. B. Fraeijs de Veubeke, A conforming finite element for plate bending, *Internat. J. of Solids Struct.* **4**, No. 1, 95–108 (1968).

42. R. Clough and C. Felippa, A refined quadrilateral element for the analysis of plate bending, *Proc. Conf. Matrix Methods Struct. Mech., 2nd, Wright-Patterson AFB, Ohio 15–17 October, 1968* (AFFDL-TR-68-150) pp. 399–440 (December 1969).

43. S. A. Fjeld, Three dimensional theory of elasticity, *in* "Finite Element Methods in Stress Analysis," pp. 333–364. Tapir Press, Trondheim, Norway, 1969.

44. J. R. Hughes and H. Allik, Finite elements for compressible and incompressible continua, *Proc. Symp. Appl. Finite Element Methods Civil Engrg.* Vanderbilt Univ., Nashville, Tennessee (November 1969).

45. R. W. Clough, Comparison of three dimensional finite elements, *Proc. Symp. Appl. Finite Element Methods Civil Engrg.* Vanderbilt Univ., Nashville, Tennessee (November 1969).

46. Y. R. Rashed, P. D. Smith, and N. Price, On further application of the finite element method of three dimensional elastic analysis, *Proc. Symp. High Speed Comput. Elastic Struct.* Univ. of Liege Press, Belgium, 1970.

47. P. Silvester, Tetrahedral finite elements for the Helmholtz equation, *Internat. J. Numer. Methods. Engrg.* **4**, No. 3, 405–413 (1972).

48. J. H. Argyris, I. Fried, and D. W. Scharpf, The TET20 and TEA8 elements for the matrix displacement method, *Aero. J.* **72**, No. 691, 618–623 (July 1968).

49. O. C. Zienkiewicz, B. M. Irons, J. Ergatoudis, S. Ahmad, and F. C. Scott, Iso-parametric and associated element families for two- and three-dimensional analysis, *in* "Finite Element Methods in Stress Analysis," pp. 383–432. Tapir Press, Trondheim, Norway, 1969.

50. I. Ergatoudis, B. M. Irons, and O. C. Zienkiewicz, Curved iso-parametric 'quadrilateral' elements for finite element analysis, *Internat. J. Solids Struct.* **4**, 31–42 (1968).

51. O. C. Zienkiewicz and B. M. Irons, Iso-parametric elements, *in* "Finite Element Techniques in Structural Analysis" (H. Tottenham and C. Brebbia, eds.), Chapter 10. Southampton Univ. Press, 1970.

52. L. T. Isaacs, A curved cubic triangular finite element for potential flow problems, *Internat. J. Numer. Methods Engrg.* **7**, No. 3, 337–344 (1973).

10

EQUATION SOLVERS
AND PROGRAMMING TECHNIQUES

For a linear equilibrium problem, a finite element formulation yields a set of linear algebraic equations of the form

$$\mathbf{AX} = \mathbf{B} \qquad (10.1)$$

where $\mathbf{A} = [a_{ij}]$ is the coefficient matrix, $\mathbf{X} = [x_j]$ is the system nodal vector of unknowns, and $\mathbf{B} = [b_i]$ is a column matrix of known values. Eigenvalue and propagation problems also yield linear algebraic equations, although of form different than in Eq. (10.1).

The techniques available for solving linear systems of equations can be classified as either direct or iterative methods. In the former category, a solution for $\mathbf{X}$ is obtained directly by a single application of a computational procedure. In contrast, with an iterative method, the solution requires the repeated application of an algorithm. An iterative procedure necessarily begins with an initial estimate of the solution, with subsequent iterations yielding an increasingly refined estimate. As a test for convergence, the latest estimate can be checked against the previous one, with the iteration process being terminated when the difference becomes less than a preassigned value.

Within the principal categories of direct and iterative methods, many variations have been developed. Some techniques, for example, take advantage of properties of the matrix $\mathbf{A}$, such as symmetry, bandedness, or sparseness, to reduce the number of computational operations and/or storage requirements. Most of the better known procedures have been programmed for computer application, and many are available as standard library subroutines. Such computer programs have become known as *equation solvers*, although sometimes the term is also used to denote the procedure on which the program is based.

This chapter outlines the more important equation solving techniques and considers their relative merits. Guidelines to aid in the choice of an equation solver for a particular problem are also presented.

10.1 SELECTION OF AN EQUATION SOLVER

The first decision to be made when choosing an equation solver is whether the method should be of the direct or the iterative type. For simple problems, where the coefficient matrix A is small, a standard library subroutine is usually used. For more complex systems, where large amounts of computation and storage are required, the computer costs become of considerable importance, and there is an incentive to search for, and if necessary develop, that procedure minimizing total computational costs. The more important criteria for selection are the number of computational operations required by the method, the difficulty of programming, the computer storage necessary, and the amount of housekeeping needed in the program. The most efficient solution may well demand core storage beyond that available, forcing the choice of a program that requires additional computation. Tradeoffs between memory, computation, capacity, time, and cost are commonly required.

In the 1950s, the core storage of computers was small and the rates for transferring data on and off magnetic tape were slow. Consequently, the direct methods of solution could be used only for simple problems. Iterative techniques, with their smaller core requirements were used for the larger, more complex problems.

Since that time, the size of core storage in computers has increased dramatically, and improved direct methods have become available. With a large present-day computer the coefficient matrix of even a moderately sized problem can be in full storage mode, allowing the solution to be obtained expeditiously using a direct method. For large problems (or even for smaller ones on minicomputers), use can be made of the fact that for direct methods, only part of the coefficient matrix need be in core at any one time. With the high transfer rates now available from peripheral storage devices, the core can be successively reloaded with portions of the matrix at great speed, allowing the solution to proceed sequentially, without taking excessive time. For very large problems, however, the iterative methods may still be found necessary, even when minimum core storage techniques are used.

In general, the following broad guidelines for selecting an equation solver can be given:

(1) For small problems, use the most convenient standard subroutine, preferably based on a direct method.

(2) For problems of moderate to large size, select a direct method, economizing on the storage modes wherever possible.

(3) For very large problems, when storage is a limitation, an iterative method should be considered.

Although the present chapter considers only linear systems of equations of the type in Eq. (10.1), resulting from elliptic differential equations, similar procedures exist for other types of problems. The finite element formulation for linear eigenvalue problems, for example, yields a system of algebraic eigenvalue equations, which can be solved by either direct or iterative means. Guidelines for the choice of method are similar to those for the equilibrium problem. Linear propagation problems, however, yield time-dependent equations for which iterative methods are more suited. For nonlinear systems of equations, it is necessary to use iterative procedures since direct methods are not available.

In the next sections, direct and iterative methods are briefly reviewed together with some associated techniques that can be used to reduce solution time and cost.

10.2 DIRECT SOLUTION METHODS

The direct solution methods most commonly used consist of two processes: first, triangularization of the system matrix by elimination or factorization, and secondly, back substitution. Elimination and factorization are both, strictly speaking, decomposition procedures and will be so regarded in this text, although decomposition is used by some authors to mean factorization only.

For finite element application, the most important direct methods at the present time are Gauss elimination, LDL^T factorization, Cholesky decomposition, and the frontal solution. The latter three methods can be regarded as variants of Gauss elimination [1]. Among the more recent direct methods, the *fast Fourier transform* and the *method of nested dissection* show particular promise [2]. Other methods are available which exploit the sparsity of the band in the matrix A (as does nested dissection) and they will be considered in Section 10.2.6. FORTRAN program listings for a number of direct methods have been compiled in convenient form by Sequi [3, 4] and those for other methods can be found elsewhere in the literature.

10.2.1 Gauss Elimination

In the method of Gauss elimination a series of row manipulations transforms the symmetric banded matrix $\mathbf{A}$ into an upper triangular matrix $\mathbf{U}$, with the column matrix $\mathbf{B}$ correspondingly changed to a column matrix $\mathbf{C}$. The procedure is outlined for a matrix $\mathbf{A}$ whose bandwidth B is 5. The upper part of the system matrix equation [Eq. (10.1)] becomes in this case

$$
\begin{bmatrix}
a_{11} & a_{12} & a_{13} \\
a_{21} & a_{22} & a_{23} & a_{24} \\
a_{31} & a_{32} & a_{33} & a_{34} & a_{35} \\
& a_{42} & a_{43} & a_{44} & a_{45} & a_{46} \\
& & a_{53} & a_{54} & a_{55} & a_{56} & a_{57} \\
& & & & \vdots
\end{bmatrix}
\begin{bmatrix}
x_1 \\ x_2 \\ x_3 \\ x_4 \\ x_5 \\ \vdots
\end{bmatrix}
=
\begin{bmatrix}
b_1 \\ b_2 \\ b_3 \\ b_4 \\ b_5 \\ \vdots
\end{bmatrix}.
\tag{10.2}
$$

In the first stage of elimination, the first row in Eq. (10.2) is multiplied by a_{i1}/a_{11}, and subtracted from the ith row, where i successively takes the values $2, 3, \ldots$, to give

$$
\begin{bmatrix}
a_{11} & a_{12} & a_{13} \\
& \bar{a}_{22} & \bar{a}_{23} & a_{24} \\
& \bar{a}_{32} & \bar{a}_{33} & a_{34} & a_{35} \\
& a_{42} & a_{43} & a_{44} & a_{45} & a_{46} \\
& & a_{53} & a_{54} & a_{55} & a_{56} & a_{57} \\
& & & & \vdots
\end{bmatrix}
\begin{bmatrix}
x_1 \\ x_2 \\ x_3 \\ x_4 \\ x_5 \\ \vdots
\end{bmatrix}
=
\begin{bmatrix}
b_1 \\ \bar{b}_2 \\ \bar{b}_3 \\ b_4 \\ b_5 \\ \vdots
\end{bmatrix}.
\tag{10.3}
$$

The bars in Eq. (10.3) indicate elements which have been modified. To illustrate, $\bar{a}_{32}$ and $\bar{b}_2$ are respectively given by

$$
\begin{aligned}
\bar{a}_{32} &= a_{32} - (a_{31}/a_{11})a_{12}, \\
\bar{b}_2 &= b_2 - (a_{31}/a_{11})b_1.
\end{aligned}
\tag{10.4}
$$

In the second elimination stage, the second row in Eq. (10.4) is multiplied by $[a_{i2}$ (or $\bar{a}_{i2})]/\bar{a}_{22}$, and subtracted from the ith row, where i takes the values $3, 4, \ldots$, resulting in

$$
\begin{bmatrix}
a_{11} & a_{12} & a_{13} \\
& \bar{a}_{22} & \bar{a}_{23} & a_{24} \\
& & \bar{\bar{a}}_{33} & \bar{a}_{34} & a_{35} \\
& & \bar{a}_{43} & \bar{a}_{44} & a_{45} & a_{46} \\
& & a_{53} & a_{54} & a_{55} & a_{56} & a_{57} \\
& & & & \vdots
\end{bmatrix}
\begin{bmatrix}
x_1 \\ x_2 \\ x_3 \\ x_4 \\ x_5 \\ \vdots
\end{bmatrix}
=
\begin{bmatrix}
b_1 \\ \bar{b}_2 \\ \bar{\bar{b}}_3 \\ \bar{b}_4 \\ b_5 \\ \vdots
\end{bmatrix}.
\tag{10.5}
$$

To illustrate, $\bar{\bar{a}}_{33}$, $\bar{a}_{43}$, and $\bar{\bar{b}}_3$ are respectively given by

$$
\begin{aligned}
\bar{\bar{a}}_{33} &= \bar{a}_{33} - (\bar{a}_{32}/\bar{a}_{22})a_{23}, \\
\bar{a}_{43} &= a_{43} - (a_{42}/\bar{a}_{22})a_{24}, \\
\bar{\bar{b}}_3 &= \bar{b}_3 - (\bar{a}_{32}/\bar{a}_{22})\bar{b}_2.
\end{aligned}
\tag{10.6}
$$

Continuing this process finally yields the system equation in the form

$$
\begin{bmatrix}
u_{11} & u_{12} & u_{13} \\
& u_{22} & u_{23} & u_{24} \\
& & u_{33} & u_{34} & u_{35} \\
& & & u_{44} & u_{45} & u_{46} \\
& & & & \ddots \\
& & & & & u_{n-1,n-1} & u_{n-1,n} \\
& & & & & & u_{nn}
\end{bmatrix}
\begin{bmatrix}
x_1 \\ x_2 \\ x_3 \\ x_4 \\ \vdots \\ x_{n-1} \\ x_n
\end{bmatrix}
=
\begin{bmatrix}
c_1 \\ c_2 \\ c_3 \\ c_4 \\ \vdots \\ c_{n-1} \\ c_n
\end{bmatrix}.
\tag{10.7}
$$

It can be seen that the elimination procedure converts the symmetric band matrix $\mathbf{A}$ in Eq. (10.1) to the upper triangular band matrix $\mathbf{U}$, resulting in

$$
\mathbf{UX} = \mathbf{C}.
\tag{10.8}
$$

Premultiplying Eq. (10.8) by the appropriate lower triangular matrix $\mathbf{L}$ can be shown to regenerate the original equation [Eq. (10.1)]. The elimination process, therefore, corresponds to the decomposition of $\mathbf{A}$ into $\mathbf{L}$ and $\mathbf{U}$, that is,

$$
\mathbf{A} = \mathbf{LU}.
\tag{10.9}
$$

Substituting Eq. (10.9) into Eq. (10.1), and premultiplying the result by the inverse of $\mathbf{L}$, yields

$$
\mathbf{UX} = \mathbf{L}^{-1}\mathbf{B} = \mathbf{C},
\tag{10.10}
$$

which is the same result as obtained above in Eq. (10.8).

The solution for the unknown x_n is given directly by the nth equation in Eq. (10.7). Substituting this value into the $(n - 1)$th equation then yields x_{n-1}. Continuing this back substitution process in Eq. (10.7) successively yields the remaining unknowns. The procedure is seen to be equivalent to premultiplying Eq. (10.10) by the inverse of $\mathbf{U}$, hence

$$
\mathbf{X} = \mathbf{U}^{-1}\mathbf{L}^{-1}\mathbf{B} = \mathbf{U}^{-1}\mathbf{C}.
\tag{10.11}
$$

For a banded matrix, close examination of the elimination method shows that during the kth stage of the elimination only the coefficients in the triangular area denoted I in Fig. 10.1a need to be held in core storage. The coefficients in triangle II of Fig. 10.1a are also needed in the process but because of symmetry can be obtained from triangle I.

Following the kth elimination stage, often called the kth reduction, the kth row in the active working area can be moved out to auxiliary storage, and a new column of data transferred into core storage as shown schematically in Fig. 10.1b. The next stage of the elimination can now proceed.

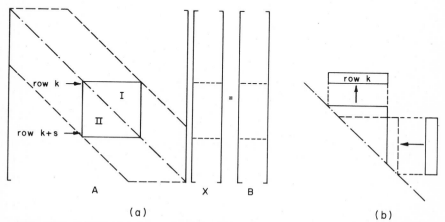

Fig. 10.1 (a) Working area at kth elimination stage; (b) change in active area from kth to $(k + 1)$th elimination stage.

10.2.2 LDL^T Factorization

The coefficient matrix **A** can be decomposed into the product of a lower triangular matrix, a diagonal matrix, and an upper triangular matrix, that is,

$$\mathbf{A} = \mathbf{LDU} \tag{10.12}$$

provided **A** and its upper-left principal submatrices are nonsingular [5, 6]. Furthermore, if **A** is symmetric, the upper triangular matrix is the transpose of the lower triangular matrix and both the lower and upper triangular matrices have 1's on their principal diagonals, hence

$$\mathbf{A} = \mathbf{LDL}^\mathsf{T}. \tag{10.13}$$

For obvious reasons, this decomposition is often termed triple factoring [7].

Using Eq. (10.13), the system matrix equation can be solved in two stages since Eq. (10.1) can be written as

$$\mathbf{LC} = \mathbf{B}, \tag{10.14a}$$

where

$$\mathbf{DL}^T\mathbf{X} = \mathbf{C}. \tag{10.14b}$$

Equation (10.14a) is first solved for $\mathbf{C}$, followed by the solution of Eq. (10.14b) for $\mathbf{X}$. The elements of the matrices $\mathbf{D}$ and $\mathbf{L}$ can be computed from

$$d_{ij} = a_{ii} - \sum_{m=1}^{i-1} l_{im}^2 d_{mm} \tag{10.15a}$$

$$l_{ii} = 1 \tag{10.15b}$$

$$l_{ij} = \frac{1}{d_{jj}}\left(a_{ij} - \sum_{m=1}^{j-1} l_{im}l_{jm}d_{mm}\right), \qquad i > j, \tag{10.15c}$$

$$l_{ij} = 0, \qquad i < j. \tag{10.15d}$$

where the summation is taken as zero if the upper limit is less than the lower limit.

The solution for the vector $\mathbf{C}$ can be obtained from

$$c_i = b_i - \sum_{n=1}^{i-1} l_{in}c_n \tag{10.16a}$$

and the solution for the vector $\mathbf{X}$ from

$$x_i = \frac{1}{d_{ii}}\left(c_i - \sum_{m=i+1}^{n} d_{ii}l_{mi}x_m\right) \tag{10.16b}$$

where $n \times n$ is the size of the coefficient matrix $\mathbf{A}$.

The $\mathbf{LDL}^T$ decomposition can be organized very effectively by calculating the final elements of $\mathbf{D}$ and $\mathbf{L}$ by columns [5]. This procedure is preferable to a simple Gauss elimination, being considerably faster.

10.2.3   Cholesky Factorization

The Cholesky decomposition, sometimes referred to as the square-root method, is possible only if the coefficient matrix $\mathbf{A}$ is symmetric and positive definite.[†] Under these conditions, the matrix $\mathbf{A}$ can be decomposed into a lower triangular matrix $\mathbf{L}$ with positive diagonal elements such that

$$\mathbf{A} = \mathbf{LL}^T. \tag{10.17a}$$

[†] If the quadratic form given by Eq. (A.43a) in Appendix A is positive for all nonzero u, the matrix $\mathbf{A}$ is termed positive definite.

Alternatively, **A** can be decomposed into an upper triangular matrix **U** satisfying

$$\mathbf{A} = \mathbf{U}\mathbf{U}^{\mathrm{T}}. \tag{10.17b}$$

For the former decomposition, substitution of Eq. (10.17a) into Eq. (10.1) results in

$$\mathbf{L}\mathbf{L}^{\mathrm{T}}\mathbf{X} = \mathbf{B}, \tag{10.18}$$

which can be written as the sequence of equations

$$\mathbf{L}\mathbf{C} = \mathbf{B}, \tag{10.19a}$$

$$\mathbf{L}^{\mathrm{T}}\mathbf{X} = \mathbf{C}. \tag{10.19b}$$

Forward substitution, using Eq. (10.19a) will yield the matrix **C**, and back substitution based on Eq. (10.19b) can then be used to obtain the required solution **X**. The triangular matrix $\mathbf{L} = [l_{ij}]$ needed in these processes can be obtained explicitly from **A** through the relationships

$$l_{ii} = \left(a_{ii} - \sum_{j=1}^{i-1} l_{ij}^2 \right)^{1/2} \qquad i = 1, \ldots, n \tag{10.20a}$$

$$l_{ij} = \frac{1}{l_{jj}} \left(a_{ij} - \sum_{m=1}^{j-1} l_{jm} l_{im} \right) \qquad i = i+1, i+2, \ldots, n; \quad j = 1, \ldots, n, \tag{10.20b}$$

$$l_{ij} = 0 \qquad i < j, \tag{10.20c}$$

where the summation is taken as zero if the upper limit is less than the lower limit.

From this algorithm, and referring to Fig. 10.2a, it is evident that the computation of an element l_{ij} requires the element a_{ij} plus the elements covered by the two solid bars in the **L** matrix. If the elements in triangle I are retained in core and if the elements a_{ij} therein are overwritten by the elements l_{ij} as they are generated, it will be seen that the necessary elements for the computation of l_{ij} (those covered by solid bars in the **L** matrix) will be available from the (overwritten) shaded-bar portions of the **A** matrix. After its determination, each new l_{ij} will overwrite the corresponding a_{ij}. Computation of the elements l_{ij} thus proceeds along BC until the lower band edge is reached. It is then necessary to transfer one row of the l_{ij} to peripheral storage from the active triangle and to transfer a new column of a_{ij} into core as shown schematically in Fig. 10.2b.

Although they have no great advantages over simple Gauss elimination insofar as storage is concerned, both **LDL**^{T} factorization and Cholesky decomposition use significantly less computation time, which is reflected in faster and cheaper solutions. The **LDL**^{T} procedure has slightly fewer operations than the Cholesky.

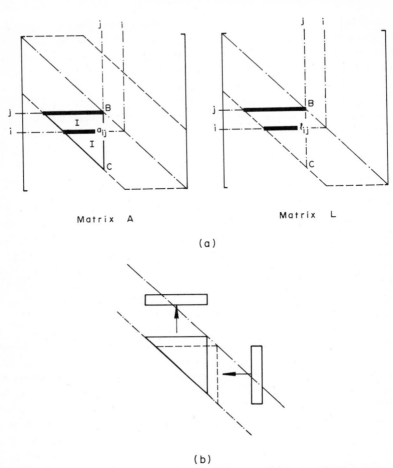

Matrix A   Matrix L

(a)

(b)

Fig. 10.2   Cholesky decomposition: (a) working areas; (b) change in active area.

10.2.4   Frontal Solution

In Section 10.2.1, it was seen that Gauss elimination can be carried out in stages with only the coefficients within the so-called active area being required in core at any stage. The back-substitution can similarly be accomplished in successive stages, with the aid of the necessary active area.

If elimination by stages is adopted, the assembly of the element **k** matrices into the system (coefficient) matrix can also proceed by stages, with only those element matrices being generated and assembled at each stage as are necessary to maintain the coefficients in the current active area as known values.

The frontal solution makes use of these concepts in an elegant fashion, with a wavefront moving across the domain of the problem in such a way that the elements are crossed by the front in the order in which they need to be assembled. An advantage of the method is that zeros of the coefficient matrix are excluded from computation. Another advantage is that midside and midface nodes are processed at little greater cost than vertex nodes.

Although the principles of the frontal solution were known earlier,[†] it was the formulation presented by Irons in 1970 that established the method firmly in finite element practice. For details of the frontal solution, this paper [8] and other descriptions in the literature [1, 5, 7, 9, 10] should be consulted.

10.2.5   Block Elimination

If the symmetric coefficient matrix $\mathbf{A}$ is partitioned into square sub-matrices[‡] $\mathbf{A}_{ij}$, and the $\mathbf{X}$ and $\mathbf{B}$ matrices are similarly partitioned into submatrices $\mathbf{X}_j$ and $\mathbf{B}_i$, the Gauss elimination procedure can be used exactly as before to solve for the $\mathbf{X}_j$. The previous formulas and algorithms remain valid except that $\mathbf{A}_{ij}$ replaces a_{ij}, $\mathbf{B}_i$ replaces b_i, and $\mathbf{X}_j$ replaces x_j. The only difference is that wherever multiplication by a_{kk}^{-1} (the reciprocal of a_{kk}) occurs in the original procedure, multiplication by the inverse of $\mathbf{A}_{kk}$ ($\mathbf{A}_{kk}^{-1}$) is the corresponding operation in the submatrix procedure. Thus, with the assistance of a subroutine to invert the symmetric $\mathbf{A}_{kk}$ submatrices, Gauss elimination can be used to convert or reduce the square matrix of submatrices $\mathbf{A}_{ij}$ to an upper triangular matrix of submatrices. Similarly, the submatrix $\mathbf{B}_i$ transforms to the corresponding submatrix $\mathbf{C}_i$. Since each of the new submatrices is explicitly known, a process of back substitution, similar to that used previously, yields the unknown $\mathbf{X}$. Examination of this process shows that only three blocks of the coefficient matrix $\mathbf{A}$ need to be in core storage at any one time to allow the elimination to proceed. Similarly, only three blocks of the matrix $\mathbf{B}$ need to be in core storage for the back substitution.

Block partitioning can also be used with other direct algorithms, such as the Cholesky procedure. Considerable use has been made of such block storage schemes in recent years to reduce core storage requirements for the direct solution of large sets of equations. In [11], for example, two blocks are used with the lower one being moved up as the upper (reduced) one is transferred out to peripheral storage and a new lower one is read in. In [12] integer strings are used to record the position of the blocks of nonzero terms. In [13], a block storage algorithm, related to nodes, is described. References

---

[†] This approach appears to have been first used by the Boeing Company, circa 1960.
[‡] Commonly called *blocks*.

[14–16] also give block elimination algorithms, with that of [16] storing only the nonzero elements. More recently, [17, 18] use an automatic determination of block size to suit a particular problem.

Despite the continuing interest in block schemes, it can be questioned whether those based on unconditional partitioning[†] offer any particular advantage for symmetric banded matrices. The original raison d'être for block storage schemes was that they still permitted a direct solution to be obtained even when the system of equations was so large that the coefficient matrix could not be wholly stored in core. Multipartitioning of the matrix and storage of only the necessary blocks at any one time does reduce the core requirement (hence allowing larger problems to be solved), but the penalty is increased computing time due to the transfers between core and peripheral storage. Additional *housekeeping* routines are required which add further demands for storage and time. While in principle, a block scheme could be designed with smaller core requirement than the triangular active area approach outlined earlier, thus allowing the solution of larger problems, the additional computing time makes other approaches preferable. When the triangular active area is too large to be stored wholly in core, procedures requiring storage of smaller portions can be resorted to [19]. Exploitation of the sparsity of the coefficient matrix, discussed in the next section, can extend the capacity of the direct methods, avoiding recourse to block methods employing unconditional partitioning. Alternatively, or additionally, conditional partitioning can be adopted to permit the solution of large problems. When these approaches reach their limits, or even before, iterative methods should be exploited.

10.2.6   Sparse Matrix Operations

The symmetric banded coefficient matrix **A** usually has many zero elements scattered along the edges of the band[‡] as well as within it. The sparsity of the nonzero terms in the band may even be such that the number of zero terms exceeds the nonzero terms. In most direct methods, in their simplest form, all the elements in the band are assumed nonzero, with storage and computation being ordered accordingly. The program may, however, use a test to prevent actual computation with zeros. Such unsophisticated approaches become more efficient, for a given problem, if the sparsity is decreased by reducing the bandwidth of the coefficient matrix. As noted previously, the bandwidth depends on the way in which the nodes are

---

[†] Where the partitioning is arbitrary. If the partitions are positioned so that the matrix elements become grouped according to some chosen criteria, the partitioning is said to be *conditional*. Reordering of the equation set may be necessary to allow the desired grouping.

[‡] That is, the *profile* is ragged.

numbered. In simple problems the sequence that will yield minimum bandwidth is usually obvious, but this may not be so in more complex problems. In such cases, subroutines that automatically renumber the nodes can be used to reduce the bandwidth (see Section 10.4.4).

For large problems, it becomes increasingly important to exploit the sparse nature of the matrix in order to reduce storage and computation. Mathematically speaking, sparse matrix algorithms "embody a very important idea: that the *graph* of a matrix is the key to its structure" [2].

Profile elimination methods are a relatively simple type of sparse matrix procedure; these take advantage of the detailed shape of the band profile.[†] Each row of the coefficient matrix is associated with markers that give the location numbers of the leftmost and rightmost nonzero elements, with the intermediate elements being considered as nonzero.

More complex are those procedures that exclude zero elements from storage since a record of the location of all nonzero coefficients must then be stored. Methods by which this can be accomplished are detailed in [20] and [21]. In the integer-string method, a string of integers giving the location numbers of the nonzero elements is kept for each row of the matrix. Another approach uses a Boolean string, sometimes referred to as a binary mask, for each row. This is a sequence of binary digits corresponding to the elements of the matrix row, with a 1 signifying a nonzero entry and a 0 denoting a zero element.

A problem that must be considered in sparse matrix methods is that some of the elements in the coefficient matrix, originally calculated as zero, become nonzero as the elimination proceeds. This is known as *fill-in*; and it is desirable that this be reduced as far as possible since each additional nonzero element created means additional computation subsequently. One approach that often reduces fill-in considerably is based on automatic reassignment of the node numbers. In the sparse Gauss elimination of [22] fill-in terms are not retained if less than a predetermined magnitude, which is another approach.

Considerable ingenuity in avoiding the unnecessary storage or computational use of zeros in the band is evident in the literature. Among the many schemes that have been proposed are the wavefront or frontal method of Irons (Section 10.2.4) and the sparse Cholesky factorization approach of Jennings and Tuff [23]. Other procedures and additional details on sparse matrices are given in [24–32].

Although housekeeping in a sparse algorithm may itself require significant core storage and involve appreciable computation, it appears from the literature that there is generally a net reduction in overall solution cost

---

[†] The boundary on either side of the band, outside of which lie only zero entries.

compared to an equivalent nonsparse algorithm, for larger problems at least. For smaller equation systems there may not be a significant advantage. (Sophistication does not usually yield benefits with small systems. For example, in discussing band elimination schemes, Birkhoff and Fix [2] suggest that "the gain in efficiency which can be achieved by using sophisticated schemes is seldom very great unless more than 2000 unknowns are involved.")

For large and complex problems, program strategy should be matched to the computer hardware and the operating system if maximum efficiency is to be achieved [33].

10.2.7 Iterative Improvement

The accuracy of a solution obtained from a direct method can be significantly increased with relatively little additional computation[†] by *iterative refinement or improvement*. Let $\bar{X}$ be the solution obtained by the direct method. Using double precision arithmetic, the residual $\mathbf{R}$ is calculated from

$$\mathbf{R} = \mathbf{B} - \mathbf{A}\bar{\mathbf{X}} \qquad (10.21)$$

A new system matrix equation

$$\mathbf{A}\mathbf{Y} = \mathbf{R}, \qquad (10.22)$$

is now defined and solved for $\mathbf{Y}$, using the triangular matrices from the decomposition of $\mathbf{A}$ in the direct method. The improved solution is then taken to be

$$\mathbf{X} = \bar{\mathbf{X}} + \mathbf{Y}. \qquad (10.23)$$

If $\bar{X}$ and $\mathbf{Y}$ were computed exactly, $\mathbf{X}$ would satisfy Eq. (10.1) exactly since from Eqs. (10.21)–(10.23)

$$\mathbf{A}\mathbf{X} = \mathbf{A}\bar{\mathbf{X}} + \mathbf{A}\mathbf{Y} = \mathbf{B} - \mathbf{R} + \mathbf{R} = \mathbf{B}. \qquad (10.24)$$

Since $\bar{X}$ and $\mathbf{Y}$ are not exact, $\mathbf{X}$ is also not exact but is more accurate than $\bar{X}$. The process can be repeated in an attempt to further increase the accuracy of the solution.

If the elements of $\mathbf{Y}$ are small in magnitude relative to the corresponding elements of $\bar{X}$ this suggests that the system Eq. (10.1) is well conditioned and that $\bar{X}$ is an accurate solution. This, however, is not *necessarily* true [34]. If $\mathbf{Y}$ is not small, this implies that the matrix $\mathbf{A}$ is ill conditioned, although $\mathbf{X}$ will usually still converge, under repeated iterative improvements, to a more accurate value if the condition number of $\mathbf{A}$ is not too large. For further information on iterative improvement, see [34].

[†] Although the storage requirements are increased since $\mathbf{A}$ and its triangular matrices need to be preserved.

10.2.8   Accuracy

In the selection and application of a solution procedure for a set of linear algebraic equations, the storage requirement[†] and computation time are important considerations but accuracy must not be overlooked. The number of computational operations in the method will influence the accuracy of the solution, through truncation or roundoff errors, as well as determining the computational time and cost.

At this point, it is worth reviewing how computational errors develop. A computer typically carries out arithmetic operations using a predetermined word length for each number. Storage of the numbers is commonly in a nondecimal form (e.g., binary, octal), but for the purposes of explanation a decimal form will be assumed. Suppose that the word length of the computer is 10 significant digits and that two of the numbers to be manipulated are 685378.9879 and 437896.4879. Simple addition of the two numbers gives 1123275.4758, but since only 10 digits can be stored, the computer records the sum as 1123275.475 or 1123275.476, according to whether the truncating or rounding option of the system has been selected. In either case, there is a loss of essential information and the error that has been introduced will increase in compound fashion in subsequent operations. Suppose that the next number 111111.1111 is to be added to the previous sum. Adding 111111.1111 to 1123275.475 gives 1234386.586, for both truncation and rounding. Similarly, adding 111111.1111 to 1123275.476 yields 1234386.587. To obtain the exact value of the sum, 111111.1111 must be added to the 11-digit form of the previous sum, which yields 1234386.5869. It will be noted that the rounding option has given an approximate result closer to the exact value than the truncating option but *both* are in error. With truncation, the tenth significant digit is in error almost by one unit. Subsequent addition, subtraction, multiplication, or division operations will cause errors successively in the ninth digit, the eighth, and so on. With roundoff, the same process will occur but at a slower rate. In general, therefore, a computed result will be in error by an amount that depends upon the word length and the number and type of operations performed. For a given set of operations, the error can be reduced by using multiple words[‡] or a longer word length. For a given word length, it can be reduced by choosing a procedure involving fewer operations.

It will be noted that a dramatic increase in error can occur when two almost equal numbers are subtracted from one another during a computational sequence. If only the last two of the digits in the result, for example,

---

[†] Core storage is the common limitation but peripheral or auxiliary storage (generally disk) can also be a limitation.

[‡] As in double-precision arithmetic.

were significant, subsequent computational errors could soon swamp this two-digit information completely. Thus, for a fixed word length, the relative magnitudes of the numbers being manipulated must also be taken into account, in addition to the number of operations and their type, when determining the rate at which errors increase. To illustrate, if the matrix **A** is ill conditioned, the relative magnitudes of the coefficients are such that errors rapidly accumulate in the solution process for **X**.

For a given direct solution method, there are two ways in which the accuracy of the solution **X** can be increased. Double-precision arithmetic can be used throughout, but this approximately doubles the core storage requirement and is thus impractical for larger problems. The execution time will also increase with double precision. Alternatively, iterative improvement can be used, provided the elements of the matrix **A** and its triangular factors are stored[†] during the elimination or decomposition process. This method has the advantage that a considerable increase in accuracy can be achieved for a relatively small increase in computation time and some increase in storage if the coefficient matrix is well conditioned.

For large problems, iterative improvement is the obvious choice if improved accuracy is desired. For smaller problems, either double-precision or iterative improvement can be used. In such cases numerical experiments [3, 4] suggest that double-precision arithmetic is the easiest way to gain accuracy without excessive execution times.

10.3   ITERATIVE METHODS

Compared with the direct methods, the iterative methods have the advantages that (a) they are generally simpler to program, (b) can cope efficiently with matrix sparsity by storing and processing only nonzero coefficients, and (c) have lower core storage requirements. Convergence is rapid for iterative methods when the diagonal terms of the coefficient matrix are dominant but for ill-conditioned problems convergence may be very slow. When iterative methods are used, assembly by nodes can be advantageous (Section 6.3.3). Iterative methods are especially suitable for finite element formulations in which the assembly of the system matrix equation and its solution are meshed (Section 3.3.8), thus allowing a further economy in core storage. For the very largest problems, therefore, where core storage is an inevitable limitation, iterative methods would appear to be preferable. There are programs designed for such problems, however, which use direct methods [19, 32].

[†] Initially in core, with subsequent transfer to disk.

It is commonly accepted that, with some exceptions, the direct methods require less computation time than the iterative procedures. This however is an oversimplification since the number of computational operations depends not only on the method but also on the problem type and problem size.[†] For a given class of problems, examination of the number of computational operations for various direct and iterative methods shows the operation count to typically depend upon some power k of the problem size [2, 35]. Assessing the total operation counts for the iterative methods as the count per iteration cycle times the number of iterations needed for accuracy equivalent to the direct methods, allows a computation time comparison of the two approaches to be made. When this is done, there appears to be an overlap in the value of the index k for the direct and iterative methods. Birkhoff and Fix [2], when considering total operation counts in second-order linear source problems show that the index k is less for the successive over-relaxation (SOR) iterative method than for standard band elimination. For such second-order problems, these authors conclude that although band elimination requires less computing time than SOR for the smaller size problem; the situation will be reversed when the problem becomes sufficiently large. Numerical experiments [36], in fact, showed this to be the case. Such considerations led to the conclusion that

> iterative methods are more advantageous than direct methods for most elliptic problems when sufficiently many unknowns are involved. Thus, they become more advantageous for large linear source problems in two or three dimensions, and for sufficiently large fourth-order three-dimensional problems. A lone exception (among the cases tested), was provided by linear fourth-order plane problems like those involving the biharmonic equation, for which there seems to be no cross-over point above which iterative methods become more efficient [2].

This assessment is qualified in the same reference by the remarks, that

> Although the preceding conclusions are supported by the experimental tests ... they must still be taken with reservations. In the first place, they are surely to some extent problem-dependent and the experimental tests ... did not cover a wide range of problems. More importantly, it is typically most effective to use a combination of direct and iterative methods in solving large linear elliptic problems, and such combinations were not studied [in the experimental work cited].

[†] As specified, for example, by the size of the coefficient matrix.

10.3.1   Jacobi and Gauss–Seidel Iterative Methods

The two classic iterative methods, the Jacobi and Gauss–Seidel algorithms have the following virtues:

(1)   simplicity;
(2)   both methods always converge[†] if the coefficient matrix **A** is symmetric and positive-definite;
(3)   core storage requirements are smaller than with Gauss elimination.

Unfortunately, these advantages are outweighed by relatively slow convergence. Compared to other iterative methods described subsequently, the Jacobi and Gauss–Seidel procedures require a larger number of iterations to achieve the same accuracy. Thus, for the larger problems, the Jacobi and Gauss–Seidel methods are quite unsuited. For smaller problems, direct methods can achieve an equivalent level of accuracy with a great deal less computation. For example, on a banded system of 100 equations, the better of two Gauss–Seidel subroutines required 25 times the execution time of the fastest direct equation solver to obtain a solution of substantially inferior accuracy [4]. Both the Gauss–Seidel procedure and the direct solver used in this test were designed for sparse matrices and both retained the coefficients of **A** necessary for solution wholly in core.

From the above, it is seen that the Jacobi and Gauss–Seidel methods should be regarded as of historical importance rather than of utilitarian value.

10.3.2   Point Iterative Methods

In iterative methods, generally, a set of approximate values for the elements x_i in the unknown matrix **X** [Eq. (10.1)] are used as input for a procedure that generates a *new* set of approximate values for these elements. Commencing with an initially assumed set of values for the x_i, the iteration process is continued until the respective differences between the last two sets of approximate values is less than some preassigned value.

In a *point iterative* method, at the kth level of iteration, the approximate solution x_i^k for the unknown x_i is determined explicitly from an equation involving the approximate values of x_i from previous levels of iteration. In the early application of these methods, each x_i was related uniquely to some point in the region of the physical problem and the description of such methods as point iterative (implying that an improved solution was obtained successively at each point) was understandable. It is not such an apt description today, when in a higher order finite element formulation several variables x_i may belong to one point.

[†] Gauss–Seidel faster than Jacobi.

A general *linear iteration* for the set of equations given in Eq. (10.1) can be defined in the form

$$\mathbf{X}^k = \mathbf{G}_k \mathbf{X}^{k-1} + \mathbf{R}_k,$$ (10.25)

where $\mathbf{X}^k$, $\mathbf{X}^{k-1}$ are the approximations for $\mathbf{X}$ at the kth and $(k - 1)$th levels of iteration, respectively, $\mathbf{G}_k$ is a matrix depending on $\mathbf{A}$ and $\mathbf{B}$, and $\mathbf{R}_k$ is a column vector. Both $\mathbf{G}_k$ and $\mathbf{R}_k$, as the subscript indicates, are, in general, different at each level of iteration k.

In the limiting case when $k \to \infty$, $\mathbf{X}^k$ converges to the exact solution

$$\mathbf{X} = \mathbf{A}^{-1}\mathbf{B},$$ (10.26)

and the iteration equation (10.25) becomes by substitution from Eq. (10.26)

$$\mathbf{A}^{-1}\mathbf{B} = \mathbf{G}_k \mathbf{A}^{-1}\mathbf{B} + \mathbf{R}_k.$$ (10.27a)

From Eq. (10.27a), $\mathbf{R}_k$ is obtained as

$$\mathbf{R}_k = (\mathbf{I} - \mathbf{G}_k)\mathbf{A}^{-1}\mathbf{B}.$$ (10.27b)

Satisfaction of Eq. (10.27b) is thus required as a consistency condition on Eq. (10.25). Equation (10.27b) can also be written as

$$\mathbf{R}_k = \mathbf{M}_k \mathbf{B},$$ (10.27c)

where $\mathbf{M}_k$ is defined as

$$\mathbf{M}_k = (\mathbf{I} - \mathbf{G}_k)\mathbf{A}^{-1}.$$ (10.28)

The general iterative scheme can now be written in the form

$$\mathbf{X}^k = \mathbf{G}_k \mathbf{X}^{k-1} + \mathbf{M}_k \mathbf{B},$$ (10.29)

where $\mathbf{R}_k$, as can be seen from Eq. (10.25), has been replaced by $\mathbf{M}_k \mathbf{B}$.

The various linear point iterative methods differ in their specification of the iteration matrices $\mathbf{G}_k$ and $\mathbf{M}_k$ and will be examined on this basis. The derivations of the methods and the accompanying limitations on convergence will not be considered, but can be found in [37–41].

Several of the methods use a partitioning of $\mathbf{A}$ into

$$\mathbf{A} = \mathbf{L} + \mathbf{D} + \mathbf{U},$$ (10.30)

where $\mathbf{L}$ is a strictly[†] lower triangular matrix, $\mathbf{D}$ is the diagonal matrix, and $\mathbf{U}$ is a strictly upper triangular matrix.

[†] Strictly means without the diagonal elements.

10.3.2.1   Jacobi Method

In this method, the iteration matrices $\mathbf{G}_k$ and $\mathbf{M}_k$ of Eq. (10.29) are the same for all levels of iteration k, and are given by

$$\mathbf{G} = -\mathbf{D}^{-1}(\mathbf{L} + \mathbf{U}), \tag{10.31a}$$

$$\mathbf{M} = \mathbf{D}^{-1}. \tag{10.31b}$$

In algebraic form, the iterative scheme can be written, from Eqs. (10.29), (10.31a), and (10.31b), as

$$x_i^{\,k} = \left[\sum_{j=1}^{n} g_{ij} x_j^{k-1} \right] + d_i, \tag{10.32}$$

where x_i, g_{ij}, and d_i are the elements of $\mathbf{X}$, $\mathbf{G}$, and $\mathbf{D}^{-1}\mathbf{B}$, respectively, and where the summation limit n is the total number of elements in $\mathbf{X}$.

10.3.2.2   Gauss–Seidel Method

For this iterative procedure, the matrices $\mathbf{G}_k$ and $\mathbf{M}_k$ are constant for all levels of iteration k and are, respectively, given by

$$\mathbf{G} = -(\mathbf{L} + \mathbf{D})^{-1}\mathbf{U}, \tag{10.33a}$$

$$\mathbf{M} = (\mathbf{L} + \mathbf{D})^{-1}. \tag{10.33b}$$

In algebraic form, the iterative algorithm becomes from Eqs. (10.29), (10.33a), and (10.33b)

$$x_i^{\,k} = \sum_{j=1}^{i-1} g_{ij} x_j^{k} + \sum_{j=i+1}^{n} g_{ij} x_j^{k-1} + d_i, \tag{10.34}$$

where the notation is the same as in Section 10.3.2.1.

10.3.2.3   Successive Overrelaxation (SOR)

This method can be regarded as an extension of the Gauss–Seidel procedure, which uses an acceleration of the convergence. The iteration matrices $\mathbf{G}_k$ and $\mathbf{M}_k$ are again constant for all levels of iteration and are, respectively, given by

$$\mathbf{G} = (\mathbf{D} + \omega\mathbf{L})^{-1}[(1 - \omega)\mathbf{D} - \omega\mathbf{U}], \tag{10.35a}$$

$$\mathbf{M} = \omega(\mathbf{D} + \omega\mathbf{L})^{-1}, \tag{10.35b}$$

where ω is the relaxation parameter or overrelaxation factor. In algebraic form, the iterative scheme becomes from Eqs. (10.29), (10.35a), and (10.35b)

$$x_i^{\,k} = (1 - \omega)x_i^{k-1} + \omega\left\{ \sum_{j=1}^{i-1} g_{ij} x_j^{k} + \sum_{j=i+1}^{n} g_{ij} x_j^{k-1} + d_i \right\}, \tag{10.36}$$

where the previously defined notation has been used.

For a symmetric, positive-definite matrix $\mathbf{A}$, the method is convergent [37] at a rate dependent on the value of the relaxation parameter ω. If the matrix $\mathbf{A}$ has a certain characteristic, *Property A*, it can be shown [39, 41] that the optimal value of ω maximizing the convergence rate can be determined directly from the elements of $\mathbf{A}$. Although matrices with Property A do not often result from finite element formulations, SOR has generally been found to give excellent results. Sometimes, a near-optimal value of ω can be selected on the basis of previous experience with similar problems. For example, "selection of $1.85 < \omega < 1.92$ provides good convergence behavior for well-posed plane-stress problems in elasticity applications" [42]. Where there is no such guide to the choice of ω, methods available [43, 44] that allow the determination of a near optimal ω.

The SOR iterative method, although still a simple procedure to program, converges much faster than the Gauss–Seidel method. When an iterative method is required, the SOR technique can be recommended as a first choice, except where the region is regular as the following indicates:

> For very irregular regions, there is considerable support in the view that the SOR method is by far the simplest and best method to program, requiring just one vector storage in use. However, for fairly regular domains, some improvement in the convergence of the method must be sought, otherwise computation times will tend to become prohibitive [22].

If the problem is so large that the computational cost of even SOR is too high, then recourse must be had to one of the methods described subsequently. For certain problems, the block iterative versions of SOR converge more rapidly than point SOR. Semi-iterative methods are also sometimes superior to the SOR procedure.

10.3.2.4   Gradient Methods

The basis for this class of iterative method is the fact that the solution $\mathbf{X}$ to Eq. (10.1) also gives a minimum value to the quadratic function

$$F = \mathbf{X}^\mathrm{T}\mathbf{A}\mathbf{X} - 2\mathbf{B}^\mathrm{T}\mathbf{X}. \tag{10.37}$$

Equation (10.37) can be used to define a family of similar ellipsoids whose common center corresponds to the minimum condition. An approximate solution $\mathbf{X}^k$ corresponds to a point on the surface of a particular ellipsoid. An *iterative gradient method* consists of successive steps from a larger to a smaller ellipsoid, with the point corresponding to the approximate solution moving progressively toward the common center. The various gradient methods differ in their choice of direction for each step.

In the *steepest descent* gradient method, the iteration step is along the inward normal to the ellipsoid and it can be shown [43, 45] that this leads to the iteration defined by Eq. (10.29) with

$$\mathbf{G}_k = \mathbf{I} - v_{k-1}\mathbf{A}, \tag{10.38a}$$

$$\mathbf{M}_k = v_{k-1}, \tag{10.38b}$$

where

$$v_{k-1} = (\mathbf{R}^{k-1})^T\mathbf{R}^{k-1}/(\mathbf{R}^{k-1})^T\mathbf{A}\mathbf{R}^{k-1}, \tag{10.39}$$

and

$$\mathbf{R}^{k-1} = \mathbf{B} - \mathbf{A}\mathbf{X}^{k-1}. \tag{10.40}$$

Convergence is relatively slow and the method is not recommended for practical use.

The *conjugate gradient method* requires each iterative step to consist of two substeps, the first along the inward normal and the second parallel to the previous iterative step, which improves the convergence rate. Although the basic method is useful [46, 47] because of the small storage required, it is the preconditioned form of the method [22] that is the more promising approach. The *conjugate Newton* method [48] which is algebraically similar to the conjugate gradient technique, uses a considerable amount of Gauss reduction in addition to the iteration, and this type of formulation would appear to be promising.

10.3.2.5   Richardson's Method

This method is based on the iterative scheme defined by Eqs. (10.29), (10.38a), and (10.38b) but uses a relaxation factor v_{k-1} selected in a different manner. There are several variants of the method, but each determines v_{k-1} by requiring that the deviation between an approximate value $\mathbf{X}^k$ and the exact solution $\mathbf{A}^{-1}\mathbf{B}$ be small in some sense. Since the SOR technique converges more rapidly than Richardson's method and requires less storage [43], its use is to be preferred.

10.3.2.6   Semi-Iterative Methods

Consider the general iteration of Eq. (10.29) in the form

$$\mathbf{X}^{k+1} = \mathbf{G}\mathbf{X}^k + \mathbf{M}\mathbf{B}. \tag{10.41}$$

At a given level of iteration N, a new approximation to the solution $\mathbf{X} = \mathbf{A}^{-1}\mathbf{B}$ denoted by $\mathbf{Y}^N$, can be constructed as

$$\mathbf{Y}^N = \sum_{k=0}^{N} \beta_{N,k}\mathbf{X}^k, \tag{10.42}$$

where the coefficients $\beta_{N,k}$ are required to satisfy

$$\sum_{k=0}^{N} \beta_{N,k} = 1, \tag{10.43}$$

and to minimize, in some sense, the error matrix

$$\mathbf{E}^N = \mathbf{Y}^N - \mathbf{X}. \tag{10.44}$$

Providing the sequence $\mathbf{X}^1, \mathbf{X}^2, \ldots, \mathbf{X}^k, \ldots$ is convergent, Eq. (10.42) with the aid of Eq. (10.43) shows that $\mathbf{Y}^N \to \mathbf{X}^N \to \mathbf{X}$ as $N \to \alpha$. The relationships [Eqs. (10.41)–(10.44)] define a semi-iterative process. With an appropriate choice of the basic iteration method [Eq. (10.41)] it is possible to derive from the preceding equations a recurrence relationship involving $\mathbf{Y}^N$, $\mathbf{Y}^{N-1}$, $\mathbf{Y}^{N-2}$ but not any of the $\mathbf{X}^k$ iterants of the basic method. This new relationship is then used to calculate successive iterates $\mathbf{Y}^N$ commencing from an initially assumed value $\mathbf{Y}^0$.

If the matrix $\mathbf{A}$ has certain properties and can be partitioned in a particular way, the cyclic Chebyshev semi-iterative method [37, 43, 49, 50] can be developed. When considered in relation to the other point SOR methods, Birkhoff and Fix [2] believe "the semi-iterative multiline cyclic Chebyshev" method to be "the one to recommend for general use." Another semi-iterative method, which appears to be of considerable promise, is the semi-iterative symmetric overrelaxation (SSOR) introduced by Sheldon [51, 52].

10.3.2.7   Preconditioning

It has been shown that the convergence rates for the basic iterative methods, with symmetric positive-definite matrices, depend inversely on the *P-condition number* of the coefficient matrix $\mathbf{A}$ [53]. One way of making this number as small as possible is to transform Eq. (10.1) by multiplication with a suitable nonsingular matrix $\mathbf{Q}$ [22, 53] defined by

$$\mathbf{Q} = (\mathbf{I} - \omega\mathbf{L})^{-1}, \tag{10.45}$$

where $\mathbf{L}$ is obtained from the decomposition of $\mathbf{A}$ into

$$\mathbf{A} = \mathbf{I} - \mathbf{L} - \mathbf{L}^\mathrm{T}. \tag{10.46}$$

It can be shown that the system matrix equation (10.1) transforms into an equivalent system given by

$$\mathbf{BY} = \mathbf{D}. \tag{10.47}$$

The P-condition number of $\mathbf{B}$ can be minimized by properly choosing the acceleration factor ω, thus maximizing the convergence rate of the iterative process chosen for the solution of Eq. (10.47). The extra computation of the preconditioning process is increasingly outweighed by the faster convergence rate in the solution of Eq. (10.47) as the size of the problem increases [22]. It is suggested in the reference cited that the conjugate gradient method when used with preconditioning becomes "an extremely attractive method."

10.3.3   Group and Block Iterative Methods

In the point iterative methods, at each level of iteration, a new approximate solution is determined separately for each unknown x_i. Moreover, the value of each iterant x_i^k at the kth level of iteration is obtained from an explicit relation involving previously computed approximate values of x_i.

In group iterative methods, new approximate values for a group of unknowns $\mathbf{X}_g$ in the matrix $\mathbf{X}$ are determined at the same time. Such methods are implicit because the new approximate values for the unknowns in a group are determined at the kth level of iteration by simultaneous solution of a system of equations involving the $\mathbf{X}_g^k$ and the previously computed values of the $\mathbf{X}_g$ from earlier levels of iteration.

If the groups are formed by simple partitioning of the vector $\mathbf{X}$, they are known as *blocks*. The group iterative method based on these partitions is a *block iterative method*. Any desired grouping of the elements of $\mathbf{X}$ can be obtained by reordering the equations in the system [Eq. (10.1)] and then partitioning appropriately. Equivalently, the reordering can be accomplished by renumbering the nodes of the finite element mesh, with the desired grouping then being obtained by partitioning. Since group iterative methods can be transformed in such ways to block methods, only the latter will be considered in the following.

For any point iterative method, an analogous block iterative method can be obtained, by first partitioning $\mathbf{X}$ into blocks, secondly, partitioning $\mathbf{A}$ and $\mathbf{B}$ in corresponding fashion, and, thirdly, constructing the iterative algorithm in the same form as for the point iteration but now replacing the matrix elements by blocks. It should be noted that an algebraic reciprocal operation in the point method becomes a matrix inversion operation in the block method. Thus, the block Gauss–Seidel iteration can be written as

$$\mathbf{X}_i^k = \mathbf{A}_{ii}^{-1}\left[\mathbf{B}_i - \sum_{j=1}^{i-1} \mathbf{A}_{ij}\mathbf{X}_j^k - \sum_{j=i+1}^{N} \mathbf{A}_{ij}\mathbf{X}_j^{k-1} \right], \tag{10.48}$$

where $\mathbf{X}$ has been partitioned into N blocks $\mathbf{X}_i$ with $\mathbf{A}$ and $\mathbf{B}$ correspondingly partitioned into the blocks $\mathbf{A}_{ij}$ and $\mathbf{B}_i$.

When the coefficient matrix $\mathbf{A}$ is banded, some of the $\mathbf{A}_{ij}$ in Eq. (10.48) may contain only zero elements. Computation will be reduced if the node numbering and the partitioning is such that the $\mathbf{X}_j$ in Eq. (10.48) are the smallest possible number of blocks that are adjacent to, and near to, the block X_i. If the domain of the problem is regular, the nodes can be chosen on a regular mesh and numbered in sequence by rows or by columns, giving a coefficient matrix A that will be block tridiagonal.[†] Then, partitioning this matrix in the obvious manner for a block iterative method will lead to the blocks X_j in Eq. (10.48) being adjacent to the block X_i. Moreover, it will be found that in this case A satisfies block Property A [22, 45], so that SOR can be used in block iterative form, with the optimal value of ω being theoretically predicted. For such regular problems, the block SOR procedure converges faster than the block Gauss–Seidel method, which in turn is faster than the block Jacobi iterative scheme.

When comparing block methods against the corresponding point methods, the additional computation in solving the implicit equations [for example, Eq. (10.48)] in the block scheme must be taken into account. Whether there is a net reduction in computation time in going from a point method to the analogous block method depends upon the problem being considered. For regular domains, the block method generally is faster than its corresponding point iterative method. For irregular domains, the block method may be faster than the point method, depending on the problem.

The recently developed alternating direction implicit (ADI) methods can be related to block processes but have significant differences. For certain problems, ADI methods can be much faster than SOR procedures, but such model problems are exceptional [2].

Since the matrices arising from finite element formulations do not often have the block tridiagonal and Property A structure that give block iterative methods their advantage, it does not appear that these methods will find much application in this field. Further information on block methods will be found in Refs. [37, 39, 43].

10.4 SPECIAL TECHNIQUES TO AID EQUATION SOLVING

The preceding sections have described the most commonly used methods of solving systems of linear equations. In finite element applications, a persistent problem has been the inadequacy of computers in relation to problem size. As the capability of computers has increased, ever larger problems have been formulated so that computer capability still continues to be a limitation. The literature contains many references to techniques for reducing the effective size of a problem, such as partitioning, condensation, substructuring, and renumbering. Although particularly useful for

---

[†] A matrix is block tridiagonal if it can be partitioned so that the nonzero blocks are on the diagonal, the adjacent subdiagonal, and the adjacent superdiagonal, and provided that the diagonal blocks (but not necessarily the sub- and superdiagonal blocks) are square submatrices.

very large problems, these methods often given substantial saving of computer time for smaller problems. A selection of these approaches is presented in the next sections and others may be found in the literature.

10.4.1 Conditional Partitioning

As indicated earlier, the system matrix equation can either be partitioned arbitrarily or conditionally. Various criteria for the partitioning are available. For example, in the large structural analysis program NASTRAN an "important aid to efficiency" [32] is the partitioning of the vector X of Eq. (10.1) into X_a and X_o, where X_a contains those unknowns which are excessively coupled. It is shown in the reference cited, that X_a is obtained by solution of an equation of the form

$$PX_a = Q, \tag{10.49}$$

and X_o by solution of

$$X_o = X_o + G_o X_a. \tag{10.50}$$

Although the matrices P, G_o, Q, and $X_o{}^\circ$ must be computed from the submatrices in Eq. (10.1), the linear equations to be finally solved [Eqs. (10.49) and (10.50)] are smaller in size than the original equation [Eq. (10.1)]. An added advantage is the increase in efficiency already noted.

Conditional partitioning can also be used to eliminate those variables that are prescribed in the following way. The system matrix equation (10.1) is first partitioned to give

$$\begin{bmatrix} A_{11} & A_{12} \\ A_{21} & A_{22} \end{bmatrix} \begin{bmatrix} X_1 \\ X_2 \end{bmatrix} = \begin{bmatrix} B_1 \\ B_2 \end{bmatrix}, \tag{10.51}$$

where the submatrix X_1 contains only those nodal values that have been prescribed. The necessary grouping can be achieved by reordering the equations or by a renumbering subroutine that assigns sequential numbers to the prescribed nodes. With this partitioning, it will be found that:

A_{11} is the identity matrix I;
A_{12} is a rectangular matrix with zero elements;
A_{21} is a sparse rectangular matrix;
A_{22} is a symmetric square matrix.

Equation (10.51) represents the two equations

$$X_1 = B_1, \tag{10.52a}$$

$$A_{21}X_1 + A_{22}X_2 = B_2. \tag{10.52b}$$

Substitution of Eq. (10.52a) into (10.52b) yields, after some rearrangement,

$$\mathbf{A}_{22}\mathbf{X}_2 = \mathbf{B}_\mathrm{o},\tag{10.53}$$

where

$$\mathbf{B}_\mathrm{o} = \mathbf{B}_2 - \mathbf{A}_{21}\mathbf{B}_1.\tag{10.54}$$

The column matrix $\mathbf{B}_\mathrm{o}$ can be computed from known values through Eq. (10.54), allowing Eq. (10.53), which is of the same general form as Eq. (10.1), to be solved in the usual way.

If the region of the problem is regular and the nonprescribed nodes are numbered[†] so as to minimize the bandwidth, it will be found that the bandwidth of $\mathbf{A}_{22}$ is not significantly different from the bandwidth of $\mathbf{A}$. The solution of Eq. (10.53) is thus simpler than Eq. (10.1), but involves the minor penalty of computing Eq. (10.54).

Other variants of conditional partitioning into two or more partitions can be found in the literature. Substructuring, discussed in Section 10.4.3, may be regarded as a form of conditional partitioning.

10.4.2   Condensation

The technique used in the previous section to eliminate prescribed nodal values from the solution can also be used, although in a modified form, to eliminate the nodal values of interior nodes from an element matrix equation.[‡] Suppose that Eq. (10.51) were the element matrix equation for an element with nodes on its boundaries (*exterior* nodes) and in its domain (*interior* nodes). The quadrilateral shown in Fig. 10.3 might be such an element. The element matrix equation could be partitioned as shown in Eq. (10.51), by grouping the exterior nodal values into the matrix $\mathbf{X}_2$ and the interior nodal values into $\mathbf{X}_1$. The two submatrix equations comprising

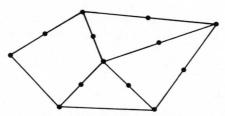

Fig. 10.3   Quadrilateral element with both exterior and interior nodes.

[†] Subsequent to sequentially numbering the prescribed nodes.
[‡] The condensation procedure can also be carried out at *element level* as shown in Chapter 5.

Eq. (10.51) can then be solved to give

$$\bar{A}X_2 = \bar{B}, \tag{10.55}$$

where

$$\bar{A} = A_{22} - A_{21}A_{11}^{-1}A_{12} \tag{10.56a}$$

and

$$\bar{B} = B_2 - A_{21}A_{11}^{-1}B_1. \tag{10.56b}$$

The element matrix equation (10.51) containing both exterior and interior nodal values has thus been condensed into Eq. (10.55), which is of similar form, but which contains only the exterior nodal variables. Applying this procedure to all the elements of the system leads, after assembly, to a system matrix equation containing only the exterior (element) nodal parameters. In contrast, assembly without condensation would have resulted in a larger system matrix equation containing both the exterior and interior (element) nodal parameters.

The above method may be used in dynamic analysis to eliminate massless degrees of freedom and thus reduce the size of the system eigenvalue matrix equation, in which case the procedure is often called *static condensation* [54]. Condensation is further described in [55–58].

10.4.3 Substructuring

For a large system matrix equation the process of substructuring can be used to condense the equation to manageable size. The region or structure is divided into two or more sections, each of which is subdivided into finite elements. The element matrix equations are assembled for each section to form the section matrix equations. By regarding a section as a large element with many interior and exterior nodes, and using the condensation process as described in the previous section, each section matrix equation can be condensed to a form involving only exterior nodal values. The condensed section matrix equations can now be assembled to form a system matrix equation of reduced size from which the section nodal vector can be obtained. These exterior values are now used as prescribed boundary values on each section, allowing the uncondensed section matrix equations to be solved for the interior nodal values.

One advantage of substructuring is that significant redesign of each section can be accomplished before the boundary loads originally determined for the section need to be recalculated. This approach is particularly useful in the design of aircraft structures. In Figs. 10.4a and 10.4b the use of substructuring for a Boeing 747 aircraft body [59] is illustrated.

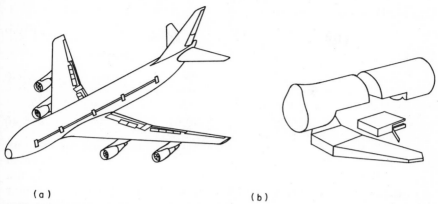

Fig. 10.4   Substructuring of Boeing 747 aircraft: (a) aircraft configuartion; (b) schematic of the substructures.

10.4.4   Node Renumbering or Relabeling

It was seen in Chapter 6 that the bandwidth of the system matrix equation is determined by the manner in which the nodes are numbered. For simple structures or regions, it is easy to label the nodes in an order that will give minimum bandwidth, but this becomes nearly impossible for larger problems. Moreover, automatic mesh generation is now commonly used; and the node numbers assigned by the mesh algorithm may give a bandwidth far from the minimum.

If a direct method of the banded type (Sections 6.1, 10.2) is being used to solve the system matrix equation, reduction in the bandwidth will result in a quicker and cheaper solution. It may even mean the difference between using an *in-core* equation solver† or having to resort to an *out-of-core* method. Where sparse matrix techniques are used in the equation solver, bandwidth reduction is generally irrelevant except in special cases.

Most of the automatic bandwith renumbering schemes allow the mesh to be initially numbered in an arbitrary way. An algorithm then renumbers the nodes to reduce the system matrix bandwidth prior to solving the system matrix equation. Often another algorithm is provided to convert the node numbers back into the original ones, after the solution has been obtained.

Automatic renumbering techniques are considered in [60–74]. A recent comparison of the various approaches [72] shows that the reverse Cuthill–McKee method [73] and the King method [68] are the best algorithms,

---

† Where the coefficient matrix **A** is stored wholly in core.

with the latter being superior for the finite element class of matrices. This comparison, however, did not include the Grooms algorithm [66]. A more recent algorithm [69] has shown excellent results when compared with the Cuthill–McKee,[†] Rosen, Akyuz–Utku, and Grooms methods.

As pointed out by Bolstad *et al.* [72], attempting to minimize the bandwidth does not necessarily result in minimal storage requirements since the ordering method influences the block size if partitioning is used. In the cited reference, these same authors also consider the interrelation of ordering and partitioning in some detail.

10.4.5 Diagnostic Routines

Commerical finite element programs usually have extensive diagnostics to detect mistakes in the input data. The processing of incorrect information can be quite expensive and, moreoever, can lead to time-consuming redesign later. For some of the useful diagnostics for equation solvers, see [47].

REFERENCES

1. E. Schrem, Computer implementation of the finite element procedures, *in* "Numerical and Computer Methods in Structural Mechanics" (S. J. Fenves, A. R. Robinson, and W. C. Schnobrich, eds.), pp. 79–121. Academic Press, New York, 1973.
2. G. Birkhoff and G. Fix, Higher order finite element methods. Tech. Rep. No. 1. Office of Naval Res. Contract N00014-67-A-0298-0015, AD-779341 (March 1974).
3. W. T. Sequi, Computer programs for the solution of systems of linear algebraic equations, *Int. J. Numer. Methods Engrg.*, **7**, No. 4, 479–490 (1973).
4. W. T. Sequi, Computer programs for the solution of systems of linear algebraic equations. NASA Contractor's Rep. CR-2173 (January 1973).
5. K-J. Bathe and E. L. Wilson, "Numerical Methods in Finite Element Analysis." Prentice-Hall, Englewood Cliffs, New Jersey, 1976.
6. J. R. Westlake, "A Handbook of Numerical Matrix Inversion and Solution of Linear Equations." Krieger, Huntington, New York, 1975.
7. Pin Tong and J. R. Rossettos, "The Finite Element Methods." MIT Press, Cambridge, Massachusetts, 1977.
8. B. M. Irons, A frontal solution program for finite element analysis, *Internat. J. Numer. Methods Engrg.* **2**, No. 1, 5–32 (1970).
9. R. J. Melosh and R. M. Bamford, Efficient solution of load-deflection equations, *Proc. ASCE, J. Struct. Div.* **94**, No. ST4, 661–676 (1969).
10. T. K. Heller, A frontal solution for finite element techniques, Rep. RD-B-N1459, Central Electricity Generating Board, Berkeley Nuclear Laboratories, England (1969).
11. R. S. Dunham and R. E. Nickell, Finite element analysis of axisymmetric solids with arbitrary loadings. Rep. 67–6. Civil Engrg. Dept. Univ. of California, Berkeley (N.T.I.S. AD-655-253) (1967).
12. D. F. Brooks and D. M. Brotton, Computer system for analysis of large frameworks, *Proc. ASCE, J. Struct. Div.* **ST-6**, 1–23 (December 1967).

[†] It was not compared with the reverse Cuthill–McKee method, however.

13. W. D. Whetstone, Computer analysis of large linear frames, *Proc. ASCE J. Struct. Div.* **ST-11**, 2401–2417 (1969).

14. G. von Fuchs and J. R. Roy, Solution of the stiffness matrix equations in ASKA/FORTRAN N Version. Rep. No. 50. Inst. für Statik und Dynamik, Univ. Stuttgart (1968).

15. G. Cantin, An equation solver of very large capacity. *Internat. J. Numer. Methods. Engrg.* **3**, No. 3, 379–388, (1971).

16. M. Svoboda and M. Sahm, Some comments on the equation block solver, *Internat. J. Numer. Methods. Engrg.* **7**, No. 2, 227–228, (1973).

17. K. J. Bathe, E. L. Wilson, and F. E. Peterson, SAP IV: A structural analysis program for static and dynamic response of linear systems, Rep. No. EERC-73-11, Earthquake Engrg. Res. Centre, Univ. of California, Berkeley (N.T.I.S. PB-221–967/3) (June 1973).

18. J. Lestingi and S. Prachuktam, A blocking technique for large scale structural analysis, *Internat. J. Comput. Struct.* **3**, No. 3, 669–714 (1973).

19. C. W. McCormick, Sparse matrix operations in NASTRAN, *Proc. 1973 Tokyo Seminar Finite Element Anal.*, pp. 611–631. Univ. of Tokyo Press, Toyko, 1973.

20. R. W. McLay, M. Kawahara, B. K. Stearns, and E. M. Buturla, Sparse matrices in finite element program development, "Theory and Practice in Finite Element Structural Analyses" (*Proc. 1973 Tokyo Seminar Finite Element Anal.*), pp. 633–650. Univ. of Tokyo Press, Tokyo, 1973.

21. U. W. Pooch and A. Nieder, A survey of indexing techniques for sparse matrices, *Comput. Surv.* **5**, No. 2, 109–133, (1973).

22. D. J. Evans, Analysis and application of sparse matrix algorithms in the finite element method, "The Mathematics of Finite Elements and Applications" (*Proc. 1972 Brunel Univ. Conf. Inst. Math. Appl.*), pp. 427–447. Academic Press, New York, 1973.

23. A. Jennings and A. D. Tuff, A direct method for the solution of large sparse symmetric simultaneous equations, *in* "Large Sparse Sets of Linear Equations" (J. K. Reid, ed.). Academic Press, New York, 1971.

24. J. K. Reid (ed.), "Large Sparse Sets of Linear Equations," Academic Press, New York, 1971.

25. D. J. Rose and R. A. Willoughby (eds.), "Sparse Matrices and Their Application." Plenum, New York, 1972.

26. F. Gustavson, W. Liniger, and R. Willoughby, Symbolic generation of an optimal Crout algorithm for sparse systems of linear equations, *J. Assoc. Comput. Machinery* **17**, No. 1, 87–109 (1970).

27. A. Jennings, A sparse matrix scheme for computer analysis of structures, *Internat. J. Comput. Math.* **2**, 1–21 (1968).

28. T. S. Chow and J. S. Kowalik, Computing with sparse matrices, *Internat. J. Numer. Methods Engrg.* **7**, No. 2, 211–223 (1973).

29. R. A. Willoughby, A survey of sparse matrix technology, *Proc. Conf. Comput. Oriented Anal. Shell Struct., Palo Alto, California, August 1971* (N.T.I.S. AD-740-547; AFFDL-TR-71-79) (June 1972).

30. R. P. Tewarson, The Crout reduction for sparse matrices, *Comput. J.* **12**, 158–159 (1969).

31. R. P. Tewarson, Computations with sparse matrices, *SIAM Rev.* **12**, No. 4, 527–543 (1970).

32. C. W. McCormick, The NASTRAN program for structural analysis, "Advances in Computational Methods in Structural Analysis and Design" (*Proc. U.S.–Jpn. Seminar on Matrix Methods Struct. Anal. Design, August 1972*), pp. 551–571. Univ. of Alabama Press, Huntsville, Alabama, 1972.

33. S. J. Fenves, Design philosophy of large interactive systems, *in* "Numerical and Computer

Methods in Structural Mechanics" (S. J. Fenves, A. R. Robinson, and W. C. Schnobrich, eds.), pp. 403–414. Academic Press, New York, 1973.

34. R. S. Martin, G. Peters, and J. H. Wilkinson, Iterative refinement of the solution of a positive definite system of equations, *Numer. Math.* **8**, 203–206, (1966).

35. J. L. Traub (ed.), "Proceedings of Symposium on Complexity of Sequential and Parallel Numerical Algorithms," p. 224. Academic Press, New York, 1973.

36. G. Fix and K. Larsen, The convergence of SOR iterations for finite element methods, *SIAM J. Numer. Anal.* **8**, 536–547, (1971).

37. R. S. Varga, "Matrix Iterative Analysis." Prentice-Hall, Englewood Cliffs, New Jersey, 1962.

38. J. F. Traub, "Iterative Methods for the Solution of Equations." Prentice-Hall, Englewood Cliffs, New Jersey, 1964.

39. D. M. Young, "Iterative Solution of Large Linear Systems." Academic Press, New York, 1971.

40. E. Wachpress, "Iterative Solution of Elliptic Systems." Prentice-Hall, Englewood Cliffs, New Jersey, 1966.

41. D. M. Young, Iterative methods for solving partial differential equations of elliptic type, *Trans. Amer. Math. Soc.* **76**, 92–111 (1954).

42. H. Martin and G. Carey, "Finite Element Analysis—Theory and Application." McGraw-Hill, New York, 1973.

43. W. F. Ames, "Numerical Methods for Partial Differential Equations." Nelson, London, 1969.

44. B. Carré, The determination of the optimum accelerating factor for successive over-relaxation, *Comput. J.* **4**, 73–78, (1961).

45. H. R. Schwarz, H. Rutishauser, and E. Stiefel, "Numerical Analysis of Symmetric Matrices." Prentice-Hall, Englewood Cliffs, New Jersey, 1973.

46. J. K. Reid, On the method of conjugate gradients for the solution of large sparse systems of linear equations, *in* "Large Sparse Systems of Linear Equations" (J. K. Reid, ed.). Academic Press, New York, 1971.

47. B. M. Irons and D. K. Y. Kan, Equation-solving algorithms for the finite element method, *in* "Numerical and Computer Methods in Structural Mechanics" (S. J. Fenves, A. R. Robinson, and W. C. Schnobrich, eds.), pp. 497–511. Academic Press, New York, 1973.

48. B. M. Irons, The conjugate Newton method, *in* Lectures to the NATO Advanced Study Institute on Finite Elements, University of Calgary, Alberta, Canada, August 1973.

49. G. Birkhoff, "Numerical Solution of Elliptic Equations." SIAM Publ., Philadelphia, Pennsylvania, 1971.

50. G. H. Golub and R. S. Varga, Chebyshev iterative methods, successive over-relaxation iterative methods, and second-order Richardson iterative methods, *Numer. Math.* **3**, 147–168 (1961).

51. J. W. Sheldon, On the numerical solution of elliptic difference equations, *Math. Tabl. Aids Comput., Nat. Res. Council, Washington* **9**, 101–112 (1955).

52. J. W. Sheldon, On the spectral norms of several iterative processes, *J. Assoc. Comput. Machinery* **6**, 494–505 (1959).

53. D. J. Evans, The use of pre-conditioning in iterative methods for solving linear equations with symmetric positive definite matrices, *J. Inst. Math. Appl.* **4**, 295–314 (1967).

54. R. W. Clough and C. A. Felippa, A refined quadrilateral element for the analysis of plate bending, *Proc. Conf. Matrix Methods Struct. Mech., 2nd, Wright-Patterson AFB, Ohio, October 1968.* (AFFDL-TR-68-150), pp. 399–440 (December 1969).

55. E. L. Wilson, The static condensation algorithm, *J. Numer. Methods Engrg.* **8**, No. 1, 198–203 (1974).

56. C. A. Felippa and R. W. Clough, The finite element method in solid mechanics, *in* "Numerical Solution of Field Problems in Continuum Physics" (*SIAM–AMS Proc.*), Vol. 2, pp. 210–252. Amer. Math. Soc., Providence, Rhode Island, 1970.

57. C. S. Desai and J. F. Abel. "Introduction to the Finite Element Method," Van Nostrand-Reinhold, Princeton, New Jersey, 1972.

58. K. H. Hueber, "The Finite Element Method for Engineers." Wiley, New York, 1975.

59. S. D. Hansen, G. L. Anderson, N. E. Connacher, and C. S. Dougherty, Analysis of the 747 aircraft wing-body intersection, *Proc. Conf. Matrix Methods Struct. Mech., 2nd, Wright-Patterson, Ohio, 15–17 October, 1968* (AFFDL-TR-68-150) (December 1969).

60. G. C. Alway and D. W. Martin, An algorithm for reducing the bandwidth of a matrix of symmetrical configuration, *Comput. J.* **8**, 264–272 (1965).

61. T. A. Akyuz and S. Utku, An automatic node-relabelling scheme for bandwidth minimization of stiffness matrices, *AIAA J.* **7**, No. 2, 380–381 (February 1969).

62. E. Cuthill, Several strategies for reducing the bandwidth of matrices, *in* "Sparse Matrices and Their Applications" (D. Rose and R. Willoughby, eds.), Plenum, New York, 1972.

63. E. Cuthill and J. McKee, Reducing the bandwidth of sparse symmetric matrices, *Proc. Nat. Conf. Assoc. Comput. Machinery, San Francisco* pp. 157–172 (1969).

64. I. Arany, W. F. Smyth, and L. Szoda, An improved method for reducing the bandwidth of sparse symmetric matrices, *in* "Information Processing 71: Proceedings of IFIP Congress." North-Holland Publ., Amsterdam, 1972.

65. J. Barlow and C. G. Marples, Comment on "An automatic node-relabelling scheme for bandwidth minimization of stiffness matrices," *AIAA J.* **7**, No. 2, 380–382 (1969).

66. H. R. Grooms, Algorithm for matrix bandwidth reduction, *J. Struct. Div. Am. Soc. Civ. Eng.* **98**, ST1, 203–214 (1972). [See also discussion: *ibid.* **98**, No. ST12, 2820–2821 (1972).]

67. R. Rosen, Matrix bandwidth minimization, *Proc. Nat. Conf. Assoc. Comp. Mach., 23rd*, pp. 585–595 (1968).

68. I. P. King, An automatic re-ordering scheme for simultaneous equations derived from network systems, *Internat. J. Numer. Methods Engrg.* **2**, No. 4, 523–533 (1970).

69. R. J. Collins, Bandwidth reduction by automatic renumbering, *Internat. J. Numer. Methods Engrg.* **6**, No. 3, 345–356 (1973).

70. K. Y. Cheng, Note on minimizing the bandwidth of sparse symmetric matrices, *Computing (Arch. Elektron. Rechnen)* **11**, 27–30 (1973).

71. E. Roberts Jr., Relabelling of Finite Element Meshes Using a Random Process. NASA TM X-2660 (October 1972).

72. J. H. Bolstad, G. K. Leaf, A. J. Lindeman, and H. G. Kaper, An empirical investigation of the reordering and data management for finite element systems of equations. Rep. ANL-8056, Argonne Nat. Lab., Argonne, Illinois (September 1973).

73. J. A. George, Computer Implementation of the Finite Element Method, Ph.D. Thesis, STAN-CS-71-208. Stanford Univ., California (1971).

74. G. Dhatt and G. Akhras, An Automatic relabeling algorithm for bandwidth minimization, *Proc. Canad. Congr. Appl. Mech., 5th, University of New Brunswick, 26–30 May 1975*, pp. 690–691.

11

SELECTED APPLICATIONS OF THE FINITE ELEMENT METHOD

In the earlier chapters of this book, the application of the finite element method to equilibrium problems was demonstrated. For simplicity, the Laplace equation was chosen to illustrate the procedure on several occasions. In the present chapter, the application of the finite element method to other problems is considered.

11.1 SOLID MECHANICS—PLANE STRAIN AND PLANE STRESS

This section considers only the simplest elasticity problem–plane stress and plane strain. For a more complete treatment of the finite element method in solid mechanics, reference should be made to finite element texts oriented to this area [1–9].

In plane strain formulation, the strain normal to the plane of loading is assumed zero. Thus the displacements and stresses in a long straight retaining wall (if uniformly loaded, lengthwise) can be derived by considering a transverse slice of the wall and assuming that the strain is zero normal to the slice. In a plane stress problem, the solid is a *thin* plate and stresses normal to the plate are assumed zero.

The finite element formulation is the same for both cases except that the elements of the elastic constant matrix are different in the two approaches. For simplicity, linear triangular elements are used in the following. However, as seen in the earlier chapters, the analysis can be modified for higher order approximations.

Consider the two-dimensional domain D enclosed by the boundary curve S shown in Fig. 11.1a as the plane view of an elastic body of thickness t. On a portion A_T of the circumferential boundary surface a distributed loading is applied, which at any point can be expressed as the applied force per unit area **T**. The loading is uniform across the thickness t and the portion S_T of the boundary curve corresponds to the portion A_T of circumferential area. If the x and y components of **T** at a point are denoted by T_x and T_y, respectively, the *surface* stress matrix **T** at that point may be written as

$$\mathbf{T} = \begin{bmatrix} T_x \\ T_y \end{bmatrix}. \tag{11.1}$$

Let the body force per unit volume **P** and the displacement vector **U** at a point be similarly given in terms of their x and y components, respectively, as

$$\mathbf{P} = \begin{bmatrix} p_x \\ p_y \end{bmatrix}, \qquad \mathbf{U} = \begin{bmatrix} u \\ v \end{bmatrix}. \tag{11.2}$$

For two-dimensional solid mechanics problems, the principle of *minimum potential energy* allows the following functional [1, 9–11] to be obtained

$$\chi = \int_V \tfrac{1}{2}\boldsymbol{\varepsilon}^{\mathsf{T}}\boldsymbol{\sigma}\, dV - \int_V \mathbf{U}^{\mathsf{T}}\mathbf{P}\, dV - \int_{A_T} \mathbf{U}^{\mathsf{T}}\mathbf{T}\, dA_T, \tag{11.3}$$

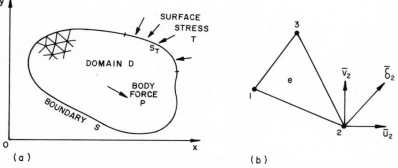

(a)   (b)

Fig. 11.1   Two-dimensional elastic body subdivided into finite elements: (a) two-dimensional region; (b) typical element e.

where ε, σ are the strain and stress vectors, respectively, given by

$$\varepsilon = \begin{bmatrix} \varepsilon_x \\ \varepsilon_y \\ \gamma_{xy} \end{bmatrix}, \qquad \sigma = \begin{bmatrix} \sigma_x \\ \sigma_y \\ \tau_{xy} \end{bmatrix}, \qquad (11.4)$$

and where V and A_T are the volume and circumferential area on which $\mathbf{P}$ and $\mathbf{T}$ act, respectively (Fig. 11.1).

In Eqs. (11.4) ε_x, ε_y and σ_x, σ_y are the normal strain and stress in the x, y directions, respectively, γ_{xy} is the shear strain,[†] and τ_{xy} is the shear stress. For the latter, the first subscript indicates the axis that is *normal* to the shear plane and the second subscript designates the axis parallel to the shear force.

From the principle of minimum potential energy, it can be shown that the displacement field that satisfies the equilibrium (and compatibility) equations for this two-dimensional elasticity problem also minimizes the functional given in Eq. (11.3).

For the subdivided region shown in Fig. 11.1a, Eq. (11.3) can be written as

$$\chi = \sum_{e=1}^{l} t_e \int_{D_e} \tfrac{1}{2} \varepsilon^T \sigma \, dD_e - \sum_{e=1}^{l} t_e \int_{D_e} \mathbf{U}^T \mathbf{P} \, dD_e - \sum_{e=1}^{l} t_e \int_{S_{T_e}} \mathbf{U}^T \mathbf{T} \, dS_{T_e}, \quad (11.5)$$

where D_e is an element subdomain, with t_e being its thickness, and where l is the total number of elements in the system. The last right-hand side term will, of course, be nonzero only for elements along the S_T boundary. For each element, the matrices ε, σ, and $\mathbf{U}$ can be expressed in terms of the element nodal displacement vector δ^e, as shown below, allowing χ to be determined as a function of the system nodal displacement vector δ. To accomplish this, the displacement $\mathbf{U}$ within an element is first written in terms of nodal displacements, using the shape function relation

$$\mathbf{U} = \mathbf{N}^e \delta^e, \qquad (11.6)$$

where $\mathbf{N}^e$ is the shape function matrix for the element and δ^e is the vector of element nodal displacements assumed here to be in the order $(\delta_1, \delta_2, \delta_3)^T \equiv (\bar{u}_1, \bar{v}_1, \bar{u}_2, \bar{v}_2, \bar{u}_3, \bar{v}_3)^T$. For convenience, the superscript e will be omitted in the remainder of this section, where the context prevents confusion.

For the (linear) triangular elements chosen for the present problem, the shape function relations for the displacements u and v are given by

$$u = \begin{bmatrix} N_1 & N_2 & N_3 \end{bmatrix} \begin{bmatrix} \bar{u}_1 \\ \bar{u}_2 \\ \bar{u}_2 \end{bmatrix}, \qquad v = \begin{bmatrix} N_1 & N_2 & N_3 \end{bmatrix} \begin{bmatrix} \bar{v}_1 \\ \bar{v}_2 \\ \bar{v}_3 \end{bmatrix}, \qquad (11.7)$$

[†] γ_{xy} is the engineering shear strain, which relates to the strain tensor shear strain ε_{xy} by $\gamma_{xy} = 2\varepsilon_{xy}$.

where $\bar{u}_1, \bar{u}_2, \bar{u}_3$ are the x components and $\bar{v}_1, \bar{v}_2, \bar{v}_3$ are the y components, of the nodal displacements $\bar{\delta}_1, \bar{\delta}_2, \bar{\delta}_3$, respectively (see Fig. 11.1b).

Using Eqs. (11.2) and (11.7), the shape function relation in Eq. (11.6) can be written as

$$
\mathbf{U} = \begin{bmatrix} u \\ v \end{bmatrix} = \begin{bmatrix} N_1 & 0 & N_2 & 0 & N_3 & 0 \\ 0 & N_1 & 0 & N_2 & 0 & N_3 \end{bmatrix} \begin{bmatrix} \bar{u}_1 \\ \bar{v}_1 \\ \bar{u}_2 \\ \bar{v}_2 \\ \bar{u}_3 \\ \bar{v}_3 \end{bmatrix}. \tag{11.8}
$$

For two-dimensional elasticity, strains relate to displacements through the standard relations

$$
\varepsilon_x = \frac{\partial u}{\partial x}, \qquad \varepsilon_y = \frac{\partial v}{\partial y}, \qquad \gamma_{xy} = \frac{\partial u}{\partial y} + \frac{\partial v}{\partial x}. \tag{11.9}
$$

Using Eq. (11.9), the strain matrix ε can be written in the form

$$
\varepsilon = \begin{bmatrix} \varepsilon_x \\ \varepsilon_y \\ \gamma_{xy} \end{bmatrix} = \begin{bmatrix} \partial/\partial x & 0 \\ 0 & \partial/\partial y \\ \partial/\partial y & \partial/\partial x \end{bmatrix} \begin{bmatrix} u \\ v \end{bmatrix} = \begin{bmatrix} \partial/\partial x & 0 \\ 0 & \partial/\partial y \\ \partial/\partial y & \partial/\partial x \end{bmatrix} \mathbf{U}. \tag{11.10}
$$

Substitution from Eq. (11.6) then gives

$$
\varepsilon = \begin{bmatrix} \partial/\partial x & 0 \\ 0 & \partial/\partial y \\ \partial/\partial y & \partial/\partial x \end{bmatrix} \mathbf{N}\boldsymbol{\delta} = \mathbf{B}\boldsymbol{\delta}, \tag{11.11}
$$

where $\mathbf{B}$ is defined within the equation. For the three-node linear element under consideration, it can simply be shown that $\mathbf{B}$ is given by

$$
\mathbf{B} = \frac{1}{2\Delta} \begin{bmatrix} y_2 - y_3 & 0 & y_3 - y_1 & 0 & y_1 - y_2 & 0 \\ 0 & x_3 - x_2 & 0 & x_1 - x_3 & 0 & x_2 - x_1 \\ x_3 - x_2 & y_2 - y_3 & x_1 - x_3 & y_3 - y_1 & x_2 - x_1 & y_1 - y_2 \end{bmatrix}. \tag{11.12}
$$

The stress matrix σ can now be obtained in terms of δ by expressing σ in terms of ε, and then substituting for ε from Eq. (11.11). For plane stress (zero stress σ_z normal to the body), standard textbooks on elasticity give the relations

$$
\varepsilon_x = \frac{\sigma_x}{E} - \frac{v\sigma_y}{E}, \tag{11.13a}
$$

$$
\varepsilon_y = \frac{-v\sigma_x}{E} + \frac{\sigma_y}{E}, \tag{11.13b}
$$

$$\gamma_{xy} = \frac{2(1 + v)}{E}\tau_{xy}. \tag{11.13c}$$

Using Eqs. (11.13), ε can be written as

$$\varepsilon = \begin{bmatrix} \varepsilon_x \\ \varepsilon_y \\ \gamma_{xy} \end{bmatrix} = \frac{1}{E}\begin{bmatrix} 1 & -v & 0 \\ -v & 1 & 0 \\ 0 & 0 & 2(1 + v) \end{bmatrix}\begin{bmatrix} \sigma_x \\ \sigma_y \\ \gamma_{xy} \end{bmatrix}, \tag{11.14}$$

where E is Young's modulus of elasticity and v is Poisson's ratio. Solving Eq. (11.14) for σ_x, σ_y, and τ_{xy} then gives

$$\sigma = \begin{bmatrix} \sigma_x \\ \sigma_y \\ \tau_{xy} \end{bmatrix} = \frac{E}{1 - v^2}\begin{bmatrix} 1 & v & 0 \\ v & 1 & 0 \\ 0 & 0 & (1 - v)/2 \end{bmatrix}\begin{bmatrix} \varepsilon_x \\ \varepsilon_y \\ \gamma_{xy} \end{bmatrix}, \tag{11.15}$$

or simply

$$\sigma = D\varepsilon, \tag{11.16}$$

where D is defined through Eq. (11.15).

For plane strain, Eq. (11.16) is also applicable but in this case the matrix D is defined by

$$D = \frac{E(1 - v)}{(1 + v)(1 - 2v)}\begin{bmatrix} 1 & v/(1 - v) & 0 \\ v/(1 - v) & 1 & 0 \\ 0 & 0 & (1 - 2v)/2(1 - v) \end{bmatrix}. \tag{11.17}$$

In passing, it should be noted from Eqs. (11.15) and (11.17), that the matrix D is symmetric.

Substitution of Eqs. (11.6), (11.11), and (11.16) into Eq. (11.5) finally yields the functional χ in the form

$$\chi = \sum_{e=1}^{l}\chi^e = \sum_{e=1}^{l}t_e\int_{D_e}\tfrac{1}{2}\delta^T B^T D B\delta\,dD_e - \sum_{e=1}^{l}t_e\int_{D_e}\delta^T N^T P\,dD_e$$
$$- \sum_{e=1}^{l}t_e\int_{S_{T_e}}\delta^T N^T T\,dS_{T_e} \tag{11.18}$$

where the matrix D is given through Eq. (11.15) or (11.17), depending on whether the problem is plane stress or plane strain.

The element matrix equation now follows by differentiating[†] an element contribution χ^e, from Eq. (11.18), with respect to δ to give

$$\frac{\partial\chi^e}{\partial\delta} = \int_{D_e}t_e B^T D B\delta\,dD_e - \int_{D_e}t_e N^T P\,dD_e - \int_{S_{T_e}}t_e N^T T\,dS_e. \tag{11.19}$$

[†] Using Eq. (B.16a) of Appendix B.

Equation (11.19) can also be written in standard form, as

$$\frac{\partial \chi^e}{\partial \boldsymbol{\delta}} = \mathbf{k}^e \boldsymbol{\delta} - \mathbf{F}_P{}^e - \mathbf{F}_T{}^e, \tag{11.20}$$

where

$$\mathbf{k}^e = t_e \int_{D_e} \mathbf{B}^T \mathbf{D} \mathbf{B} \, dD_e, \tag{11.21a}$$

$$\mathbf{F}_P{}^e = t_e \int_{D_e} \mathbf{N}^T \mathbf{P} \, dD_e, \tag{11.21b}$$

$$\mathbf{F}_T{}^e = t_e \int_{S_T} \mathbf{N}^T \mathbf{T} \, dS_{T_e}. \tag{11.21c}$$

and where the superscript e on $\boldsymbol{\delta}$ in Eq. (11.20) (also, Eqs. (11.1), (11.18), (11.19)) has been ommitted.

The matrices in Eqs. (11.21) are constant with respect to the integration and can be taken outside the integral. The integral remaining evaluates as Δ_e, where Δ_e is the area of the triangle, so that the element stiffness matrix $\mathbf{k}^e$ becomes

$$\mathbf{k}^e = \mathbf{B}^T \mathbf{D} \mathbf{B} \Delta_e t_e. \tag{11.22}$$

Since $\boldsymbol{\delta}^e$ is a 6×1 matrix [see Eqs. (11.6) and (11.8)], $\mathbf{k}^e$ is 6×6 in size. From Eq. (11.22) the elements of $\mathbf{k}^e$ can be determined as functions of E, v, and the nodal coordinates of the element. Thus, for plane stress,

$$k_{11}^e = \frac{t_e}{4\Delta_e} \left\{ \frac{E}{1 - v^2}(y_2 - y_3)^2 + \frac{E}{2(1 + v)}(x_3 - x_2)^2 \right\}, \tag{11.23}$$

with similar expressions being obtained for the other elements of $\mathbf{k}^e$.

Substituting $\mathbf{N}^T$ from Eq. (11.8) and $\mathbf{P}$ from Eqs. (11.2) into Eq. (11.21b), allows the column vector $\mathbf{F}_P{}^e$ to be written as

$$\mathbf{F}_P{}^e = t_e \int_{D_e} \begin{bmatrix} N_1 p_x \\ N_1 p_y \\ N_2 p_x \\ N_2 p_y \\ N_3 p_x \\ N_3 p_y \end{bmatrix} dD_e. \tag{11.24}$$

If both p_x and p_y are assumed constant over an element (for example, at their average values), the integration in Eq. (11.24) can be easily carried out. Thus, for the second element of $\mathbf{F}_P{}^e$,

$$(F_P{}^e)_2 = t_e \int_{D_e} N_1 p_y \, dD_e = p_y{}^e t_e \int_{D_e} N_1 \, dD_e = \frac{t_e \Delta_e}{3} p_y{}^e. \tag{11.25}$$

where $p_y{}^e$ is the constant or average value of p_y over the element.

More generally, it can be shown that the pairs of elements in the matrix $\mathbf{F}_P{}^e$ corresponding to the node identifiers $i = 1, 2, 3$ can be written as

$$(\mathbf{F}_P{}^e)^i = \frac{t_e \Delta_e}{3} \begin{bmatrix} p_x{}^e \\ p_y{}^e \end{bmatrix}. \tag{11.26}$$

From Eq. (11.21c), the boundary load matrix $\mathbf{F}_T{}^e$ can be evaluated in a form identical to Eq. (11.26), but with T_x, T_y replacing p_x, p_y, and with the integration being over S_{T_e} instead of D_e. From this relationship, if i is a boundary node as shown in Fig. 11.2, the corresponding pair of elements in the matrix $\mathbf{F}_T{}^e$ can be obtained as

$$(\mathbf{F}_T{}^e)^i = \int_{S_{T_e}} N_i \begin{bmatrix} T_x \\ T_y \end{bmatrix} dS_{T_e} = \int_0^{L_{ij}^e} N_i \begin{bmatrix} T_x \\ T_y \end{bmatrix}^e ds, \tag{11.27}$$

where L_{ij}^e is the length of the element side along S_T. In deriving Eq. (11.27), the applied loads T_x and T_y have been assumed constant along the element boundary (for example, at their average values).

Since the trial function for this element is linear, the shape function N_i can be written (see Fig. 11.2) as the linear relation

$$N_i = \left(1 - \frac{s}{L_{ij}^e}\right), \tag{11.28}$$

along s. Substituting Eq. (11.28) into Eq. (11.27) and integrating yields $(\mathbf{F}_T{}^e)^i$ as

$$(\mathbf{F}_T{}^e)^i = \frac{L_{ij}^e}{2} \begin{bmatrix} T_x{}^e \\ T_y{}^e \end{bmatrix}. \tag{11.29}$$

For the node j of the element (see Fig. 11.2), the relation $N_j = s/L_{ij}$ replaces Eq. (11.28), but it can be easily verified that the subsequent integration yields the same expression as given on the right-hand side of Eq. (11.29). Hence, Eq. (11.29) is applicable to any node of the element that lies on the boundary S_T. For a node that is not on the boundary, the right-hand side of Eq. (11.29) should, of course, be replaced by zero. Thus, if nodes 2, 3 are on the boundary S_T but node 1 is not, the form of $\mathbf{F}_T{}^e$ will be

$$\mathbf{F}_T{}^e = L_{23}^e \begin{bmatrix} 0 \\ 0 \\ T_x{}^e \\ T_y{}^e \\ T_x{}^e \\ T_y{}^e \end{bmatrix}. \tag{11.30}$$

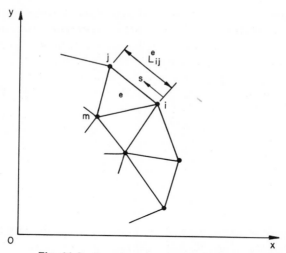

Fig. 11.2   Boundary elements of the domain.

For a problem in plane strain or plane stress, therefore, the choice of three-node triangular elements with linear trial functions allows the element matrix equations to be determined in the form of Eq. (11.20) with relative ease. Assembly of these element equations into the system matrix equation and insertion of the prescribed displacements then follows in the usual way. Solution of this equation yields the nodal displacements, with the stresses and strains then being obtained through Eqs. (11.11) and (11.16).

For the three-node triangular element used in this section, the linear trial function corresponds to a linear distribution of displacement over the element. The strain distributions however are constant as Eq. (11.10) shows. For this reason, the element is often known as the constant strain triangle (CST). Other elements, of course, can be used if the formulation is modified accordingly.

11.2   THREE-DIMENSIONAL STRESS ANALYSIS

In this section the extension of the plane strain and plane stress formulation to general three-dimensional problems in solid mechanics is briefly considered. For a more complete treatment, finite element texts in structural engineering and solid mechanics should be consulted [1–9].

The functional corresponding to the principle of minimum potential energy in the three-dimensional case is identical in form to Eq. (11.3), except that A_T is now that portion of the three-dimensional bounding surface on which tractions are prescribed. The number of elements in each of the matrices ε, σ, **F**, **P**, and **T** should, of course, be appropriate to the three-dimensional situation. Thus, ε contains the six components necessary to describe the state of stress at a point and σ contains a similar number of components to specify the strain. Using the previous subscript convention for stress and strain components, the ε and σ matrices can be written as

$$
\varepsilon = \begin{bmatrix} \varepsilon_x \\ \varepsilon_y \\ \varepsilon_z \\ \gamma_{xy} \\ \gamma_{yz} \\ \gamma_{zx} \end{bmatrix}, \qquad \sigma = \begin{bmatrix} \sigma_x \\ \sigma_y \\ \sigma_z \\ \tau_{xy} \\ \tau_{yz} \\ \tau_{zx} \end{bmatrix}, \qquad (11.31)
$$

If consideration is restricted to a Lagrangian element with s nodes, the displacement matrix δ can be written, either as

$$
\delta = [\bar{u}_1, \bar{v}_1, \bar{w}_1; \bar{u}_2, \bar{v}_2, \bar{w}_2; \ldots; \bar{u}_s, \bar{v}_s, \bar{w}_s]^{\mathrm{T}} \qquad (11.32)
$$

or as

$$
\delta = [\bar{u}_1, \bar{u}_2, \ldots, \bar{u}_s; \bar{v}_1, \bar{v}_2, \ldots, \bar{v}_s; \ldots; \bar{w}_1, \bar{w}_2, \ldots, \bar{w}_s]^{\mathrm{T}}, \qquad (11.33)
$$

where $\bar{u}, \bar{v}, \bar{w}$ are the nodal displacements in the x, y, z directions, respectively, and the subscripts denote the node identifiers of the element. The matrices **P** and **T** each have three elements that correspond to the x, y, and z components, respectively.

The finite element formulation is generally similar to that of the previous section, with appropriate modifications for the three-dimensional situation. For example, the matrix **D** becomes a 6×6 symmetric matrix of elastic constants and **B** is evaluated from strain-displacement relations analogous to Eqs. (11.10) and 11.11).

If initial strains ε_0 are present, an additional force matrix $\mathbf{F}_{\varepsilon_0}^e$, given by

$$
\mathbf{F}_{\varepsilon_0}^e = \int_{V_e} \mathbf{B}^{\mathrm{T}} \mathbf{D} \varepsilon_0 \, dV_e, \qquad (11.34)
$$

needs to be subtracted from the right-hand side of Eq. (11.20). Temperature strains ε_t similarly require a matrix identical in form to Eq. (11.34) to be subtracted from the right-hand side of Eq. (11.20).

The principle of minimum potential energy, from which the preceding displacement finite element method was derived, is not the only variational principle available in solid mechanics. The principle of minimum complementary energy yields a functional allowing variations of stress, and its corresponding finite element formulation is thus based on an assumed stress field, instead of a displacement field. Reissner's principle allows both displacement and stress fields to be assumed over the region. Further details of these and other principles may be found in the textbooks cited earlier.

For purposes of illustration, only Lagrangian elements have been considered in this chapter. Other elements, such as the Hermitian, can also be used, with the procedure being modified accordingly. The adoption of a local coordinate system for the element, in general, will be found to simplify the formulation. For particular types of problems, such as plate bending and shell structures, specialized approaches have been developed, and for details of these the literature should be consulted.

Exercise 11.1   For the constant strain triangle (CST), derive explicit relations for the entries of the element **k** matrix in terms of the elastic constants and nodal coordinates, for both plane strain and plane stress.

Exercise 11.2   For the linear rectangular Lagrangian element shown in Fig. 11.3, show that the matrix **B** is given by

$$
\mathbf{B} =
\begin{bmatrix}
-\dfrac{1}{a}+\dfrac{y}{ab} & 0 & -\dfrac{y}{ab} & 0 & \dfrac{1}{a}-\dfrac{y}{ab} & 0 & \dfrac{y}{ab} & 0 \\[2mm]
0 & -\dfrac{1}{b}+\dfrac{x}{ab} & 0 & \dfrac{1}{b}-\dfrac{x}{ab} & 0 & -\dfrac{x}{ab} & 0 & \dfrac{x}{ab} \\[2mm]
-\dfrac{1}{b}+\dfrac{x}{ab} & -\dfrac{1}{a}+\dfrac{y}{ab} & \dfrac{1}{b}-\dfrac{x}{ab} & -\dfrac{y}{ab} & -\dfrac{x}{ab} & \dfrac{1}{a}-\dfrac{y}{ab} & \dfrac{x}{ab} & \dfrac{y}{ab}
\end{bmatrix},
$$

$$(11.35)$$

and hence obtain the entries in the element **k** matrix in terms of elastic constants and nodal coordinates for both plane strain and plane stress.

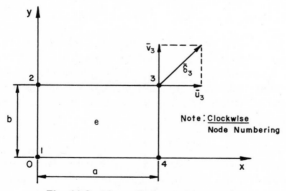

Fig. 11.3   Linear displacement rectangle.

Project 11-1   Using CST elements, develop a computer program to solve the plane stress problem shown in Fig. 11.4. Select appropriate dimensions and loads for the problem. From the results, estimate the stress concentration factor and compare this with published data. Run the program with two different sizes of element and compare the results.

Project 11-2   Modify the computer program of Project 11.1 to solve a plane strain problem similar to that shown in Fig. 11.4, where the hole is rectangular instead of circular. Use the linear rectangular Lagrangian elements shown in Fig. 11.3.

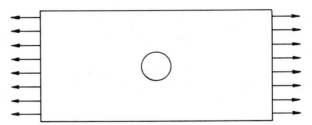

Fig. 11.4   Hole in tension-loaded bar.

Project 11-3   Rework either Project 11.1 or 11.2 using previous CST elements around the hole and rectangular elements elsewhere.

11.3   ACOUSTICS, ELECTROMAGNETIC WAVES, AND SEICHE MOTION

Periodic wave phenomena, of the free vibration type, are commonly governed by the Helmholtz equation

$$\frac{\partial}{\partial x}\left(k_x\frac{\partial\phi}{\partial x}\right) + \frac{\partial}{\partial y}\left(k_y\frac{\partial\phi}{\partial y}\right) + \frac{\partial}{\partial z}\left(k_z\frac{\partial\phi}{\partial z}\right) + \lambda^2\phi = 0, \qquad (11.36)$$

where ϕ is a scalar variable, k_x, k_y, k_z are the properties of the medium in the directions of the principal coordinate axes x, y, z, respectively, and λ is a frequency parameter. For a medium that is isotropic and homogeneous, $k_x = k_y = k_z = k$, where k is constant throughout the region. In this case, k can be absorbed within the frequency parameter λ and the Helmholtz equation reduces to the form

$$\nabla^2\phi + \lambda^2\phi = 0, \qquad (11.37)$$

where ∇^2 is the Laplace operator. Often, a Dirichlet boundary condition is specified on part of the boundary and a homogeneous Neumann condition applies on the remaining part.

For an acoustic field within a closed volume, the governing Helmholtz equation has the form

$$\nabla^2\phi + \left(\frac{\omega}{c}\right)^2 p = 0, \tag{11.38}$$

where p is the pressure change from ambient, ω is the wave frequency, and c is the velocity of sound. The propagation of electromagnetic waves through a hollow wave guide filled with homogeneous dielectric is also governed by the Helmholtz equation, in this case written as

$$\nabla^2\phi + (\omega^2\mu_\circ\varepsilon_\circ\varepsilon_d)\phi = 0, \tag{11.39}$$

where ϕ is a component of the magnetic field strength vector **H** or the electric field vector **E**, ω is the wave frequency, $\mu_\circ$ is the permeability of free space, and $\varepsilon_\circ$, ε_d are the permittivities of free space and the dielectric, respectively. Standing waves on an enclosed mass of water in a lake or harbor can be represented by the same basic equation in the following form

$$\frac{\partial}{\partial x}\left(h\frac{\partial z}{\partial x}\right) + \frac{\partial}{\partial y}\left(h\frac{\partial z}{\partial y}\right) + \left(\frac{4\pi^2}{gT^2}\right)z = 0, \tag{11.40}$$

where z is the standing wave elevation measured from the mean water level, h is the depth from mean water level to the harbor bed, g is the acceleration due to gravity, and T is the period of oscillation. For these and other phenomena governed by the Helmholtz equation, a variational finite element solution can be developed, as shown in the remainder of this section.

Consider the Helmholtz equation in the form given by Eq. (11.37). Using the variational calculus (Chapter 7), it can be shown that the solution ϕ to this equation also minimizes the functional

$$\chi = \int_D \left\{ \left(\frac{\partial\phi}{\partial x}\right)^2 + \left(\frac{\partial\phi}{\partial y}\right)^2 + \left(\frac{\partial\phi}{\partial z}\right)^2 - \lambda^2\phi^2 \right\} dD. \tag{11.41}$$

Trial functions $\hat{\phi}$ to be used in Eq. (11.41) must belong to the class of admissible functions, which in this case requires that they be continuous and have piecewise continuous first derivatives in $D + S$. Moreover, the trial functions must satisfy the Dirichlet boundary conditions of the problem. A homogeneous Neumann boundary condition

$$\partial\phi/\partial n = 0,$$

applying on a portion of the boundary, is a natural boundary condition and will be satisfied automatically.

The first part of the finite element formulation closely follows that for a Laplace problem since the Helmholtz functional has only one additional term $-\lambda^2\phi^2$. Subdividing the domain D into l finite elements, substituting the element trial functions $\hat{\phi}^e$ into the element contributions χ^e, and differentiating with respect to the element nodal vector $\boldsymbol{\phi}^e$ in the usual way yields the element matrix equation in the form

$$\frac{\partial \chi^e}{\partial \boldsymbol{\phi}^e} = \mathbf{k}^e \boldsymbol{\phi}^e - \lambda^2 \mathbf{h}^e \boldsymbol{\phi}^e. \tag{11.42}$$

The elements of the matrices $\mathbf{k}^e$ and $\mathbf{h}^e$ can be shown to be

$$k_{\alpha\beta}^e = \int_{D_e}\left(\frac{\partial N_\alpha}{\partial x}\frac{\partial N_\beta}{\partial x} + \frac{\partial N_\alpha}{\partial y}\frac{\partial N_\beta}{\partial y} + \frac{\partial N_\alpha}{\partial z}\frac{\partial N_\beta}{\partial z}\right)dD_e \tag{11.43}$$

and

$$h_{\alpha\beta}^e = \int_{D_e} N_\alpha N_\beta \, dD_e, \tag{11.44}$$

where N_α, N_β are the shape functions and α, β represent the node identifiers $1, 2, \ldots$ of the element.

For four-node tetrahedral elements with linear trial functions, the shape function relation can be written as

$$\hat{\phi}^e = \sum_{k=1}^{4} N_k \bar{\phi}_k. \tag{11.45}$$

In Chapter 9 it was shown that the shape functions N_i for these elements are identical to the volume coordinates L_i. Making this substitution and differentiating with respect to x, y, and z then yields

$$\frac{\partial L_i}{\partial x} = \frac{b_i}{6V} \quad, \quad \frac{\partial L_i}{\partial y} = \frac{c_i}{6V} \quad, \quad \frac{\partial L_i}{\partial z} = \frac{d_i}{6V} \quad, \tag{11.46}$$

where the b_i, c_i, and d_i have been defined previously in Section 9.3.3. Hence, for these elements, Eq. (11.43) can be written as

$$k_{\alpha\beta} = \frac{1}{36V}(b_\alpha b_\beta + c_\alpha c_\beta + d_\alpha d_\beta). \tag{11.47}$$

Similarly, Eq. (11.44) with the aid of Eq. (9.55),[†] can be obtained as

$$h_{\alpha\beta} = \begin{cases} V/10, & \alpha = \beta, \\ V/20, & \alpha \neq \beta. \end{cases} \tag{11.48}$$

Assembly of the element matrix equations then follows in the usual way to yield a system matrix equation of the form

$$\mathbf{K}\boldsymbol{\phi} - \lambda^2 \mathbf{H}\boldsymbol{\phi} = \mathbf{0}, \tag{11.49a}$$

[†] Note that $0! = 1$.

where ϕ is the system nodal vector and $\mathbf{0}$ is the null matrix. It will be assumed that the Dirichlet boundary conditions are of the form $\phi_i = 0$. Insertion of these into Eq. (11.49a) allows these ϕ_i to be eliminated and the set of equations represented by Eq. (11.49a) to be condensed to

$$\mathbf{K}^*\boldsymbol{\phi}^* - \lambda^2\mathbf{H}^*\boldsymbol{\phi}^* = \mathbf{0}, \tag{11.49b}$$

where $\boldsymbol{\phi}^*$ is the reduced system nodal vector (that is, $\boldsymbol{\phi}$ with the ϕ_i deleted) and $\mathbf{K}^*$, $\mathbf{H}^*$ are the reduced $\mathbf{K}$, $\mathbf{H}$ matrices. Equation (11.49b) can be written as

$$\mathbf{K}^*\boldsymbol{\phi}^* = \lambda^2\mathbf{H}^*\boldsymbol{\phi}^* \tag{11.50}$$

which is an eigenvalue or characteristic value equation. In general, the number of eigenvalues λ that satisfy Eq. (11.50) is the same as the order of the square matrices in the equation. Corresponding to each eigenvalue is a particular solution for the column vector $\boldsymbol{\phi}^*$, known as an eigenfunction, which defines a corresponding distribution (mode) of $\boldsymbol{\phi}^*$ across the region. Since the set of equations represented by Eq. (11.50) is homogeneous, the eigenfunctions cannot be obtained uniquely, although for any eigenfunction the ratios of its elements can be determined. It is often convenient to assign the value unity to the largest element of an eigenfunction and to determine the remaining elements explicitly in relation to this arbitrary datum.

There are both direct and iterative methods available for solving eigenvalue equations such as Eq. (11.50). Since in many physical eigenvalue problems, the vibration modes decrease in amplitude as the frequency increases, often only the first few eigenvalues λ_i and their corresponding modal vectors $\boldsymbol{\phi}_i$ are required. In this case, iterative solution methods will generally be found preferable. For details of eigenvalue solution procedures, the reader is directed to the literature [12–15].

11.4   TIME-DEPENDENT PROBLEMS

In reality, problems of the Helmholtz type are time dependent but the periodicity of the motion allows the governing equations to be reduced to the time-independent eigenvalue form shown earlier. Problems such as transient vibration, which retain a time dependence in the governing equations, are considerably more difficult to solve. The finite element approach is still a valuable one for such problems, but since in most cases there is no variational principle, an alternative formulation such as the Galerkin must be used.

It can be shown [1, 8, 11] that the finite element method for a time-dependent problem yields a system matrix equation of the form

$$\mathbf{K}\boldsymbol{\phi} + \mathbf{C}\dot{\boldsymbol{\phi}} + \mathbf{M}\ddot{\boldsymbol{\phi}} = \mathbf{P}(t) \tag{11.51}$$

where **K** is the system **K** matrix, **C** is a damping (or capacitance) matrix, **M** is an inertia matrix (most commonly, a mass matrix), and the matrix **P**(t) is the discretized forcing function for the problem. The matrices $\boldsymbol{\phi}$, $\dot{\boldsymbol{\phi}}$, and $\ddot{\boldsymbol{\phi}}$ are the system nodal vectors for ϕ, $\partial\phi/\partial t$, and $\partial^2\phi/\partial t^2$, respectively. In this formulation, the function ϕ is approximated over the spatial region by the trial function $\hat{\phi}$. An alternative approach uses a trial function that approximates ϕ over both space and time, but this will not be considered in this book.

The solution procedure to be adopted for Eq. (11.51) will depend, among other factors, on the precise characteristics of the equation, the type of forcing function, and the number of system nodal parameters. Mode superposition and time-marching approaches are widely used, but there are also specialized procedures that have been developed for particular problems. In all transient problems, the possibility of instability in the computation is an ever-present danger.

Since a detailed treatment of time-dependent problems is beyond the scope of this book, the reader is directed to the literature for further information.

11.5 OTHER APPLICATIONS

The finite element method has been applied to a wide variety of physical and engineering problems. Reviews such as [16, 17] indicate the scope of the method and finite element formulations for specific problems can be identified through a recent bibliography [18]. For details of particular approaches, the literature should be consulted.

REFERENCES

1. O. C. Zienkiewicz, "The Finite Element Method in Engineering Science." McGraw-Hill New York, 1971.
2. O. Ural, "The Finite Element Method." Intext Educational Publ., New York, 1973.
3. H. C. Martin and G. F. Carey, "Introduction to Finite Element Analysis." McGraw-Hill, New York, 1973.
4. R. H. Gallagher, "Finite Element Analysis." Prentice-Hall, Englewood Cliffs, New Jersey, 1975.
5. J. H. Robinson, "Integrated Theory of Finite Element Methods." Wiley (Interscience), New York, 1973.
6. K. C. Rockey, H. R. Evans, D. W. Griffiths, and D. A. Nethercot, "The Finite Element Method." Crosby Lockwood Staples, London, 1975.
7. C. A. Brebbia and J. J. Connor, "Fundamentals of Finite Element Techniques." Halsted Press, Wiley, New York, 1974.
8. C. S. Desai and J. F. Abel, "Introduction to the Finite Element Method." Van Nostrand-Reinhold, Princeton, New Jersey, 1972.
9. R. D. Cook, "Concepts and Applications of Finite Element Analysis." Wiley, New York, 1974.

10. D. H. Norrie and G. de Vries, "The Finite Element Method." Academic Press, New York, 1973.

11. K. H. Huebner, "The Finite Element Method for Engineers." Wiley, New York, 1975.

12. F. S. Acton, "Numerical Methods That Work." Harper, New York, 1970.

13. J. H. Wilkinson, "The Algebraic Eigenvalue Problem." Oxford Univ. Press, London and New York, 1965.

14. A. Ralston and H. S. Wilf, "Mathematical Methods for Digital Computers," Vol. 2. Wiley, New York, 1967.

15. C. A. Brebbia, H. Tottenham, G. B. Warburton, J. Wilson, and R. Wilson, "Vibrations of Engineering Structures." Computational Mechanics, Southampton, England, 1976.

16. O. C. Zienkiewicz, From intuition to generality, *Appl. Mech. Rev.*, **23**, 249–256 (1970).

17. D. H. Norrie and G. de Vries, A survey of finite element applications in fluid mechanics, Rep. No. 83, Dept. of Mech. Eng., Univ. of Calgary, Alberta, Canada (December 1976).

18. D. H. Norrie and G. de Vries, "Finite Element Bibliography." Plenum, New York, 1976.

12

OTHER FINITE ELEMENT METHODS

It was noted previously that for linear self-adjoint problems, the conditions under which the functional has a stationary value can be used in a finite element formulation to determine the solution to the governing equation. For other types of problem where a functional of this kind does not exist, a variational finite element formulation may still be possible if a quasi-variational or restricted variational principle can be found. In such cases, however, it is generally preferable [1] to use a nonvariational approach.

The most popular of the other finite element formulations is the Galerkin method, which is a particular form of the residual finite element method, as is also the least-squares procedure. Another method, which has also been applied to a wide range of problems, is variously known as the direct, energy balance, global balance, or control volume finite element method.

12.1 RESIDUAL FINITE ELEMENT METHODS

Consider a problem for which the governing equation in the domain D involves one dependent variable u and its derivatives, and several independent variables $x_1, x_2, \ldots, x_n$ collectively denoted by x_i. Let the governing equation be written in the general form

$$f_D(u; x_i) = 0. \tag{12.1}$$

Substitution of an approximate solution $\hat{u}$ into Eq. (12.1) will not in general result in

$$f_D(\hat{u}; x_i) = 0, \tag{12.2}$$

and an equation error or residual R can thus be formed as

$$R = f_D(u; x_i) - f_D(\hat{u}; x_i). \tag{12.3}$$

The closer the approximate solution $\hat{u}$ approaches the exact solution u, the nearer will the residual R approach toward zero. Substitution of Eq. (12.2) into Eq. (12.3) allows the simplified residual form

$$R = -f_D(\hat{u}; x_i) \tag{12.4}$$

to be obtained.

In residual trial function methods, the residual R is required to satisfy some condition that forces it to be small. For a residual finite element method it is the weighted integral over the domain, $\int_D Wf(R)dD$, which is required to satisfy the smallness criterion, where W is a weighting function. The choice of W and $f(R)$ defines the particular method, thus

Galerkin: $\int_D W_p R\, dD = 0, \qquad p = 1, 2, \ldots, n, \tag{12.5}$

where the W_p are domain interpolation functions to be defined subsequently and n is the total number of nodal parameters.

Least Squares: $\int_D R^2\, dD = \text{minimum}. \tag{12.6}$

It is evident that in Eq. (12.5) there are as many equations as there are unknown nodal parameters, thus allowing a solution to be obtained. For the least-squares case, substitution of the finite element interpolation into R, via Eq. (12.4), converts the integrand into a function of all the system nodal parameters. The conditions for R to be a minimum are the same as used previously, namely, that the derivative with respect to each nodal parameter respectively equal zero. A set of n equations in terms of the n unknown nodal values is thus obtained.

So far, only residuals arising from the governing domain equations have been considered. If the trial function does not satisfy the boundary conditions exactly, there will be nonzero boundary residuals[†] formed analogously

---

[†] In propagation (time-dependent) problems, the initial conditions also allow *initial residuals* to be defined.

to the domain residuals which must be also considered. The residual solution method is then based on both sets of residuals. It is usual, however, to require the trial function to satisfy the boundary conditions exactly and the boundary residuals, then being zero, do not require further consideration.

12.2 GALERKIN FINITE ELEMENT METHOD

For a Galerkin residual method, Eq. (12.5) applies, where in general, the W_p are interpolation functions in a domain trial function $\hat{u}$ of the form

$$\hat{u} = \sum_{p=1}^{n} W_p \bar{u}_p \qquad \text{in } D, \tag{12.7}$$

In Eq. (12.7), the $\bar{u}_p$ are the system nodal parameters and n is the total number of nodal parameters. By analogy with Eq. (9.1) for an element trial function, it will be seen that the W_p can be described as domain shape functions and to emphasize this, the W_p will hereafter be denoted as the $\underline{N}_p$. Thus, Eqs. (12.5) and (12.7) can, respectively, be written as

$$\int_D \underline{N}_p R dD = 0, \qquad p = 1, 2, \ldots, n, \tag{12.8}$$

$$\hat{u} = \sum_{p=1}^{n} \underline{N}_p \bar{u}_p. \tag{12.9}$$

It can be shown that each domain shape function $\underline{N}_p$ is a sum of element shape functions, as follows. For any element e, the trial function within the element is expressed in terms of its element shape functions, $N_k{}^e$, as

$$\hat{u}^e = \sum_{k=1}^{m} N_k{}^e \bar{u}_k \qquad \text{in } e, \tag{12.10}$$

where m is defined by

$$m = s \cdot q. \tag{12.11}$$

In Eq. (12.11), s is the total number of nodes of element e and q is the number of degrees of freedom per node. The subscripts of the nodal parameters $\bar{u}_k$ in Eq. (12.10) refer to the node identifiers and not to the system node numbers (see Section 9.2).

The trial function in Eq. (12.10) can be written in a more general form as

$$\hat{u}^e = \sum_{p=1}^{n} N_p{}^e \bar{u}_p \qquad \text{in } e, \tag{12.12a}$$

where it is to be understood that $N_p{}^e = 0$ if p does not correspond to a nodal parameter belonging to element e. In Eq. (9.12a), m of the subscripts of the nodal parameter $\bar{u}_p$ are node numbers corresponding to the node identifiers of Eq. (12.10).

In matrix form Eq. (12.12a) becomes

$$\hat{u}^e = \underline{N}^e \mathbf{u} \qquad \text{in } e, \tag{12.12b}$$

where $\mathbf{u}$ is the system nodal vector and $\underline{N}^e$ is the expanded shape function matrix for element e.

The set of trial functions derived by applying Eq. (12.12b) to the l elements of the system allow the interpolation over the domain to be written as

$$\hat{u} = \sum_{e=1}^{l} \hat{u}^e = \sum_{e=1}^{l} \underline{N}^e \mathbf{u} \qquad \text{in } D. \tag{12.13}$$

Writing Eq. (12.9) in matrix form gives

$$\hat{u} = \mathbf{N}\mathbf{u}. \tag{12.14}$$

where $\underline{N}$ is defined from Eq. (12.9). Comparison of Eqs. (12.13) and (12.14) leads to the conclusion that

$$\underline{N} = \sum_{e=1}^{l} \underline{N}^e. \tag{12.15a}$$

and that a matrix element $\underline{N}_p$ of $\underline{N}$ is given by

$$\underline{N}_p = \sum_{e=1}^{l} N_p^{\,e} \tag{12.15b}$$

as was to be proven. The set of equations represented by Eq. (12.8) can be put in the matrix from

$$\int_D \mathbf{N}R\,dD = \mathbf{0}. \tag{12.16}$$

Substitution of Eqs. (12.4) and (12.15a) into Eq. (12.16) results in

$$\int_D \left\{ \sum_{e=1}^{l} \underline{N}^e \right\} f_D(\hat{u}; x_i)\,dD = \mathbf{0}. \tag{12.17a}$$

The summation in Eq. (12.17) can be taken outside the integral to give

$$\sum_{e=1}^{l} \left\{ \int_{D_e} \underline{N}^e f_D(\hat{u}^e; x_i)\,dD_e \right\} = \mathbf{0}. \tag{12.17b}$$

Equation (12.17b) represents the set of system equations. Consider the pth equation of the set, which is

$$\sum_{e=1}^{l} \int_{D_e} N_p^{\,e} f_{D_e}(\hat{u}^e; x_i)\,dD_e = 0. \tag{12.18}$$

It can be seen from Eq. (12.18) that an element contribution X^e can be defined as

$$X_p{}^e = \int_{D_e} N_p{}^e f_D(\hat{u}^e; x_i)\, dD_e, \qquad (12.19a)$$

For linear problems, substitution of the chosen shape function interpolation for $\hat{u}^e$ into Eq. (12.19) and subsequent integration will be found to yield $X_p{}^e$ in the form

$$X_p{}^e = \mathbf{k}_p{}^e \mathbf{u}^e + F_p{}^e, \qquad (12.19b)$$

where the element nodal vector $\mathbf{u}^e$ lists the nodal values of element e according to node number subscripts and where $\mathbf{k}_p{}^e$ is a row matrix.

Assembling the element contributions given by Eq. (12.19b) into Eq. (12.18) gives the pth system equation as

$$\sum_{e=1}^{l} X_p{}^e = \mathbf{K}_p \mathbf{u} + F_p = 0, \qquad (12.20)$$

where the row matrix $\mathbf{K}_p$ is formed by expanding the row matrices $\mathbf{k}_p{}^e$ to system size and adding the expanded $\mathbf{k}_p{}^e$ matrices, and where F_p is formed as the sum of the $F_p{}^e$ terms.

The set of system equations, of which Eq. (12.20) is the pth member, can be written as

$$\mathbf{Ku} + \mathbf{F} = \mathbf{0}. \qquad (12.21)$$

In Eq. (12.21), the elements of the matrices $\mathbf{K}$ and $\mathbf{F}$ can be shown to be given respectively by

$$K_{pq} = \sum_{e=1}^{l} k_{pq}^e, \qquad F_p = \sum_{e=1}^{l} F_p{}^e. \qquad (12.22a,b)$$

where p and q are node number subscripts and k_{pq}^e is an element of the row matrix $\mathbf{k}_p{}^e$. The assembly process used above corresponds to assembly by nodes.

For an element e, the set of equations of which Eq. (12.19b) is the pth member, has the matrix form

$$\mathbf{X}^e = \mathbf{k}^e \mathbf{u}^e + \mathbf{F}_p{}^e. \qquad (12.23)$$

Assembling these element equations, according to Eq. (12.17b), corresponds to assembly by elements and again yields the system matrix equation (12.21). It is easily shown that Eqs. (12.22a, b) remain valid for this latter case.

It had been assumed implicitly that the trial function $\hat{u}$ satisfies the boundary conditions exactly, thus giving zero boundary residuals. For Dirichlet boundary conditions this is easily achieved by correcting the system matrix

K in the usual way. The procedure for other boundary conditions is illustrated by example subsequently.

Illustrative Example 12.1   Consider the two-dimensional region shown in Fig. 12.1 in which heat is generated internally and which loses heat by convection across portion of the boundary.

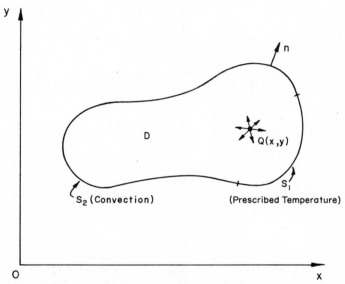

Fig. 12.1   Conduction in a two-dimensional region with heat generation.

For a homogeneous and isotropic medium, the governing equation is Poisson's equation

$$\frac{\partial^2 T}{\partial x^2} + \frac{\partial^2 T}{\partial y^2} + Q = 0 \qquad \text{in } D, \tag{12.24}$$

where $T(x, y)$ is the temperature at a point and $Q(x, y)$ is the internal (local) heat generation. It is assumed that the temperature is prescribed along a portion S_1 of the boundary, that is,

$$T = g(x, y) \qquad \text{on } S_1, \tag{12.25a}$$

and that there is heat loss along the remaining portion S_2 due to convection, so that

$$\frac{\partial T}{\partial n} + h(T - T_a) = 0 \qquad \text{on } S_2, \tag{12.25b}$$

where n is in the direction of the unit outward normal to S_2, T_a is the ambient temperature of the surrounding fluid, h is the convection heat transfer coefficient, and $S = S_1 + S_2$ encloses D; see Fig. 12.1.

Let the domain be subdivided into l finite elements of the Lagrangian type so that the nodal parameters are temperatures only. The trial function $\hat{T}$ within any element e can be written in shape function form as

$$\hat{T}^e = \mathbf{N}^e \mathbf{T}^e, \tag{12.26}$$

Rewriting Eq. (12.24) in terms of the trial function $\hat{T}^e$ and substituting the result into the element contribution [Eq. (12.19a)] yields

$$X_p^e = \int_{D_e} N_p^e \left(\frac{\partial^2 \hat{T}^e}{\partial x^2} + \frac{\partial^2 \hat{T}^e}{\partial x^2} + Q \right) dD_e. \tag{12.27}$$

Substitution of Eq. (12.26) into Eq. (12.27) then gives

$$X_p^e = \int_{D_e} N_p^e \left[\frac{\partial^2 \mathbf{N}^e}{\partial x^2} + \frac{\partial^2 \mathbf{N}^e}{\partial y^2} \right] \mathbf{T}^e dD_e + \int_{D_e} N_p^e Q \, dD_e. \tag{12.28}$$

It will be noted that this equation is of the same form as Eq. (12.19b). Although a similar form is obtained from a variational formulation, the derivatives in the element contribution in the present case are of the same order as those in the governing domain equation, whereas for the variational formulation the derivatives in the element contribution will be of a lower order (see Section 2.1). Thus, for the Galerkin approach, it appears that a higher order trial function representation is required than for the corresponding variational formulation. It is shown in the following, however, that this requirement can be averted. The Dirichlet boundary condition [Eq. (12.25a)] can be imposed in the usual way but it is not obvious how the Cauchy boundary condition [Eq. (12.25b)] can be introduced. The means by which this can be accomplished is also considered below. Green's theorem for two variables u and v over the subdomain D_e of an element e can be written as

$$\int_{D_e} \nabla u \cdot \nabla v \, dD_e + \int_{D_e} u \nabla^2 v \, dD_e = \int_{S_e} u \frac{\partial v}{\partial n} dS_e, \tag{12.29}$$

where S_e is the boundary around D_e. Replacing u by N_p^e and v by $\hat{T}^e$ in Eq. (12.29)[†] allows the first domain integral in Eq. (12.28) to be converted into a domain integral (with lower order derivatives) and a surface integral,

---

[†] Provided N_p^e and $\hat{T}^e$ satisfy the required continuity conditions of Green's theorem [See Eq. (7.37)].

as shown below

$$
X_p^e = \int_{D_e} N_p^e \left[\frac{\partial^2 \hat{T}^e}{\partial x^2} + \frac{\partial^2 \hat{T}^e}{\partial y^2} \right] dD_e + \int_{D_e} N_p^e Q \, dD_e
$$

$$
= \int_{S_e} N_p^e \left[\frac{\partial \hat{T}^e}{\partial n} \right] dS_e - \int_{D_e} \left(\frac{\partial N_p^e}{\partial x} \frac{\partial \hat{T}^e}{\partial x} + \frac{\partial N_p^e}{\partial y} \frac{\partial \hat{T}^e}{\partial y} \right) dD_e
$$

$$
+ \int_{D_e} N_p^e Q \, dD_e. \tag{12.30}
$$

Substitution of Eq. (12.26) into the domain integrals of Eq. (12.30) then yields

$$
X_p^e = -\int_{D_e} \left[\frac{\partial N_p^e}{\partial x} \frac{\partial \mathbf{N}^e}{\partial x} + \frac{\partial N_p^e}{\partial y} \frac{\partial \mathbf{N}^e}{\partial y} \right] \mathbf{T}^e \, dD_e + \int_{D_e} N_p^e Q \, dD_e
$$

$$
+ \int_{S_e} N_p^e \left[\frac{\partial \hat{T}^e}{\partial n} \right] dS_e. \tag{12.31}
$$

The term $\partial \hat{T}^e / \partial n$ appearing in the surface integral of Eq. (12.31) can be used to impose the Cauchy boundary condition [Eq. (12.25b)] on the formulation. Using Eq. (12.26), the Cauchy condition Eq. (12.25b) can be substituted into the surface integral of Eq. (12.31) to give the element contribution in the form

$$
X_p^e = -\int_{D_e} \left[\frac{\partial N_p^e}{\partial x} \frac{\partial \mathbf{N}^e}{\partial x} + \frac{\partial N_p^e}{\partial y} \frac{\partial \mathbf{N}^e}{\partial y} \right] \mathbf{T}^e \, dD_e + \int_{D_e} N_p^e Q \, dD_e
$$

$$
- \int_{S_{2e}} N_p^e [h \mathbf{N}^e \mathbf{T}^e - h T_a] \, dS_{2e}. \tag{12.32}
$$

The element contribution for any element is given by Eq. (12.32), but the surface integral exists only for those elements along the S_2 boundary.

Each element contribution as computed using Eq. (12.32) will be of the form shown in Eq. (12.19b). Assembly of these contributions into the system matrix equation can thus follow in the usual way. After insertion of the Dirichlet conditions, the system matrix equation can be solved by any standard technique and the nodal parameters obtained explicitly. A survey of Galerkin methods, including the finite element formulation, can be found in [2].

Exercise 12.1   Solve the two-dimensional heat conduction problem given in Fig. 2.1 by the Galerkin finite element method, dividing the region into the triangular mesh shown in Fig. 2.4 and using three-node triangular elements with linear trial functions. Evaluate the element contributions from Eq. (12.32) and assemble these by nodes into the system matrix equation. Repeat using assembly by elements. Correct the system matrix equations for the Dirichlet boundary conditions and compare the resulting equations with Eq. (2.36) obtained from the variational finite element solution. (For linear self-adjoint problems, such as considered here, it can be shown [3] that the variational finite element method and the Galerkin finite element method yield identical system matrix equations if the same mesh is used.)

12.3 LEAST-SQUARES FINITE ELEMENT METHOD

The basis for the least-squares method was earlier shown to be

$$\int_D W R^2 \, dD = \text{minimum}, \tag{12.33}$$

where $W(x_i)$ is a (positive) weighting function. Commonly, W is chosen as unity and the smallness criterion [Eq. (12.33)] then reduces to

$$\int_D R^2 \, dD = \text{minimum}. \tag{12.34}$$

To illustrate the formulation, consider a two-dimensional region for which the governing field equation is

$$Au = f \quad \text{in } D, \tag{12.35}$$

subject to the boundary condition

$$Bu = g \quad \text{on } S. \tag{12.36}$$

In Eqs. (12.35) and (12.36), A and B are linear differential operators with f and g being prescribed functions of x and y, and S encloses D. Using the element trial function $\hat{u}^e$, the residual R_e within an element is seen, from Eq. (12.35), to be

$$R_e = A\hat{u}^e - f \quad \text{in } e. \tag{12.37}$$

Expressing Eq. (12.34) as a sum of element integrals and then substituting Eq. (12.37) yields

$$\int_D R^2 \, dD = \sum_{e=1}^{l} \left[\int_{D_e} R_e^{\,2} \, dD_e \right] = \sum_{e=1}^{l} \left[\int_{D_e} (A\hat{u}^e - f)^2 \, dD_e \right]. \tag{12.38}$$

The element trial function $\hat{u}^e$ can be written in terms of the element shape function matrix $\mathbf{N}^e$ as

$$\hat{u}^e = \mathbf{N}^e \mathbf{u}^e, \tag{12.39}$$

where $\mathbf{u}^e$ is the element nodal vector. An alternate form of Eq. (12.39) is

$$\hat{u}^e = \underline{\mathbf{N}}^e \mathbf{u}, \tag{12.40}$$

where $\underline{N}^e$ is the expanded element shape function matrix and $\mathbf{u}$ is the system nodal vector. Substitution of Eq. (12.40) into Eq. (12.38) then gives

$$\int_D R^2\, dD = \sum_{e=1}^{l}\left[\int_{D_e} (A\underline{N}^e\mathbf{u} - f)^2 D_e\right]. \qquad (12.41)$$

Since the right-hand side of Eq. (12.41) is a function of the system nodal parameters, the minimization required by Eq. (12.34) can be accomplished by differentiating this expression with respect to each nodal parameter, successively, and setting each resultant to zero. The set of equations so obtained will be the system equations. These can alternatively be derived directly in matrix form by differentiating the right-hand side of Eq. (12.41) with respect to the system nodal vector $\mathbf{u}$ and equating the resulting expression to zero as follows

$$\frac{\partial}{\partial \mathbf{u}} \sum_{e=1}^{l} \int_{D_e} (A\underline{N}^e\mathbf{u} - f)^2\, dD_e = \mathbf{0}. \qquad (12.42)$$

Taking the differentiation into the integral converts Eq. (12.42) into

$$\sum_{e=1}^{l} \int_{D_e} \frac{\partial}{\partial \mathbf{u}} (A\underline{N}^e\mathbf{u} - f)^2\, dD_e = \mathbf{0}, \qquad (12.43)$$

which, by evaluating the squared term, can be written in the form

$$\sum_{e=1}^{l} \int_{D_e} \frac{\partial}{\partial \mathbf{u}} (\mathbf{u}^T\mathbf{A}^T\mathbf{A}\mathbf{u} - 2\mathbf{u}^T\mathbf{A}^Tf + f^2)\, dD = \quad , \qquad (12.44)$$

where

$$\mathbf{A} = A\underline{N}^e. \qquad (12.45)$$

Carrying out the differentiation[†] in Eq. (12.44) results in

$$2\left[\sum_{e=1}^{l}\left(\int_{D_e} \mathbf{A}^T\mathbf{A}\mathbf{u}\, dD_e - \int_{D_e} \mathbf{A}^Tf\, dD_e\right)\right] = \mathbf{0}. \qquad (12.46)$$

By noting that $\mathbf{u}$ is not a function of x and y, Eq. (12.46) can be written as

$$\sum_{e=1}^{l} \mathbf{k}^e\mathbf{u} + \mathbf{F}^e = \mathbf{0}, \qquad (12.47)$$

where

$$\mathbf{k}^e = \int_{D_e} \mathbf{A}^T\mathbf{A}\, dD_e, \qquad \mathbf{F}^e = -\int_{D_e} \mathbf{A}^Tf\, dD_e. \qquad (12.48)$$

[†] Using Eqs. (B.16a) and (B.17) of Appendix B.

As derived, $\mathbf{k}_e$ and $\mathbf{F}_e$ are in expanded form, that is, are of system size, so that Eq. (12.47) is the expanded element matrix equation. The relations above can also be written in element size, which is left as an exercise.

Assembly of the element matrix equations, either by nodes or by elements yields the system matrix equation

$$\mathbf{Ku} + \mathbf{F} = \mathbf{0}. \tag{12.49}$$

Equation (12.49) must now be corrected, by inserting the Dirichlet boundary conditions, in the usual way, to obtain the final system matrix equation.

From Eq. (12.48) it is seen that the element matrix $\mathbf{k}^e$ is symmetric and positive definite. Consequently, the system $\mathbf{K}$ matrix for the least-squares method is also symmetric and positive definite, prior to the insertion of the Dirichlet boundary conditions. This is in contrast to the Galerkin method, where symmetric matrices are obtained only for self-adjoint problems.

For further information on the least-squares method, [4–14] can be consulted.

12.4   DIRECT FINITE ELEMENT METHOD

To develop a finite element formulation, an equation embodying the physical laws governing the phenomena must be obtained in a form applying over the region. The functional, corresponding to a variational principle, and the smallness criterion in a residual method, are examples of such regional relationships. The governing equations in their usual differential form are not adequate since they apply to a point not to the region. As Oden pointed out [15], however, forms often exist for the governing equations which can be used as the basis of a finite element method. For example, in continuum mechanics the energy balance for the region can be written in a global or control volume form. Similarly, continuity equations can be obtained in control volume form. Discretization of the domain into finite elements and direct insertion of the approximation functions into such integral forms of the governing equation comprise a finite element procedure, as shown below, which is known as the direct, the global balance, or the energy balance method. Oden et al. [15–20] have applied the method successfully to a variety of equilibrium and propagation problems.

In the following, the equilibrium problem only is considered. The integral form of the domain equation in such a case can be represented as

$$I(u; x_i) = 0, \tag{12.50}$$

where u is the dependent variable and the x_i are the independent variables. In general, the integral I over the region can be written as the sum of the element contributions I^e, that is,

$$I = \sum_{e=1}^{l} I^e = 0. \tag{12.51}$$

Substitution of the element trial function into each element contribution in Eq. (12.51) will convert the right-hand side into an expression involving the nodal parameters of the element, the element shape functions, and the independent variables x_i. The resulting element equations can be rearranged into the standard form of element matrix equations and assembled, via Eq. (12.51), into the system matrix equation of the form

$$\mathbf{Ku} + \mathbf{F} = \mathbf{0}. \tag{12.52}$$

Dirichlet boundary conditions can be introduced by correcting the system matrix equation, with other boundary conditions being inserted via appropriate surface integrals

REFERENCES

1. B. A. Finlayson, L. E. Scriven, On the search for variational principles, *Internat. J. Heat Mass Transfer* **10**, 799–821 (1967).
2. C. A. J. Fletcher, The Galerkin method: An introduction, WRE- TM-1632 (WR&D), Weapons Research Establishment, Dept. of Defense, Adelaide, South Australia (June 1976).
3. B. A. Finlayson and L. E. Scriven, The method of weighted residuals—a review, *Appl. Mech. Rev.* **19**, No. 9, 735–748 (September 1966).
4. P. P. Lynn and S. K. Arya, Use of least squares criterion in the finite element formulation, *Internat. J. Numer. Methods in Engrg.* **6**, 75–88 (1973).
5. P. P. Lynn and S. K. Arya, Finite elements formulated by the weighted least squares method, *Internat. J. Numer. Methods in Engrg.* **8**, 71–90 (1974).
6. O. C. Zienkiewicz, D. R. J. Owen, and K. N. Lee, Least-square finite element for elastostatic problems—use of 'reduced' integration, *Internat. J. Numer. Methods in Engrg.* **8**, 341–358 (1974).
7. P. P. Lynn, Least-squares finite element analysis of laminar boundary layer flows, *Internat. J. Numer. Methods in Engrg.* **8**, 865–876 (1974).
8. J. E. Akin, A least-squares finite element solution of non-linear operations, *in* "Mathematics of Finite Elements and Applications" (J. R. Whiteman, ed.). Academic Press, New York, 1973.
9. K. N. Lee, A Note on least square residues in linear elasticity, *J. Appl. Mech.* **96**, Ser. E, 553–554, (June 1974).
10. M. P. Rossow, The least-squares variational principle for finite element applications, *J. Appl. Mech.* **97**, 900–901 (December 1975).
11. J. F. M. Ithorpe and G. P. Steven, On the least-squares approach to the integration of the Navier-Stokes equations, Preprints of the Second Symposium on Finite Element Methods in Flow Problems, St. Margherita, Italy, 14–18 June 1976, pp. 71–82.
12. G. P. Steven, Dynamics of a fluid subject to thermal and gravity diffusion, *Proc. Internat. Conf. Finite Element Methods in Engrg., Univ. of Adelaide, South Australia*, 6–8 December 1976, pp. 17.1–17.16.

13. G. de Vries, T. E. Labrujere, and D. H. Norrie, A least squares finite element solution for potential flow, Rep. No. 86, Dept. of Mech. Engrg., Univ. of Calgary, Alberta, Canada, (December 1976).

14. D. H. Norrie and G. de Vries, "Finite Element Bibliography." Plenum, New York, 1976.

15. J. T. Oden, A general theory of finite elements, *Internat. J. Numer. Methods Engrg.* (Part I) **1**, 202–221; (Part 2) **1**, 247–259 (1969).

16. J. T. Oden, "Finite Elements of Non-Linear Continua." McGraw-Hill, New York, 1972.

17. J. T. Oden, Finite element analogue of Navier–Stokes equations, *Proc. ASCE, J. Engrg. Mech. Div.* **96**, No. EM4, 529–534 (1970).

18. J. T. Oden and D. Somogyi, Finite element applications in fluid dynamics, *Proc. ASCE, J. Engrg. Mech. Div.* **95**, No. EM3, 821–826 (1969).

19. G. Aguirre-Ramirez and J. T. Oden, Finite element technique applied to heat conduction in solids with temperature dependent thermal conductivity. Paper 69-WA/HT-34, ASME Winter Annual Meeting, Los Angeles, (16–20 November 1969).

20. J. T. Oden and B. E. Kelley, Finite element formulation of general thermoelasticity problems, *Internat. J. Numer. Methods Engrg.* **3**, 161–179 (1971).

APPENDIX A

MATRIX ALGEBRA

A.1 MATRIX DEFINITIONS

A *matrix* is defined as a rectangular array of symbols or numbers, arranged in rows and columns, which can be represented in the following manner:

$$
\mathbf{A} = [A] = [a_{ij}] =
\begin{bmatrix}
a_{11} & a_{12} & \cdots & a_{1j} & \cdots & a_{1n} \\
a_{21} & a_{22} & \cdots & a_{2j} & \cdots & a_{2n} \\
\vdots & \vdots & & \vdots & & \vdots \\
a_{i1} & a_{i2} & \cdots & a_{ij} & \cdots & a_{in} \\
\vdots & \vdots & & \vdots & & \vdots \\
a_{m1} & a_{m2} & \cdots & a_{mj} & \cdots & a_{mn}
\end{bmatrix},
\tag{A.1}
$$

where the subscripts i and j, in the first and second positions, respectively, refer to the ith row and the jth column.

The matrix shown in Eq. (A.1) has m rows and n columns, with a typical element a_{ij} being in the ith row and the jth column. Consequently, this matrix $\mathbf{A}$ is said to be of order $m \times n$.

A row matrix is of order $1 \times n$ and is represented as

$$
\mathbf{A} = [a_{11} \quad a_{12} \quad \cdots \quad a_{1j} \quad \cdots \quad a_{1n}],
\tag{A.2}
$$

whereas a column matrix is of order $m \times 1$ and is represented as

$$\mathbf{A} = \begin{bmatrix} a_{11} \\ a_{21} \\ \vdots \\ a_{i1} \\ \vdots \\ a_{m1} \end{bmatrix}. \tag{A.3}$$

A single boldfaced symbol, such as $\mathbf{A}$, is used to represent a matrix throughout this text.

A *null* or *zero matrix* has all its elements equal to zero and is written as $\mathbf{0}$.

A *square matrix* has an equal number of rows and columns and is thus of the form

$$\mathbf{A} = \begin{bmatrix} a_{11} & a_{12} & \cdots & a_{1j} & \cdots & a_{1n} \\ a_{21} & a_{22} & \cdots & a_{2j} & \cdots & a_{2n} \\ \vdots & \vdots & & \vdots & & \vdots \\ a_{i1} & a_{i2} & \cdots & a_{ij} & \cdots & a_{in} \\ \vdots & \vdots & & \vdots & & \vdots \\ a_{n1} & a_{n2} & \cdots & a_{nj} & \cdots & a_{nn} \end{bmatrix}. \tag{A.4}$$

Particular types of square matrices are the scalar matrix, identity or unit matrix, diagonal matrix, band matrix, triangular matrix, symmetric matrix, and skew-symmetric matrix, each of which is defined below:

1. A *scalar matrix* is a matrix for which the elements, in terms of a scalar a, are defined by

$$a_{ij} = \begin{cases} a, & i = j, \\ 0, & i \neq j. \end{cases} \tag{A.5}$$

For example,

$$\mathbf{A} = \begin{bmatrix} a & 0 & 0 \\ 0 & a & 0 \\ 0 & 0 & a \end{bmatrix}. \tag{A.6}$$

2. The *identity* or *unit matrix* is generally denoted by $\mathbf{I}$ and has as its elements

$$I_{ij} = \begin{cases} 1, & i = j, \\ 0, & i \neq j. \end{cases} \tag{A.7}$$

The identity matrix of order 4×4, for example, is

$$\mathbf{I} = \begin{bmatrix} 1 & 0 & 0 & 0 \\ 0 & 1 & 0 & 0 \\ 0 & 0 & 1 & 0 \\ 0 & 0 & 0 & 1 \end{bmatrix}. \tag{A.8}$$

3.   A *diagonal matrix* has zero elements except along its *principal diagonal*, which runs from the upper left to the lower right-hand corner of the matrix. The elements of a diagonal matrix are thus

$$d_{ij} = \begin{cases} d_{ii}, & i = j, \\ 0, & i \neq j. \end{cases} \tag{A.9}$$

A diagonal matrix can, therefore, be written as

$$\mathbf{D} = \begin{bmatrix} d_{11} & 0 & \cdots & 0 \\ 0 & d_{22} & \cdots & 0 \\ \vdots & \vdots & & \vdots \\ 0 & 0 & & d_{nn} \end{bmatrix}. \tag{A.10}$$

4.   A *band matrix* has nonzero elements in a band that runs along the principal diagonal, with zero elements outside this band. The following is an example of a band matrix:

$$\mathbf{A} = \begin{bmatrix} a_{11} & a_{12} & 0 & 0 & 0 & \cdots & 0 & 0 & 0 \\ a_{21} & a_{22} & a_{23} & 0 & 0 & \cdots & 0 & 0 & 0 \\ 0 & a_{32} & a_{33} & a_{34} & 0 & \cdots & 0 & 0 & 0 \\ \vdots & \vdots & \vdots & \vdots & \vdots & & \vdots & \vdots & \vdots \\ 0 & 0 & 0 & 0 & 0 & \cdots & a_{n-1,n-2} & a_{n-1,n-1} & a_{n-1,n} \\ 0 & 0 & 0 & 0 & 0 & \cdots & 0 & a_{n,n-1} & a_{nn} \end{bmatrix}. \tag{A.11}$$

5.   A *triangular matrix* has all the elements above the principal diagonal equal to zero (in which case it is called a *lower triangular matrix*) or all the elements equal to zero below the principal diagonal (when it is referred to as an *upper triangular matrix*). The upper and lower triangular matrices are usually denoted by $\mathbf{U}$ and $\mathbf{L}$, respectively. Thus,

$$\mathbf{U} = \begin{bmatrix} u_{11} & u_{12} & u_{13} & \cdots & u_{1n} \\ 0 & u_{22} & u_{23} & \cdots & u_{2n} \\ 0 & 0 & u_{33} & \cdots & u_{3n} \\ \vdots & \vdots & \vdots & & \vdots \\ 0 & 0 & 0 & \cdots & u_{nn} \end{bmatrix}, \tag{A.12a}$$

$$\mathbf{L} = \begin{bmatrix} l_{11} & 0 & 0 & \cdots & 0 \\ l_{21} & l_{22} & 0 & \cdots & 0 \\ l_{31} & l_{32} & l_{33} & \cdots & 0 \\ \vdots & \vdots & \vdots & & \vdots \\ l_{n1} & l_{n2} & l_{n3} & \cdots & l_{nn} \end{bmatrix}. \qquad (A.12b)$$

6. A *symmetric matrix* is a square matrix that has equal elements placed symmetrically about the principal diagonal, that is,

$$a_{ij} = a_{ji} \qquad \text{for all } i \text{ and } j. \qquad (A.13)$$

7. A *skew-symmetric matrix* has elements that are located symmetrically about the principal diagonal but which are of opposite sign. Consequently, the principal diagonal elements are zero. Thus, for a skew-symmetric matrix,

$$a_{ij} = \begin{cases} 0 & \text{for } i = j, \\ -a_{ji} & \text{for } i \neq j. \end{cases} \qquad (A.14)$$

8. A *skew matrix* is a skew-symmetric matrix for which the principal diagonal elements are not all zero.

9. A *partitioned matrix* is one which is subdivided into *submatrices*. For example,

$$\mathbf{A} = \begin{bmatrix} a_{11} & a_{12} & a_{13} & a_{14} \\ a_{21} & a_{22} & a_{23} & a_{24} \\ a_{31} & a_{32} & a_{33} & a_{34} \\ \hline a_{41} & a_{42} & a_{43} & a_{44} \end{bmatrix} = \begin{bmatrix} \mathbf{A}_{11} & \mathbf{A}_{12} & \mathbf{A}_{13} \\ \mathbf{A}_{21} & \mathbf{A}_{22} & \mathbf{A}_{23} \end{bmatrix}, \qquad (A.15)$$

where the submatrices $\mathbf{A}_{11}, \mathbf{A}_{12}, \mathbf{A}_{13}, \mathbf{A}_{21}, \mathbf{A}_{22}, \mathbf{A}_{23}$ are given by

$$\mathbf{A}_{11} = \begin{bmatrix} a_{11} \\ a_{21} \\ a_{31} \end{bmatrix}, \qquad \mathbf{A}_{12} = \begin{bmatrix} a_{12} \\ a_{22} \\ a_{32} \end{bmatrix}, \qquad \mathbf{A}_{13} = \begin{bmatrix} a_{13} & a_{14} \\ a_{23} & a_{24} \\ a_{33} & a_{34} \end{bmatrix}. \qquad (A.16)$$

$$\mathbf{A}_{21} = a_{41}, \qquad \mathbf{A}_{22} = a_{42}, \qquad \mathbf{A}_{23} = \begin{bmatrix} a_{43} & a_{44} \end{bmatrix}.$$

A.2   MATRIX ALGEBRA

A.2.1   Addition of Matrices

Addition is possible only for matrices of the same order and is defined by the procedure of adding, respectively, corresponding elements, that is,

$$\mathbf{C} = \mathbf{A} + \mathbf{B}, \qquad (A.17)$$

where the elements of $\mathbf{C}$ are given by

$$c_{ij} = a_{ij} + b_{ij}. \tag{A.18}$$

For example,

$$\begin{bmatrix} 1 & 0 & 3 \\ 2 & 1 & 6 \\ 8 & 0 & 3 \end{bmatrix} + \begin{bmatrix} 2 & 1 & 2 \\ 0 & 1 & 3 \\ 1 & 0 & 5 \end{bmatrix} = \begin{bmatrix} 3 & 1 & 5 \\ 2 & 2 & 9 \\ 9 & 0 & 8 \end{bmatrix}. \tag{A.19}$$

A.2.2 Subtraction of Matrices

Matrix subtraction follows the same rule as for matrix addition, except that corresponding elements are subtracted instead of added, that is,

$$\mathbf{C} = \mathbf{A} - \mathbf{B}, \tag{A.20}$$

where the elements of $\mathbf{C}$ are given by

$$c_{ij} = a_{ij} - b_{ij}. \tag{A.21}$$

A.2.3 Commutativity and Associativity

The commutative law applies to matrices, that is, they can be added or subtracted in any order, for example,

$$\mathbf{A} + \mathbf{B} = \mathbf{B} + \mathbf{A}. \tag{A.22}$$

Matrices also obey the associative law and can be added or subtracted in various combinations. For example,

$$(\mathbf{A} + \mathbf{B}) + \mathbf{C} = \mathbf{A} + (\mathbf{B} + \mathbf{C}). \tag{A.23}$$

A.2.4 Transposition of Matrices

If in a matrix, all the rows are interchanged with their corresponding columns the *transposed matrix* is obtained. The transpose of a matrix $\mathbf{A}$ is denoted by $\mathbf{A}^{\mathrm{T}}$. For example, if

$$\mathbf{A} = \begin{bmatrix} a_{11} & a_{12} \\ a_{21} & a_{22} \\ a_{31} & a_{32} \end{bmatrix}, \tag{A.24a}$$

then the transpose of $\mathbf{A}$ is given by

$$\mathbf{A}^{\mathrm{T}} = \begin{bmatrix} a_{11} & a_{21} & a_{31} \\ a_{12} & a_{22} & a_{32} \end{bmatrix}. \tag{A.24b}$$

For a partitioned matrix, the transpose is obtained by changing each submatrix to its transpose and then interchanging the rows and columns of the partitioned matrix. For example, referring to Eq. (A.15), if

$$\mathbf{A} = \begin{bmatrix} \mathbf{A}_{11} & \mathbf{A}_{12} & \mathbf{A}_{13} \\ \mathbf{A}_{21} & \mathbf{A}_{22} & \mathbf{A}_{23} \end{bmatrix}, \tag{A.25a}$$

then

$$\mathbf{A}^\mathrm{T} = \begin{bmatrix} \mathbf{A}_{11}^\mathrm{T} & \mathbf{A}_{21}^\mathrm{T} \\ \mathbf{A}_{12}^\mathrm{T} & \mathbf{A}_{22}^\mathrm{T} \\ \mathbf{A}_{13}^\mathrm{T} & \mathbf{A}_{23}^\mathrm{T} \end{bmatrix}. \tag{A.25b}$$

A symmetric matrix is identical to its transpose, that is,

$$\mathbf{A}^\mathrm{T} = \mathbf{A}. \tag{A.26}$$

For a skew-symmetric matrix, the transpose is the negative of itself, thus

$$\mathbf{A}^\mathrm{T} = -\mathbf{A}. \tag{A.27}$$

A.2.5   Multiplication of Matrices

To multiply a matrix $\mathbf{A}$ by a scalar c, every element of the matrix is multiplied by c. For example, if

$$\mathbf{A} = \begin{bmatrix} a_{11} & a_{12} \\ a_{21} & a_{22} \\ a_{31} & a_{32} \end{bmatrix}, \tag{A.28a}$$

then

$$c\mathbf{A} = \begin{bmatrix} ca_{11} & ca_{12} \\ ca_{21} & ca_{22} \\ ca_{31} & ca_{32} \end{bmatrix}. \tag{A.28b}$$

Multiplication of two matrices $\mathbf{A}$ and $\mathbf{B}$ requires that the number of columns in $\mathbf{A}$ be equal to the number of rows in $\mathbf{B}$; the matrices satisfying this condition are said to be conformable. The product $\mathbf{C}$ of two conformable matrices $\mathbf{A}$ and $\mathbf{B}$ of order $m \times k$ and $k \times n$, respectively, is that matrix of order $m \times n$, for which the ijth element is given by

$$c_{ij} = \sum_{q=1}^{k} a_{iq}b_{qj}, \qquad i = 1, 2, \ldots, m, \quad j = 1, 2, \ldots, n. \tag{A.29a}$$

Adopting the summation convention[†] allows Eq. (A.29a) to be written as

$$c_{ij} = a_{iq}b_{qj}, \qquad i = 1, 2, \ldots, m, \quad j = 1, 2, \ldots, n. \tag{A.29b}$$

[†] That a repeated index means a summation over that index.

Consider, as an application of Eq. (A.29), the following example. Let

$$\mathbf{A} = \begin{bmatrix} -1 & 3 & 4 \\ 2 & 1 & 6 \end{bmatrix}, \qquad \mathbf{B} = \begin{bmatrix} 2 & 1 & 3 \\ -4 & 0 & 2 \\ 5 & 1 & 0 \end{bmatrix}. \qquad \text{(A.30a,b)}$$

The product $\mathbf{C} = \mathbf{AB}$ is then given by

$$\mathbf{C} = \mathbf{AB} = \begin{bmatrix} -1 & 3 & 4 \\ 2 & 1 & 6 \end{bmatrix} \begin{bmatrix} 2 & 1 & 3 \\ -4 & 0 & 2 \\ 5 & 1 & 0 \end{bmatrix} = \begin{bmatrix} 6 & 3 & 3 \\ 30 & 8 & 8 \end{bmatrix}. \qquad \text{(A.30c)}$$

The operation of multiplication can be extended to products of more than two matrices, provided they are of the proper orders. The matrix product

$$\mathbf{D} = \mathbf{ABC} \qquad \text{(A.31)}$$

is thus possible if the orders of $\mathbf{ABC}$ are $m \times p$, $p \times q$, $q \times n$, respectively, in which case the order of $\mathbf{D}$ is $m \times n$.

If the sequence of matrices being multiplied is not disturbed, the associative and distributive laws apply also to matrix multiplication. For example,

$$(\mathbf{AB})\mathbf{C} = \mathbf{A}(\mathbf{BC}) = \mathbf{ABC} \qquad \text{(A.32a)}$$

and

$$\mathbf{A}(\mathbf{B} + \mathbf{C} + \mathbf{D}) = \mathbf{A}(\mathbf{B} + \mathbf{C}) + \mathbf{AD} = \mathbf{AB} + \mathbf{A}(\mathbf{C} + \mathbf{D})$$
$$= \mathbf{AB} + \mathbf{AC} + \mathbf{AD}. \qquad \text{(A.32b)}$$

In general, multiplication of matrices is not commutative, thus

$$\mathbf{AB} \neq \mathbf{BA}. \qquad \text{(A.33)}$$

Partitioned matrices are multiplied in exactly the same way as ordinary matrices.

A.2.6   Inversion of a Matrix

The inverse of a square matrix is that matrix, which, when multiplied by the original matrix, gives the unit matrix of the same order. The inverse for the matrix $\mathbf{A}$ is symbolized by $\mathbf{A}^{-1}$, hence by definition

$$\mathbf{A}^{-1}\mathbf{A} = \mathbf{A}\mathbf{A}^{-1} = \mathbf{I}. \qquad \text{(A.34)}$$

It can be shown that the inverse of the matrix $\mathbf{A}$ exists only if the determinant of $\mathbf{A}$ (see Appendix C) is nonzero, that is

$$\det \mathbf{A} \neq 0. \tag{A.35}$$

The inversion operation is illustrated in the following, for a matrix $\mathbf{A}$ given by

$$\mathbf{A} = \begin{bmatrix} a_{11} & a_{12} & a_{13} \\ a_{21} & a_{22} & a_{23} \\ a_{31} & a_{32} & a_{33} \end{bmatrix}. \tag{A.36}$$

The inverse $\mathbf{A}^{-1}$, if it exists, will be of the same order as $\mathbf{A}$, hence $\mathbf{A}^{-1}$ can be written as

$$\mathbf{A}^{-1} = \begin{bmatrix} b_{11} & b_{12} & b_{13} \\ b_{21} & b_{22} & b_{23} \\ b_{31} & b_{32} & b_{33} \end{bmatrix}. \tag{A.37}$$

Substituting Eqs. (A.36) and (A.37) into Eq. (A.34) gives

$$
\begin{aligned}
\mathbf{A}^{-1}\mathbf{A} &= \begin{bmatrix} b_{11} & b_{12} & b_{13} \\ b_{21} & b_{22} & b_{23} \\ b_{31} & b_{32} & b_{33} \end{bmatrix} \begin{bmatrix} a_{11} & a_{12} & a_{13} \\ a_{21} & a_{22} & a_{23} \\ a_{31} & a_{32} & a_{33} \end{bmatrix} \\[2mm]
&= \begin{bmatrix} b_{11}a_{11} + b_{12}a_{21} + b_{13}a_{31} & b_{11}a_{12} + b_{12}a_{22} + b_{13}a_{32} & b_{11}a_{13} + b_{12}a_{23} + b_{13}a_{33} \\ b_{21}a_{11} + b_{22}a_{21} + b_{23}a_{31} & b_{21}a_{12} + b_{22}a_{22} + b_{23}a_{32} & b_{21}a_{13} + b_{22}a_{23} + b_{23}a_{33} \\ b_{31}a_{11} + b_{32}a_{21} + b_{33}a_{31} & b_{31}a_{12} + b_{32}a_{22} + b_{33}a_{32} & b_{31}a_{13} + b_{32}a_{23} + b_{33}a_{33} \end{bmatrix} \\[2mm]
&= \begin{bmatrix} 1 & 0 & 0 \\ 0 & 1 & 0 \\ 0 & 0 & 1 \end{bmatrix}. \tag{A.38}
\end{aligned}
$$

A set of equations is obtained for b_{ij}, $i, j = 1, 2, 3$, by simply equating the corresponding elements in Eq. (A.38). For example, from

$$
\begin{aligned}
b_{11}a_{11} + b_{12}a_{21} + b_{13}a_{31} &= 1, \\
b_{11}a_{12} + b_{12}a_{22} + b_{13}a_{32} &= 0, \\
b_{11}a_{13} + b_{12}a_{23} + b_{13}a_{33} &= 0,
\end{aligned} \tag{A.39}
$$

there is obtained[†]

$$
\begin{aligned}
b_{11} &= (a_{22}a_{33} - a_{32}a_{33})/\det \mathbf{A}, \\
b_{12} &= (a_{32}a_{13} - a_{12}a_{33})/\det \mathbf{A}, \\
b_{13} &= (a_{12}a_{23} - a_{32}a_{13})/\det \mathbf{A},
\end{aligned} \tag{A.40}
$$

[†] Note that if the denominator in Eqs. (A.40) and also in the equations for the remaining b_{ij} is nonzero, the condition given in Eq. (A.35) is satisfied.

where

$$\det \mathbf{A} = a_{11}(a_{22}a_{33} - a_{23}a_{33}) - a_{21}(a_{12}a_{33} - a_{13}a_{32})$$
$$+ a_{31}(a_{12}a_{23} - a_{13}a_{22}). \tag{A.41}$$

Similar results can be derived for the remaining b_{ij}.

A matrix for which the determinant vanishes is said to be *singular*. Consequently, only a nonsingular matrix has an inverse.

Further information on matrices can be obtained from standard textbooks [1, 2].

A.3 QUADRATIC AND LINEAR FORMS

A quadratic form [3] is an expression containing a sum of quadratic terms, which for the variables $u_1, u_2, \ldots, u_n$ can be represented as

$$F(u_1, u_2, \ldots, u_n) = a_{11}u_1{}^2 + a_{12}u_1u_2 + \cdots + a_{1d}u_1u_d$$
$$+ a_{21}u_2u_1 + a_{22}u_2{}^2 + \cdots + a_{2d}u_2u_d + \cdots \tag{A.42}$$
$$+ a_{n1}u_nu_1 + a_{n2}u_nu_2 + \cdots + a_{nn}u_n{}^2.$$

In matrix form Eq. (A.42) can be written as

$$F(u_1, u_2, \ldots, u_n) = \mathbf{U}^T\mathbf{A}\mathbf{U}, \tag{A.43a}$$

where

$$\mathbf{U} = \begin{bmatrix} u_1 \\ u_2 \\ \vdots \\ u_n \end{bmatrix} \tag{A.43b}$$

and

$$\mathbf{A} = \begin{bmatrix} a_{11} & a_{12} & \cdots & a_{1n} \\ a_{21} & a_{22} & \cdots & a_{2n} \\ \vdots & \vdots & & \vdots \\ a_{n1} & a_{n2} & \cdots & a_{nn} \end{bmatrix}. \tag{A.43c}$$

Without loss of generality [4], one may assume the matrix $\mathbf{A}$ to be symmetric, that is, $a_{ij} = a_{ji}$.

A quadratic functional [5] in $\phi = \phi(x, y, z)$ has a volume (or surface, or line) integral of F of the form

$$\chi = \int_V F(u_1, u_2, \ldots, u_n)dV, \tag{A.44}$$

where F is given by Eq. (A.42), or alternatively Eq. (A.43a), and where the $u_1, u_2, \ldots, u_n$ represent ϕ and its various spatial derivatives, $\phi_x, \phi_y, \phi_z, \phi_{xy}, \phi_{xz}, \phi_{yz}, \ldots$. In this case, the a_{ij} of Eq. (A.42) are functions of position.

A *linear form* comprised of variables $u_1, u_2, \ldots, u_n$ is defined as the linear combination

$$c_1 u_1 + c_2 u_2 + \cdots + c_n u_n. \tag{A.45}$$

In matrix form, Eq. (A.45) can be written as

$$c_1 u_1 + c_2 u_2 + \cdots + c_n u_n = \mathbf{U}^{\mathrm{T}}\mathbf{C} = \mathbf{C}^{\mathrm{T}}\mathbf{U}, \tag{A.46}$$

where

$$\mathbf{C} = \begin{bmatrix} c_1 \\ c_2 \\ \vdots \\ c_n \end{bmatrix}, \qquad \mathbf{U} = \begin{bmatrix} u_1 \\ u_2 \\ \vdots \\ u_n \end{bmatrix}. \tag{A.47}$$

The coefficients $c_1, c_2, \ldots, c_n$ of the linear combination in Eq. (A.45) are generally functions of position.

REFERENCES

1. A. C. Aitken, "Determinants and Matrices." Univ. Math. Tests, Oliver & Boyd, Edinburgh, 1964.
2. W. W. Sawyer, "An Engineering Approach to Linear Algebra." Cambridge Univ. Press, London and New York, 1972.
3. M. J. Forray, "Variational Calculus in Science in Engineering." McGraw-Hill, New York, 1968.
4. J. W. Dettman, "Mathematical Methods in Physica and Engineering." McGraw-Hill, New York, 1962.
5. P. W. Berg, Calculus of variations, *in* "Handbook of Engineering Mechanics" (W. Flügge, ed.), Chapter 16. McGraw-Hill, New York, 1962.

APPENDIX B

MATRIX CALCULUS

B.1 DIFFERENTIATION OF MATRICES

Differentiation of a matrix is accomplished by differentiating every element [1]. Thus, if $\mathbf{A} = \mathbf{A}(t)$ is the $p \times q$ matrix

$$\mathbf{A}(t) = \begin{bmatrix} a_{11}(t) & a_{12}(t) & \cdots & a_{1q}(t) \\ a_{21}(t) & a_{22}(t) & \cdots & a_{2q}(t) \\ \vdots & \vdots & & \vdots \\ a_{p1}(t) & a_{p2}(t) & \cdots & a_{pq}(t) \end{bmatrix}, \tag{B.1}$$

then

$$\frac{d\mathbf{A}}{dt} = D(\mathbf{A}) = \begin{bmatrix} da_{11}/dt & da_{12}/dt & \cdots & da_{1q}/dt \\ da_{21}/dt & da_{22}/dt & \cdots & da_{2q}/dt \\ \vdots & \vdots & & \vdots \\ da_{p1}/dt & da_{p2}/dt & \cdots & da_{pq}/dt \end{bmatrix}. \tag{B.2}$$

Integration, similarly, is defined by

$$\int [\mathbf{A}(t)] \, dt = \begin{bmatrix} \int a_{11} \, dt & \int a_{12} \, dt & \cdots & \int a_{12} \, dt \\ \int a_{21} \, dt & \int a_{22} \, dt & \cdots & \int a_{2q} \, dt \\ \vdots & \vdots & & \vdots \\ \int a_{p1} \, dt & \int a_{p2} \, dt & \cdots & \int a_{pq} \, dt \end{bmatrix} \tag{B.3}$$

If the matrices $\mathbf{A}(t)$, $\mathbf{B}(t)$, and $\mathbf{C}(t)$ are conformable in addition, or multiplication, as appropriate, the following can be shown [2, 3] to apply:

(1) $D(\mathbf{A} + \mathbf{B}) = D\mathbf{A} + D\mathbf{B}$. \hfill (B.4)

(2) $D(\mathbf{AB}) = (D\mathbf{A})\mathbf{B} + \mathbf{A}(D\mathbf{B})$, \hfill (B.5)

(3) $D(\mathbf{ABC}) = (D\mathbf{A})\mathbf{BC} + \mathbf{A}(D\mathbf{B})\mathbf{C} + \mathbf{AB}(D\mathbf{C})$. \hfill (B.6)

(4) $D(\mathbf{A}^{-1}) = -\mathbf{A}^{-1}(D\mathbf{A})\mathbf{A}^{-1}$. \hfill (B.7)

B.2   PARTIAL DIFFERENTIATION OF MATRICES

Partial differentiation of a matrix is defined in the same way as total differentiation. If $\mathbf{X}$ and $\mathbf{Y}$ are column and row matrices respectively, that is,

$$\mathbf{X} = \begin{bmatrix} x_1 \\ x_2 \\ \vdots \\ x_n \end{bmatrix} \quad \text{and} \quad \mathbf{Y} = [y_1 \quad y_2 \quad \cdots \quad y_m], \tag{B.8a,b}$$

then the partial derivative $\partial \mathbf{Y}/\partial x_1$ is given by

$$\partial \mathbf{Y}/\partial x_1 = [\partial y_1/\partial x_1 \quad \partial y_2/\partial x_1 \quad \cdots \quad \partial y_m/\partial x_1]. \tag{B.9}$$

The set of quantities

$$\begin{bmatrix} \partial \mathbf{Y}/\partial x_1 \\ \partial \mathbf{Y}/\partial x_2 \\ \vdots \\ \partial \mathbf{Y}/\partial x_n \end{bmatrix} = \begin{bmatrix} \partial y_1/\partial x_1 & \partial y_2/\partial x_1 & \cdots & \partial y_m/\partial x_1 \\ \partial y_1/\partial x_2 & \partial y_2/\partial x_2 & \cdots & \partial y_m/\partial x_2 \\ \vdots & \vdots & & \vdots \\ \partial y_1/\partial x_n & \partial y_2/\partial x_n & \cdots & \partial y_m/\partial x_n \end{bmatrix} \tag{B.10}$$

can be written without ambiguity as $\partial \mathbf{Y}/\partial \mathbf{X}$. Similarly, there is no ambiguity in writing $\partial \mathbf{X}/\partial \mathbf{Y}$, which expands as

$$\partial \mathbf{X}/\partial \mathbf{Y} = [\partial \mathbf{X}/\partial y_1 \quad \partial \mathbf{X}/\partial y_2 \quad \cdots \quad \partial \mathbf{X}/\partial y_m], \tag{B.11}$$

which further expands to

$$\frac{\partial \mathbf{X}}{\partial \mathbf{Y}} = \begin{bmatrix} \partial x_1/\partial y_1 & \partial x_1/\partial y_2 & \cdots & \partial x_1/\partial y_m \\ \partial x_2/\partial y_1 & \partial x_2/\partial y_2 & \cdots & \partial x_2/\partial y_m \\ \vdots & \vdots & & \vdots \\ \partial x_n/\partial y_1 & \partial x_n/\partial y_2 & \cdots & \partial x_n/\partial y_m \end{bmatrix}. \quad (B.12)$$

Expressions such as $\partial \mathbf{A}/\partial \mathbf{Y}$, where $\mathbf{A}$ is a rectangular matrix, and $\partial \mathbf{X}/\partial \mathbf{Z}$ are difficult to assign meaning to, and their use should be avoided by making use of the standard relations below. Subject to this restriction, the rules given earlier for (total) differentiation of sums and products can be used for partial differentiation.

If $\mathbf{X}$, $\mathbf{Y}$, $\mathbf{Z}$, $\mathbf{C}$ are the matrices

$$\mathbf{X} = \begin{bmatrix} x_1 \\ x_2 \\ \vdots \\ x_n \end{bmatrix}, \qquad \mathbf{Y} = [y_1 \quad y_2 \quad \cdots \quad y_m],$$

$$\mathbf{Z} = \begin{bmatrix} z_1 \\ z_2 \\ \vdots \\ z_n \end{bmatrix}, \qquad \mathbf{C} = \begin{bmatrix} c_1 \\ c_2 \\ \vdots \\ c_n \end{bmatrix}, \qquad (B.13)$$

and if the matrices $\mathbf{A}$ and $\mathbf{B}$ are conformable in multiplication or addition as required, the following standard relations can be proven by expanding before carrying out the differentiation:

(1) $\partial \mathbf{Y}/\partial \mathbf{X} = [\partial \mathbf{Y}^{\mathsf{T}}/\partial \mathbf{X}^{\mathsf{T}}]^{\mathsf{T}}.$ (B.14)

(2) $\partial \mathbf{Z}^{\mathsf{T}}/\partial \mathbf{Z} = \partial \mathbf{Z}/\partial \mathbf{Z}^{\mathsf{T}} = \partial \mathbf{Y}/\partial \mathbf{Y}^{\mathsf{T}} = \partial \mathbf{Y}^{\mathsf{T}}/\partial \mathbf{Y} = \mathbf{I}.$ (B.15)

(3)   For the quadratic form $\mathbf{X}^{\mathsf{T}}\mathbf{A}\mathbf{X}$, where $\mathbf{A} \neq f(x_i)$,

(a) $\partial(\mathbf{X}^{\mathsf{T}}\mathbf{A}\mathbf{X})/\partial \mathbf{X} = 2\mathbf{A}\mathbf{X},$ (B.16a)

(b) $\partial(\mathbf{X}^{\mathsf{T}}\mathbf{A}\mathbf{X})/\partial \mathbf{X}^{\mathsf{T}} = 2\mathbf{X}^{\mathsf{T}}\mathbf{A}.$ (B.16b)

(4) $\partial(\mathbf{X}^{\mathsf{T}}\mathbf{B})/\partial \mathbf{X} = \mathbf{B},$ where $\mathbf{B} \neq f(x_i).$ (B.17)

(5)   For the linear combination $\mathbf{Y}\mathbf{X}$ (also known as the scalar product or inner product), and where the elements of $\mathbf{Y}$ and the elements of $\mathbf{Z}$ are functions of the elements of $\mathbf{X}$, that is, $y_i, z_j = f(x_k)$,

$$\frac{\partial[\mathbf{Y}\mathbf{Z}]}{\partial \mathbf{X}} = \frac{\partial[\mathbf{Z}\mathbf{Y}]}{\partial \mathbf{X}} = \frac{\partial \mathbf{Z}^{\mathsf{T}}}{\partial \mathbf{X}}\mathbf{Y}^{\mathsf{T}} + \frac{\partial \mathbf{Y}}{\partial \mathbf{X}}\mathbf{Z}, \quad (B.18a)$$

$$\frac{\partial[\mathbf{Y}\mathbf{Z}]}{\partial \mathbf{X}^{\mathsf{T}}} = \frac{\partial[\mathbf{Z}\mathbf{Y}]}{\partial \mathbf{X}^{\mathsf{T}}} = \mathbf{Y}\frac{\partial \mathbf{Z}}{\partial \mathbf{X}^{\mathsf{T}}} + \mathbf{Z}^{\mathsf{T}}\frac{\partial \mathbf{Y}^{\mathsf{T}}}{\partial \mathbf{X}^{\mathsf{T}}}. \quad (B.18b)$$

REFERENCES

1.   D. T. Finkbeiner, "Matrices and Linear Transformations," 2nd ed. Freeman, San Francisco, California, 1965.
2.   A. D. Michal, "Matrix and Tensor Calculus." Wiley, New York, 1947.
3.   E. H. Thompson, "Introduction to the Algebra of Matrices with Some Applications." Hilger, London, 1969.

APPENDIX C

DETERMINANTS

A *determinant* is the number obtained from a square matrix using a defined procedure. The determinant of a matrix $\mathbf{A}$

$$\mathbf{A} = \begin{bmatrix} 1 & -2 & 4 \\ -8 & 6 & 2 \\ 4 & 3 & -4 \end{bmatrix} \tag{C.1}$$

is written as

$$\det \mathbf{A} = |\mathbf{A}| = \begin{vmatrix} 1 & -2 & 4 \\ -8 & 6 & 2 \\ 4 & 3 & -4 \end{vmatrix}. \tag{C.2}$$

A minor of a determinant is obtained by deleting an equal number of rows and columns from that determinant. Thus,

$$\begin{vmatrix} 1 & -2 \\ -8 & 6 \end{vmatrix}, \quad \begin{vmatrix} 6 & 2 \\ 3 & -4 \end{vmatrix}, \quad 2 \tag{C.3}$$

are minors of $|\mathbf{A}|$.

The cofactor of an element a_{ij} in a determinant $|\mathbf{A}|$ is defined as the product of $(-1)^{i+j}$ and the minor obtained by deleting the row and column containing a_{ij}.

The procedure for evaluating a determinant is as follows. The determinant of a square matrix is equal to the sum of the products of the elements in any row (or column) with their respective cofactors. Hence, the determinant of Eq. (C.1) is given by

$$|\mathbf{A}| = 1\begin{vmatrix} 6 & 2 \\ 3 & -4 \end{vmatrix} - (-2)\begin{vmatrix} -8 & 2 \\ 4 & -4 \end{vmatrix} + 4\begin{vmatrix} -8 & 6 \\ 4 & 3 \end{vmatrix}$$

$$= 1\{(6 \times -4) - (2 \times 3)\} + 2\{(-8 \times -4) - (2 \times 4)\} + 4\{(-8 \times 3) - (6 \times 4)\}$$

$$= -174. \tag{C.4}$$

Properties of Determinants:

(1)   If any row or column consists entirely of zeros, the determinant is zero.

(2)   If any two rows (or two columns) are interchanged, the determinant changes sign.

(3)   If any two rows (or two columns) are identical, the determinant is zero.

(4)   Multiplying every element in a row (or column) by a scalar λ multiplies the whole determinant by λ.

INDEX

DATE DUE